Multirate Filtering for Digital Signal Processing:
MATLAB Applications

Ljiljana Milić
University of Belgrade, Serbia

INFORMATION SCIENCE REFERENCE

Hershey · New York

Director of Editorial Content:	Kristin Klinger
Director of Production:	Jennifer Neidig
Managing Editor:	Jamie Snavely
Assistant Managing Editor:	Carole Coulson
Typesetter:	Jeff Ash
Cover Design:	Lisa Tosheff
Printed at:	Yurchak Printing Inc.

Published in the United States of America by
Information Science Reference (an imprint of IGI Global)
701 E. Chocolate Avenue, Suite 200
Hershey PA 17033
Tel: 717-533-8845
Fax: 717-533-8661
E-mail: cust@igi-global.com
Web site: http://www.igi-global.com

and in the United Kingdom by
Information Science Reference (an imprint of IGI Global)
3 Henrietta Street
Covent Garden
London WC2E 8LU
Tel: 44 20 7240 0856
Fax: 44 20 7379 0609
Web site: http://www.eurospanbookstore.com

Library of Congress Cataloging-in-Publication Data

Milic, Ljiljana.

Multirate filtering for digital signal processing : MATLAB applications / Ljiljana Milic.

     p. cm.

Includes bibliographical references and index.

Summary: "This book covers basic and the advanced approaches in the design and implementation of multirate filtering"--Provided by publisher.

ISBN 978-1-60566-178-0 (hardcover) -- ISBN 978-1-60566-179-7 (ebook)

1. Signal processing--Digital techniques--Data processing. 2. Signal processing--Digital techniques--Mathematics. 3. Electric filters, Bandpass--Computer simulation. 4. Multiplexing--Computer simulation. 5. MATLAB. I. Title.

TK5102.9.M545 2009

621.382'2--dc22

                         2008031503

British Cataloguing in Publication Data
A Cataloguing in Publication record for this book is available from the British Library.

To my grandchildren
Sara and Rajko
for the new happiness they brought to my life

Table of Contents

Foreword

In 1969 and in the mid 1970s the landmark books on digital signal processing (DSP) were published (Gold, Rader, Oppenheim, Schafer, and Rabiner). These books made it easier for individuals and practicing engineers to learn DSP and become active in the field. In the 1970s, the scientists and engineers began using the real-time DSP computers. It was the time of the first use of integrated circuits for digital signal processing. Another boost to the DSP field was the finding of widespread use of the algorithm for straightforward calculation of the discrete Fourier transform. Technical achievements and rapid technological changes revealed new horizons and significantly contributed to the signal processing revolution.

At the beginning of signal processing era, analog processing was limited by hardware (vacuum tubes and RLC circuits) while the limitation of digital signal processing was the complexity of algorithms and computational speed. The algorithms were just a mathematical curiosity to most practicing engineers. Nowadays, the largest repository of algorithms is contained within MATLAB so individuals, students, and engineers need only to learn how to apply algorithms. Ready-to-use filter design programs, code-generation tools, tools for system integration and debugging, high-level programming languages, optimizing compilers, software simulators, and hardware emulators, make digital signal processing is getting to a point where it is almost every place. There are more than 50,000 engineers who regard DSP as their specialty and much more who are relying on signal processing.

Digital signal processing has changed dramatically for over half a century. In the 1960s the main interests were audio technologies, particularly electroacoustics, speech communications and electronic music. During the past sixty years many scientists, researches, and engineers worked on interdisciplinary problems and the synergistic effect changed social, political, and economic conditions. The advances of integrated circuit technology increased the density of components on chips, increased the speed, and reduced the IC power dissipation. Combined mass production with availability of low-cost software support brought down the overall cost of digital signal processing and caused the explosive growth of DSP applications.

Despite of an extraordinary number of published papers, and many published books in the field, scientists, researches, and engineers rely on heavy use of computer. It becomes so easy to do so much using computers and available software by a single click without thinking and understanding the theory. Fabrication of such results without thoroughly understanding the assumptions that underlie the software tools may lead to a dead-end. This is especially dangerous for novice DSP users or students without excellent mathematical background. The multirate filtering is a field in which it is so easy to become unsuccessful in problem solving if one is unaware of the underling theory. On the other hand, multirate filtering is one of the most powerful approaches for exploiting all benefits brought on by signal processing chips and programmable circuits.

The book *Multirate Filtering for Digital Signal Processing*, authored by Ljiljana Milić, is the missing link offering advances and applications of multirate filtering techniques. The intent of the author is to present both theory and applications of multirate filtering in an accessible format using MATLAB. The material is well suited for researches and students who have a working knowledge of basic digital signal processing. There are very few books on this topic that can make it easier for students and practicing engineers to learn and understand multirate filtering without extensive use of strict mathematical derivations. This is a unique book in the field and the level of mathematical maturity that is required of the reader is reduced by avoiding exhaustive derivations.

The book starts from a basic level and takes the reader to advanced concepts without making use of heavy mathematics. The most important strength of the book is the clarity of presentation and its style, making the text easy to follow. The MATLAB programs provide a bridge between the theory and practical multirate filter design that certainly will appeal to practicing engineers. The text is also suitable for the final year of MSc studies and PhD studies. Also, this book can be used as a complementary text in a course of multirate systems for graduate students.

The prerequisite of readers is a basic course on digital signal processing. Hence, the number of potential readers is very large – practicing engineers and engineers in industry (such as communication companies) and also undergraduate/graduate students. Target readers are researchers, but the final year students for MSc, as well as PhD students can use this book as a valuable resource for their innovative work. Many engineers have been in situations in the field where a problem cannot be solved with the basic knowledge of signal processing. Multirate filtering could be used when improved filtering is needed, such as the reduction of the number of operations per input sample. The growth of DSP applications, and the appearance of various new ones, will result in increasing the potential readers in many other fields such as computer graphics or medical instruments.

The book is well organized in 12 chapters, which are written in considerable detail. The majority of the chapters are read without any assumption of the previous ones. Each chapter in the text contains a list of the appropriate and up-to-date references. This makes the text more readable.

Most of the books on multirate filtering do not give illustrative examples using a powerful software tool like MATLAB. This book focuses on the theory of multirate filtering, which is exemplified by means of MATLAB applications. One of the distinctive contributions of the book is the significant number of carefully selected MATLAB examples. This gives the reader a better understanding of the theory. The book covers the topic very well and there is no similar book on the market with a reasonable number of instructive examples for every topic discussed in the text. Each chapter contains a MATLAB exercise list.

This book on multirate digital signal processing will become a landmark reference for researchers and practicing engineers as well as for MSc and PhD students.

Miroslav D. Lutovac

Miroslav D. Lutovac *was born in Skopje, FYRM, in 1957. He received the Dipl.-Eng., MSc and DSc degrees from the University of Belgrade in 1981, 1985 and 1991, respectively, all in electrical engineering. He has been with Automatics Institute and Telecommunication & Electronics Institute Institute, Belgrade. His research objectives are to automate the design and real-time implementation of analog and digital signal processing systems, to apply symbolic computation in the optimization of communication and control systems, and to transfer design tools and methodologies to industry. He was appointed associate professor (2001) in Digital Signal Processing at the School of Electrical Engineering and professor (2008) in Computer Science and Electrical Engineering at the University of Novi Pazar. He is elected for Principal Research Fellow by Ministry of Science of Serbia in 1999. Dr. Lutovac is author or coauthor of more than 150 scientific papers, mainly in the field of digital signal processing, and the coauthor of the basic monograph* Filter Design for Signal Processing Using MATLAB and Mathematica *(2000) published by Prentice Hall. The book was translated into Chinese (2004). He is coauthor of the book chapter*

Efficient Multirate Filtering. *Dr. Lutovac is a senior member of IEEE, corresponding member of Academy of Engineering Sciences of Serbia, and advisory member of* Journal IEICE Transactions on Fundamentals of Electronics, Communications and Computer Sciences – *Japan. His algorithms are deployed in software FilterCAD (Linear Technology) and WIPL-D Microwave. Miroslav Lutovac and Dejan Tosic have released software SchematicSolver that is distributed by Wolfram Research Inc. (2002-2008). For more details visit http://kondor.etf.rs/~lutovac/*

Preface

Multirate signal processing techniques are widely used in many areas of modern engineering such as communications, image processing, digital audio, and multimedia. The main advantage of a multirate system is the substantial decrease of computational complexity, and consequently, the cost reduction. The computational efficiency of multirate algorithms is based on the ability to use simultaneously different sampling rates in the different parts of the system.

The sampling rate alterations generate the unwanted effects through the system: spectral aliasing in the sampling rate decrease, and spectral images in the sampling rate increase. As a consequence, the multirate processing might produce unacceptable derogations in the digital signal. The crucial role of multirate filtering is to enable the sampling rate conversion of the digital signal without significantly destroying the signal components of interest. The multirate filtering makes the general concept of multirate signal processing applicable in practice.

This book is focussed on multirate filters, the essential processing algorithm in multirate systems. The mission of the book is to bridge the existing gap between the multirate filter theory and practice. This book deeply introduces MATLAB® functions and commands in presenting and explaining various aspects of multirate filtering. MATLAB® is chosen as the most popular software widely used at universities, in research laboratories, and in industry.

A multirate filter can be defined as a digital filter in which the sampling rate of the input signal is changed in one or more intermediate points. Multirate techniques are used in filters for sampling rate conversion where the input and output rates are different, and also in constructing filters with equal input and output rates. The basic roles of multirate filtering in modern signal processing systems go in three main directions.

Firstly, the multirate filtering is used whenever two digital systems with different sampling rates have to be connected. Filtering is used to suppress aliasing in decimation, and to remove imaging in interpolation. The use of an appropriate filter enables one to convert a digital signal of a specified sampling rate into another signal with a target sampling rate without destroying the signal components of interest.

Secondly, the multirate filtering is one of the best approaches for solving complex filtering problems when a single filter operating at a fixed sampling rate is of significantly high order and suffers from output noise due to multiplication round-off errors and from the high sensitivity to variations in the filter coefficients. Various multirate design techniques provide that the overall filtering characteristic is shared between several simplified subfilters operating at the lowest possible sampling rate. Design constraints for subfilters are relaxed if compared to a single-rate overall filter. As a consequence of the reduced design constraints, the effects of quantization in subfilters and in the overall multirate filter are decreased. Multirate filters provide a practical solution for digital filters with stringent spectral characteristics that are very difficult to solve otherwise.

Third, multirate filtering is used in constructing multirate filter banks.

For multirate filters, FIR (finite impulse response) or IIR (infinite impulse response) transfer functions can be used for generating the overall system. The selection of the filter type depends on the criteria at hand. An FIR filter easily achieves a strictly linear-phase response, but requires a larger number of operations per output sample when compared with an equal magnitude response IIR filter. The linear-phase FIR filter is an adequate choice when the waveform of the signal has to be preserved. An advantage of the multirate design approach is the ability of improving significantly the efficiency of FIR filters thus making them very desirable in practice.

The multirate signal processing and multirate filtering have been attracted many researchers during the last several decades. The rapid development of the new algorithms and new design methods has been influenced by the advances in computer technology and software development. Although the existing literature on the subject is very large, the multirate signal processing is an open area of research.

The multirate filtering is an area of interest for many researchers and practicing engineers. Efficient and sophisticated design in the field of multirate filtering needs a high-level software tool such as MAT-LAB®. The adequate software enables one to use the built-in functions and algorithms and concentrate on his/her one task (or research problem).

This book presents the theory and applications of multirate filtering with the extensive use of MAT-LAB including the *Signal Processing, Filter Design, and Wavelet Toolboxes*. The material in the text is supported by examples solved in MATLAB aimed to provide experiments that illustrate and verify the underlying theory. The solved MATLAB examples given through the book and the MATLAB exercises given at the end of each chapter enable the reader to develop deeper understanding of the multirate filtering problems.

The benefit of this book is a convenient access to the theory, design and implementation of multirate filters.

The book is divided in 12 chapters.

Chapter I presents the background review of the single-rate discrete-time signals and systems. A concise review of the time-domain and the transform-domain characterization of discrete-time signals and systems is given. First, we discuss the representation of a discrete-time signal as a sequence of numbers, and explain the operations on sequences. Then, the definition and properties of discrete-time systems are given with the emphasis to the linear-time-invariant (LTI) systems. The representation in the transform domain comprises the discrete-time Fourier transform (DTFT), the discrete Fourier Transform (DFT), and the z-transform. The definitions of the discrete-time system transfer function and the frequency responses are given. The basic realization structures for FIR and IIR systems are briefly described. Finally, the relations between continuous-time and discrete-time signals are given.

Chapter II is devoted to the basics of multirate systems. This chapter considers the basic sampling rate alterations when changing the sampling rate by an integer factor. The time-domain representations of down-sampling and up-sampling operations are introduced with the emphasis to the linearity and time-dependence properties. The z-domain and frequency-domain representations of down-sampled and up-sampled signals are developed. The spectrum of the down-sampled signal is analyzed, and the concept of aliased spectra is introduced. The spectrum of the up-sampled signal has been analysed too, and the appearance of images in the signal spectrum is explained. At this point, the essential importance of filtering has been observed. The concept of decimation and interpolation that include filtering as an integral part of a sampling rate alteration operation has been explained next. The description of Six Identities that enable the reductions in computational complexity of multirate systems is given. Then, the effects of the sampling rate conversion with the phase offset are described. The polyphase decomposition of the sequence and the representation of polyphase components are explained in detail. Finally, the concept of multistage multirate system is presented.

Chapter III considers the general role of filters in multirate systems. The spectral characteristics of decimators and interpolators are discussed first. The effects of aliasing and imaging are illustrated by means of examples. Following the discussion on aliasing and imaging, the problem of proper filter specifications that could ensure the suppression of the aliased spectra and the removal of images has been underlined. Three commonly used types of filter specifications are described. In the sequel, it is shown by means of numerous examples how the existing MATLAB functions for FIR and IIR filter design can be used to meet the typical specifications. The special attention has been focussed to the computation of the residual aliasing, which is inevitably left after filtering. It is shown how the aliasing characteristics of the decimation filter can be computed. The sampling rate alteration of bandpass signals is also discussed in this chapter.

The design and implementation of FIR filters for sampling rate conversion is presented in **Chapter IV**. The implementation structures of decimators and interpolators that are based on FIR filtering are considered in this chapter. First, the application of the FIR filter direct implementation forms in constructing decimators and interpolators are analyzed. The central part of the chapter is devoted to the description of the efficient polyphase implementation of decimators and interpolators. The use of MATLAB for the verification and simulation of the decimator/interpolator structure is demonstrated. The operation of those structures is illustrated by means of example decimators and interpolators. Also, the polyphase memory-saving structures for decimators and interpolators are shown. In this chapter, the computational efficiency of FIR decimators and interpolators is discussed in order to demonstrate the significant computational savings achieved in FIR multirate filtering.

Chapter V is devoted to IIR filters for sampling rate conversion. In this chapter, the direct implementation structure for IIR decimators and interpolators has been considered first. The computational efficiency of IIR decimators and interpolators when implemented in the direct form has been presented. The application of the polyphase decomposition in constructing efficient IIR decimators and interpolators has been considered. The advantage of the solutions based on all-pass polyphase components has been underlined and illustrated by means of an example. The role of extra filter in constructing high-performance IIR decimator and interpolator is explained and illustrated. In this chapter, the particular attention has been paid to the solutions which use the implementation structures based on the parallel connection of two all-pass subfilters. It is shown that extremely efficient IIR decimators and interpolators can be achieved when using the cascade of halfband IIR filters followed by the factor-of-two down-samplers. The application of elliptic minimal Q factors (EMQF) filters in the systems for sampling rate alterations has been shown.

Chapter VI considers the sampling rate conversion by a fractional factor, sometimes called a fractional sampling rate conversion. It is shown first how the MATLAB functions can be used to convert the sampling rate of the signal by a rational factor. The technique for constructing efficient sampling rate conversion by a rational factor based on FIR filters and polyphase decomposition is presented. In the sequel, we consider the sampling rate alteration with an arbitrary conversion factor. We present the polynomial-based approximation of the impulse response of a hybrid analog/digital model, and the implementation based on the Farrow structure. We also consider the construction of fractional delay filters. MATLAB examples illustrate the applications.

Chapter VII is devoted to the theory and design of Lth-band filters and particularly to the halfband filters, the most important subclass of Lth-band filters. This chapter starts with the linear-phase Lth-band FIR filters. We introduce the main definitions and present by means of examples the efficient polyphase implementation of Lth-band FIR filters. We discuss the properties of the separable linear-phase transfer functions, and construct the minimum-phase and maximum-phase FIR transfer functions. The minimum-phase (maximum-phase) transfer function is considered as a spectral factor of the separable (factorisable)

FIR filter transfer function. In sequel, we present the design and efficient implementation of the halfband FIR filters. A halfband filter can be considered as a special class of the Lth-band filter obtained for $L = 2$. The class of IIR Lth-band and halfband filters is presented next. The particular attention is addressed to the design and implementation of IIR halfband filters.

In **Chapter VIII** we present the complementary filter pairs. First, we review the definitions of delay-complementary, all-pass complementary, power-complementary and magnitude complementary properties. The generation of a highpass filter (FIR and IIR) from the complementary lowpass filter is shown. Then, the definitions of the analysis and synthesis lowpass/highpass filter pairs are given. In the sequel, we present the design and implementation of FIR filter pairs comprising: delay-complementary, power-complementary, and magnitude complementary FIR filter pairs. The design and implementation of three classes of IIR filter pairs satisfying the allpass-complementary/power-complementary, power-complementary, and allpass-complementary/magnitude-complementary properties are presented in this chapter. We demonstrate the high-performance complementary IIR filter pairs, which benefit the advantages of FIR and IIR filter properties.

In **Chapter IX** we present the application of multirate techniques in filter design and implementation. The chapter considers filters with equal input and output sampling rate, with narrow transition bandwidths that are very difficult to be implemented by using classical single-rate techniques. Employing the multirate techniques with multistage filtering and the complementary filter pairs, one achieves to construct the overall high-order filter by combining several low-order subfilters. In this way, the overall filtering task is shared between subfilters of significantly lower order. In this chapter, we consider the application of multistage filtering to design the narrowband filters. Extremely efficient solutions are achieved when using halfband decimation and interpolation subfilters. The wideband filters with sharp transition bands are considered, as well. The solutions are based on the complementary multirate filtering and multistage design.

Chapter X considers the applications of frequency-response masking techniques in constructing digital filters with sharp transition bands. The concept of model and masking filters is introduced and the design and implementation of narrowband FIR and IIR filters is discussed. In the sequel, the frequency-response masking approach in designing filters with the arbitrary bandwidths is considered. The concept of frequency-response masking technique based on the model complementary filter pair and two masking filters that is suitable for synthesizing the arbitrary bandwidth filters is presented. The general characteristics of the model complementary filter pair and that of two masking filters are shown. The synthesis of FIR and IIR wideband filters with sharp transition bands is illustrated by means of examples. A solution that uses the halfband filter as one of the masking filters is also given in this chapter.

Chapter XI is devoted to the design and realization of the comb-based filters for decimators and interpolators. In this chapter, we first introduce the concept of the basic cascade integrator-comb (CIC) filter and discuss its properties. Then, we present the structures of the CIC-based decimators and interpolators, discuss the corresponding frequency responses, and demonstrate the overall two-stage decimator constructed as the cascade of a CIC decimator and an FIR decimator. In the next section, we expose the application of the polyphase implementation structure, which is aimed to reduce the power dissipation in the comb-based decimators and interpolators. We consider techniques for sharpening the original comb filter magnitude response and emphasize an approach that modifies the filter transfer function in a manner to provide a sharpened filter operating at the lowest possible sampling rate. Finally, we give a brief description of the modified comb filter based on the zero-rotation approach. We discuss the improvements achieved with modified comb-filter transfer function and sharpening techniques.

The final chapter, **Chapter XII**, illustrates by means of examples the applications of multirate filters in constructing multirate filter banks. First, we give a brief review of the properties of the two-channel analysis and synthesis filter banks with the condition for elimination of aliasing. The perfect-reconstruction and nearly perfect-reconstruction properties are discussed, and solutions based on FIR and IIR QMF banks and the orthogonal two-channel filter banks are shown. In the sequel, the tree-structured multichannel filter banks are considered including the uniform filter banks and nonuniform filter banks with the special emphasis to the octave filter banks. The process of signal decomposition and reconstruction is illustrated by means of examples. The application of some MATLAB functions for signal decomposition and reconstruction (from the *Wavelet Toolbox*) that are based on the octave filter banks has been also demonstrated in this Chapter.

Finally, at the end of each chapter, except Chapter I, numerous MATLAB exercises are provided, with the intention to help the reader in developing various individual solutions. Some of the exercises require only the modifications of the existing programs given in the text. However, some of the exercises are more demanding.

The material exposed in this book range in difficulty from very simple applications of multirate techniques and multirate filtering to more elaborate and demanding multirate processing algorithms.

The MATLAB examples are extensively used through the chapters. In the first chapters, the script files in the form of demo programs are given in details. Later on, the MATLAB applications are shown with the essential code fragments only. Using the given code fragments, the reader can easily complete his/her own m-file and generate the computations and figures of interest.

The majority of examples use the existing MATLAB functions from the *Signal Processing* and *Filter Design Toolboxes* in order to exploit the power of MATLAB for the easier access to the main subject of this book. In the last chapter, some functions from the *Wavelet Toolbox* are utilized.

Although the MATLAB programs in this book are written in a simple intuitive way, it is expected that the reader possess some basic knowledge in MATLAB programming.

MATLAB® is a registered trademark of The MathWorks, Inc. and is used with permission. The MathWorks does not warrant the accuracy of the text or exercises in this book. This book's use or discussion of MATLAB® software or related products does not constitute endorsement or sponsorship by The MathWorks of particular pedagogical approach or particular use of the MATLAB® software.

For product information, please contact:

The MathWorks, Inc.
3 Apple Hill Drive
Natick, MA 01760-2098 USA
Tel; 508-647-7000
Fax: 508-647-7001
E-mail: info@mathworks.com
Web: www.mathworks.com

Acknowledgment

The author would like to acknowledge the help of all involved in the developing and review process of this book, without whose support the project could not have been satisfactorily completed. The deep appreciation is due to the Serbian Ministry of Science for the financial support, to the MathWorks™ for the grant of the MATLAB software, and to Mihajlo Pupin Institute for providing the technical support to this project.

Author's appreciations and special thanks go to Prof. Jaakko Astola and Prof. Tapio Saramäki for the fruitful visits to the Tampere International Center for Signal Processing, Tampere University of Technology, Finland.

The author is indebted to the colleagues who read and reviewed the first drafts of book chapters and gave their suggestions and comments with a sincere wish to contribute to the quality of the book. My appreciations go to Prof. Miroslav Lutovac of the University of Belgrade and to Prof. Gordana Jovanović-Doleček from Institute INAOE, Puebla, Mexico. My special thanks go to my former students: Jelena Ćertić for the inspiring discussions and thoughtful comments; Sanja Damjanović for the assistance in developing some of the examples, and improving the details of presentation; Jovanka Gajica for careful reading of the entire manuscript; to Irena Janković, Marko Nikolić and Milenko Ćirić for reading and commenting some of the chapters; and to Valentina Timčenko for the technical assistance. My appreciations and sincere thanks go to Gordana Marković of Mihajlo Pupin Institute for the editorial work.

Special thanks go to the publishing team at IGI Global, whose contributions throughout the whole process have been invaluable. In particular, to Julia Mosemann whose continuous e-mail communication and an excellent assistance were of great help for keeping the project in schedule, and to Jamie Sue Snavely for an excellent collaboration we have had during the preparation of the final version of this publication.

In closing, I wish to express the gratitude to my family for their support and encouragement during the whole period it took to complete this book.

Ljiljana Milić

Chapter I
Single–Rate Discrete–Time Signals and Systems:
Background Review

INTRODUCTION

This chapter is a concise review of time-domain and transform-domain representations of single-rate discrete-time signals and systems. We consider first the time-domain representation of discrete-time signals and systems. The representation in transform domain comprises the discrete-time Fourier transform (DTFT), the discrete Fourier transform (DFT), and the z-transform. The basic realization structures for FIR and IIR systems are briefly described. Finally, the relations between continuous and discrete signals are given.

DISCRETE-TIME SIGNALS

A signal is a function of at least one independent variable. In this book, we assume that the independent variable is time even in cases where the independent variable is a quantity other than time.

We define a *continuous-time signal*, $x_c(t)$, as a signal that exists at every instant of time t. A continuous-time signal with a continuous amplitude is also called an *analog signal*. The independent variable t is a continuous variable and $x_c(t)$ can assume any value over a continuous range of numbers.

A *discrete-time signal* is a sequence of numbers denoted as $\{x[n]\}$, where n is said to be the *time index*, and $x[n]$ denotes the value of the nth element in the sequence. A discrete-time signal is called a *discrete signal*. The quantity $x[n]$ is also called the *sample value*, and its time index n is called the *sample index*. The quantity $x[n]$ can take any value over some continuous range of numbers, $x_{\min} \leq x[n] \leq x_{\max}$.

Discrete signals can be defined only for integer values of n from an interval $N_1 \leq n \leq N_2$. When the sample values of the sequence $\{x[n]\}$ are represented as binary numbers using a final number of bits, the signal $\{x[n]\}$ is a *digital signal*.

The *length of the sequence* is defined as $N \leq N_2 - N_1 + 1$. The sequence $\{x[n]\}$ is a *finite-length sequence* if N is of a finite length; otherwise, $\{x[n]\}$ is an *infinite-length sequence*.

For the purpose of the analysis, it is useful to represent signals as the combination of basic sequences. The frequently used basic sequences are included in Table 1.1.

In many applications, the discrete-time signal $\{x[n]\}$ is generated by sampling the continuous-time signal $x_c(t)$ at uniform time intervals:

$$x[n] = x_c(t)_{|t=nT} = x_c(nT) \tag{1.1}$$

A time interval T is called a *sampling interval* or a *sampling period*, and the reciprocal value,

$$F_T = \frac{1}{T} \tag{1.2}$$

is a *sampling frequency* or a *sampling rate*. In general, the unit of sampling frequency is cycles per second, and when T is given in seconds [s], F_T can be expressed in hertz [Hz].

Operations on Sequences

Processing a sequence means performing certain operations on the sequence. Generally, the processing algorithm is composed of basic operations such as addition, multiplication and scalar multiplication, time-shifting, down-sampling and up-sampling. Figure 1.1 shows a schematic representation of basic operations.

DISCRETE-TIME SYSTEMS

A *discrete-time system*, or shortly a *discrete system*, is an algorithm or physical device that converts one sequence (called *input*) into another sequence (called *output*). The input-output relation of the system can be expressed mathematically in the form

Table 1.1. Basic sequences

Sequence	Description
Unit-Sample	$\delta[n] = \begin{cases} 1, & n = 0 \\ 0, & n \neq 0 \end{cases}$
Unit- Step	$u[n] = \begin{cases} 1, & n \geq 0 \\ 0, & n < 0 \end{cases}$
Real-Valued Exponential	$x[n] = a^{bn}$
Sinusoidal	$x_s[n] = a\, \sin(2\pi f n + \varphi)$
Complex-Exponential	$x_e[n] = a e^{j(\omega n + \phi)}$

Figure 1.1. Basic operations on sequences. (a) Addition. (b) Modulation. (c) Scalar multiplication. (d) Time-shift (delay). (e) Down-sampling. (f) Up-sampling.

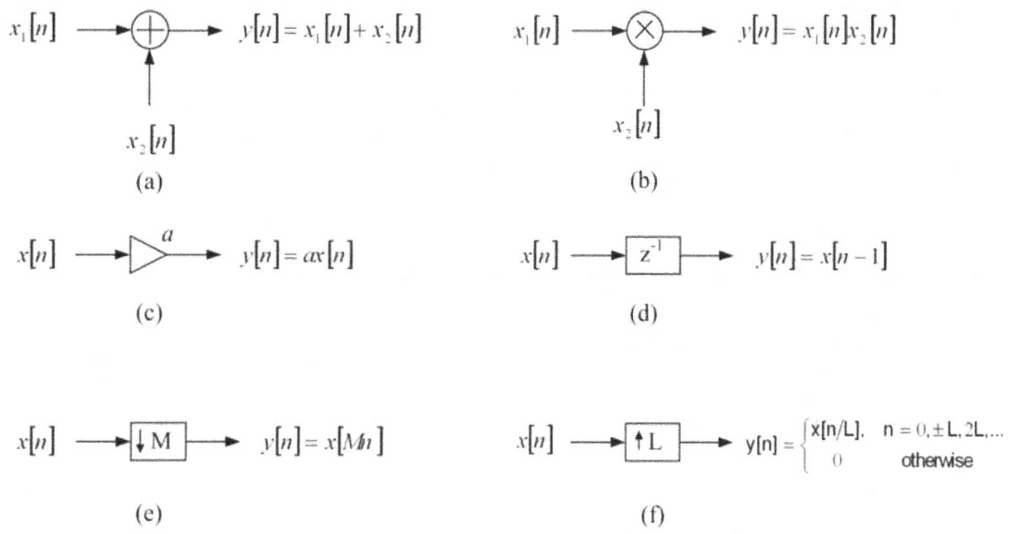

$$\{y[n]\} = \Phi\big(\{x[n]\}\big),\qquad(1.3)$$

where operator Φ represents the rule that is used to produce the output signal $\{y[n]\}$ from the input signal $\{x[n]\}$.

A discrete system is *stable* if any bounded input sequence produces a bounded output sequence. Only stable systems are of practical interest.

A discrete system is *causal* if the output depends only on the present and the past values of the input. If $y[n_0]$ is the output for the time index n_0, then $y[n_0]$ depends only on the input samples $x[n]$ for values $n \le n_0$.

Linear Time-Invariant Systems

Linear time-invariant (LTI) systems are stable systems that are linear and time (shift) invariant. The response of a system to the unit sample sequence $\{\delta[n]\}$ is the *unit-sample response* or *impulse response* and is denoted by $\{h[n]\}$,

$$h[n] = \Phi\big(\{\delta[n]\}\big).\qquad(1.4)$$

An LTI system is completely characterized by $\{h[n]\}$ since the sequence on the output of the system can be expressed as a *convolution* of the input sequence and the impulse response of the system,

$$y[n] = \sum_{k=-\infty}^{\infty} x[k]\,h[n-k], \quad \text{or alternatively } y[n] = \sum_{k=-\infty}^{\infty} h[k]\,x[n-k].\qquad(1.5)$$

3

The above convolution is referred to as a *linear convolution* and can be expressed in the compact form

$$\{y[n]\} = \{x[n]\} * \{h[n]\}.$$ (1.6)

An LTI system is said to be stable if its impulse response satisfies

$$\sum_{n=-\infty}^{\infty} |h[n]| < \infty.$$ (1.7)

An LTI system is *causal* if its impulse response $\{h[n]\}$ is a causal sequence,

$$h[n] = 0, \quad for \quad n < 0.$$ (1.8)

An LTI system is *anticausal* if its impulse response $\{h[n]\}$ is an anticausal sequence,

$$h[n] = 0, \quad for \quad n > 0.$$ (1.9)

The LTI systems are divided into two basic classes:

1. *Finite Impulse Response (FIR)* systems
2. *Infinite Impulse Response (IIR)* systems

For an FIR system, since $\{h[n]\}$ is of a finite length, the input-output relation is expressed as the finite convolution sum. Usually, we work with causal systems, that is

$$y[n] = \sum_{k=0}^{N-1} h[k]x[n-k]$$ (1.10)

where N is the length of the sequence $\{h[n]\}$.

For an IIR system, since $\{h[n]\}$ is of an infinite length, the input-output relation is an infinite convolution sum. Therefore, for a causal IIR system we write

$$y[n] = \sum_{k=0}^{\infty} h[k]x[n-k].$$ (1.11)

From the practical point of view, a class of LTI systems that can be described by a constant-coefficients difference equation is very important. For this class of systems, input-output relation is expressed in the form

$$\sum_{k=0}^{N} a_k y[n-k] = \sum_{k=0}^{M} b_k x[n-k],$$ (1.12)

where $\{x[n]\}$ and $\{y[n]\}$ are input and output of the system, and $\{a_k\}$ and $\{b_k\}$ are constants.

The output of the system defined by (1.12) can be computed recursively. If the system is causal, we can express $y[n]$ in terms of the current sample and M previous samples of the input sequence, and from N previous output samples

$$y[n] = \sum_{k=0}^{M} b_k x[n-k] - \sum_{k=1}^{N} a_k y[n-k].$$

(1.13)

Here, we assume that $a_0 = 1$.

A difference equation (1.13) gives the unique solution if N initial conditions are specified. Those initial conditions might consist of specifying fixed values of $y[n]$ for fixed values of n. This is easily achieved with causal systems since $y[n] = 0$, for $n < n_0$, where n_0 denotes the instant of excitation.

For an FIR system, the difference equation is nonrecursive, and coefficients $\{b_k\}$ are identical with those of the impulse response of the system. Sometimes FIR systems are also called *nonrecursive systems*.

For an IIR system, the difference equation is recursive. Sometimes IIR systems are called *recursive systems*.

DISCRETE-TIME FOURIER TRANSFORM

The *discrete-time Fourier transform* (DTFT) represents the discrete-time sequence in terms of the exponential sequence $\{e^{-j\omega n}\}$ where ω is the real frequency variable. Sometimes the shorter term *Fourier Transform* is used to denote DTFT. The frequency variable ω is called the *angular frequency*, and sometimes *frequency* for short. For the sequence $\{x[n]\}$, the discrete-time Fourier transform $X(e^{j\omega})$ is defined by

$$X\left(e^{j\omega}\right) = \sum_{n=-\infty}^{\infty} x[n] e^{-j\omega n}.$$

(1.14)

The discrete-time Fourier transform $X(e^{j\omega})$ is a *continuous* function of the frequency variable ω. The necessary and sufficient condition for the DTFT $X(e^{j\omega})$ to exist is that the sequence $\{x[n]\}$ is an *absolutely summable* sequence, i.e.

$$\sum_{n=-\infty}^{\infty} |x[n]| < \infty.$$

(1.15)

The complex function $X(e^{j\omega})$ is expresible in the rectangular form

$$X\left(e^{j\omega}\right) = X_R\left(e^{j\omega}\right) + jX_I\left(e^{j\omega}\right),$$

(1.16)

where $X_R(e^{j\omega})$ and $X_I(e^{j\omega})$ are real functions representing real and imaginary parts of $X(e^{j\omega})$, respectively. Alternatively, $X(e^{j\omega})$ can be expressed in the polar form

$$X\left(e^{j\omega}\right) = \left|X\left(e^{j\omega}\right)\right| e^{j\phi(\omega)},$$

(1.17)

where $|X(e^{j\omega})|$ is the *magnitude function* defined by

$$\left|X\left(e^{j\omega}\right)\right| = \sqrt{X_R^2\left(e^{j\omega}\right) + X_I^2\left(e^{j\omega}\right)},$$

(1.18)

and $\varphi(\omega)$ is the *phase function*

$$\phi(\omega) = \arg\left[X\left(e^{j\omega}\right)\right] = \tan^{-1}\frac{X_I\left(e^{j\omega}\right)}{X_R\left(e^{j\omega}\right)}. \tag{1.19}$$

Magnitude and phase functions $|X(e^{j\omega})|$ and $\varphi(\omega)$ are real functions of ω.

The sequence $\{x[n]\}$ can be computed from the transform $X(e^{j\omega})$ by using the *inverse discrete-time Fourier transform* (IDTFT) defined by

$$x[n] = \frac{1}{2\pi}\int_{-\pi}^{\pi} X\left(e^{j\omega}\right)e^{j\omega n}\,d\omega. \tag{1.20}$$

The discrete-time Fourier transform and the inverse discrete-time Fourier transform defined by (1.14) and (1.20), respectively constitute the *discrete-time Fourier transform pair*.

Since the DTFT $X(e^{j\omega})$ is a periodic function in ω with the period of 2π, i.e.,

$$X\left(e^{j(\omega+2k\pi)}\right) = X\left(e^{j\omega}\right), \tag{1.21}$$

the DTFT is completely represented in the range of 2π.

The majority of sequences used in practice are real sequences. When $\{x[n]\}$ is a real sequence, the DTFT $X(e^{j\omega})$ exhibits the conjugate symmetry property

$$X\left(e^{-j\omega}\right) = X^*\left(e^{j\omega}\right), \tag{1.22}$$

where "*" is used to denote the complex conjugate function. Since the DTFT is periodic with the period of 2π, the range of π is sufficient to represent the DTFT of a real sequence.

For signal processing applications, some properties of the discrete-time Fourier transform are very practical. The general properties of DTFT that will be used in this book are listed in Table 1.2.

Table 1.2. Some important properties of the discrete-time Fourier transform

Property	Sequence	DTFT
	$\{x[n]\}$	$X(e^{j\omega})$
	$\{h[n]\}$	$H(e^{j\omega})$
Linearity	$a\{x[n]\} + b\{h[n]\}$	$a\,X(e^{j\omega}) + b\,H(e^{j\omega})$
Time-shifting	$\{x[n-n_0]\}$	$e^{-j\omega n_0}X\left(e^{j\omega}\right)$
Frequency-shifting	$\{e^{j\omega_0 n}x[n]\}$	$X\left(e^{j(\omega-\omega_0)}\right)$
Convolution	$\{x[n]\}*\{h[n]\}$	$X\left(e^{j\omega}\right)H\left(e^{j\omega}\right)$
Modulation	$\{x[n]\}\{h[n]\}$	$\dfrac{1}{2\pi}\int_{-\pi}^{\pi}X\left(e^{j\theta}\right)H\left(e^{j(\omega-\theta)}\right)d\theta$

Spectrum of Discrete-Time Signal

When the sequence $\{x[n]\}$ represents the discrete-time signal, its discrete-time Fourier transform $X(e^{j\omega})$ defined in (1.14) represents the *spectrum* of the signal. From the polar representation of DTFT given in (1.17), we define the magnitude spectrum and the phase spectrum of the signal $\{x[n]\}$.

The *magnitude spectrum* is the magnitude function $|X(e^{j\omega})|$ defined by (1.18), and the *phase spectrum* is the phase function $\varphi(\omega)$ defined by (1.19). For real signals, the magnitude function is an even function of ω, and the phase spectrum is an odd function of ω.

In MATLAB, the function freqz can be used to compute the spectrum of the signal. The following example illustrates the application.

```
% Program demo_1_1
% Computation of Discrete-Time Fourier Transform
% signal {x[n]}
clear all, close all
x = [0.3,0.2,0.1,0.15,0.18,0.20,0.5,0.6,0.4,0.3,0.2,0.1,0.15]; % Test signal
L = length(x);
N = 256;
[X,w] = freqz(x,1,N);          % Computation of the signal spectrum
mag = abs(X);                   % Magnitude spectrum
phase = angle(X);               % Phase spectrum
figure(1)
subplot(3,1,1), stem(0:L-1,x),
legend('Signal {x[n]}'), xlabel('Time index n'), ylabel('x[n]')
subplot(3,1,2), plot(w/pi,mag),
legend('Magnitude Spectrum'), xlabel('Normalized frequency \omega/\pi'), ylabel('|X(e^{j\omega})|')
subplot(3,1,3), plot(w/pi,unwrap(phase)),
legend('Phase Spectrum'), xlabel('Normalized frequency \omega/\pi'), ylabel('\phi(\omega)')
```

Figure 1.2 plots signal $\{x[n]\}$, magnitude spectrum, and phase spectrum of the signal.

Frequency Response of Discrete-Time LTI System

When the impulse response of the discrete-time LTI system $\{h[n]\}$ satisfies the stability condition (1.7), the discrete-time Fourier transform $H(e^{j\omega})$

$$H\left(e^{j\omega}\right) = \sum_{n=-\infty}^{\infty} h[n] e^{-j\omega n} \qquad (1.23)$$

represents the *frequency response* of the discrete-time system. The frequency response describes the stable LTI system in frequency domain.

From the polar representation of DTFT we define the magnitude response and the phase response of the system,

$$H\left(e^{j\omega}\right) = \left|H\left(e^{j\omega}\right)\right| e^{j\phi(\omega)}, \qquad (1.24)$$

Figure 1.2 Signal $\{x[n]\}$, *magnitude spectrum* $|X(e^{j\omega})|$, *phase spectrum* $\arg\{X(e^{j\omega})\}$

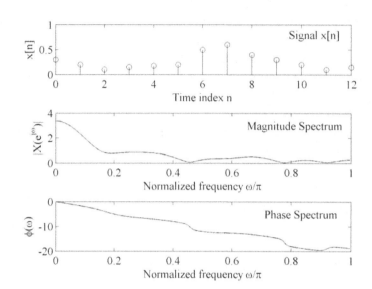

where $|H(e^{j\omega})|$ is the *magnitude response*, and $\phi(\omega)$ is the *phase response*. For the systems with real coefficients, the magnitude response is an even function of ω, and the phase response is an odd function of ω.

The quantity that expresses the magnitude response in decibels is called the *gain function*

$$g(\omega) = 20\log_{10}\left(\left|H\left(e^{j\omega}\right)\right|\right) \tag{1.25}$$

The *attenuation function* is a negative of the gain function,

$$a(\omega) = -g(\omega) = -20\log_{10}\left(\left|H\left(e^{j\omega}\right)\right|\right). \tag{1.26}$$

The *group delay function* of a discrete-time system is defined by the expression

$$\tau_g(\omega) = -\frac{d\phi(\omega)}{d\omega}, \tag{1.27}$$

and the *phase delay* is given by

$$\tau_f(\omega) = -\frac{\phi(\omega)}{\omega}. \tag{1.28}$$

DISCRETE FOURIER TRANSFORM

The *Discrete Fourier Transform* (DFT) is defined for finite-length sequences. For a given L-length sequence $\{x[n]\}$, $0 \leq n < L-1$, the discrete Fourier transform is the sequence obtained by uniformly sampling the discrete-time Fourier transform $X(e^{j\omega})$ on the ω-axis in the range $0 \leq \omega < 2\pi$. Hence, the *frequency samples* $X[k]$ are the values of $X(e^{j\omega})$ at the points $\omega_k = 2\pi k/N$, $0 \le k \le N-1$. Using the definition of $X(e^{j\omega})$ as given in (1.14), the sequence $\{X[k]\}$ is given by

$$X[k] = X\left(e^{j\omega}\right)\Big|_{\omega = 2\pi k/N} = \sum_{n=0}^{L-1} x[n]e^{-j2\pi kn/N}, \quad 0 \le k \le N-1. \tag{1.29}$$

where $N \ge L$, and $\{x[n]\}$ being the sequence of a finite length is zero valued outside the interval $0 \le n < L-1$. The sequence $\{X[k]\}$ is called the discrete Fourier transform (DFT) of the finite length sequence $\{x[n]\}$. Sometimes $\{X[k]\}$ is called the *DFT sequence*, and the sample $X[k]$ is called *DFT coefficient*. Note that the sequence $\{X[k]\}$ is a complex sequence. Introducing the commonly used notation

$$W_N = e^{-j2\pi/N}, \tag{1.30}$$

the discrete Fourier transform (1.29) is presented in the form

$$X[k] = \sum_{n=0}^{N-1} x[n]W_N^{nk}, \quad 0 \le k \le N-1. \tag{1.31}$$

The *inverse discrete Fourier transform* (IDFT) is given by

$$x[n] = \frac{1}{N}\sum_{n=0}^{N-1} X[k]W_N^{-nk}, \quad 0 \le n \le N-1. \tag{1.32}$$

Using DFT, the finite-length sequence $\{x[n]\}$ is described in frequency domain by the finite-length sequence $\{X[k]\}$. In many applications, this representation is more practical than DTFT $X(e^{j\omega})$, which is a continuous function of ω.

High popularity of DFT is due to the efficient algorithms known under the name *Fast Fourier Transforms* (FFT). The description of FFT algorithms can be found in many signal processing books; see for example (Mitra, 2006; Oppenheim and Schafer, 1989; Proakis and Manolakis, 1996).

The MATLAB function fft computes DFT for the finite-length sequences. Next, we illustrate the computation of DFT coefficients on the example of the discrete signal composed of three sinusoidal components.

```
% Program demo_1_2
% Computation of Discrete Fourier Transform (DFT)
n = 0:63;           % Time index n
x=sin(2*pi*4*n/64)+0.6*sin(2*pi*8*n/64) + 0.8*sin(2*pi*18.5*n/64);
figure (1)
subplot(2,1,1), stem(n,x), xlabel('Time Index n'), ylabel('x[n]'), axis([0,63,-3,3])
X = fft(x);         % Computation of DFT
k = n;              % Frequency index k
subplot(2,1,2), stem(k,abs(X)), xlabel('Frequency index k'), ylabel('|(X[k])|'), axis([0,63,0,40])
```

Figure 1.3 plots the signal $\{x[n]\}$ and the absolute values of the DFT coefficients, which represent the spectral components in the magnitude spectrum of the given finite-length signal $\{x[n]\}$.

THE *z*-TRANSFORM

The *z*-transform of a sequence $\{x[n]\}$ is defined as the power series

$$X(z) = \sum_{n=-\infty}^{\infty} x[n] z^{-n}, \tag{1.33}$$

where z is a complex variable.

Using the *z*-transform, we represent the time-domain signal $\{x[n]\}$ in the complex plane as a function of the complex variable z. Generally, the *z*-transform is the sum of an infinite power series and therefore it exists only for those values of z for which this series converges. The *region of convergence* (ROC) of $X(z)$ is a set of values of z for which $X(z)$ has a finite value. The region of convergence for $X(z)$ includes the regions of the complex z plane where

$$|X(z)| < \infty. \tag{1.34}$$

The *zeros* of a z-transform $X(z)$ are the values of z for which $X(z) = 0$. The *poles* of a z-transform $X(z)$ are the values of z for which $X(z) = \infty$. The region of convergence cannot contain poles.

The discrete-time Fourier transform of the sequence $\{x[n]\}$ can be considered as the z-transform of the sequence evaluated on the unit circle of the z-plane. The discrete-time Fourier transform is developed by evaluating the z-transform on the unit circle under the condition that the unit circle belongs to the region of convergence,

$$X(z)\Big|_{z=e^{j\omega}} = X(e^{j\omega}) = \sum_{n=-\infty}^{\infty} x[n] e^{-j\omega n}. \tag{1.35}$$

Figure 1.3. Discrete signal $\{x[n]\}$ composed of three sinusoids – upper subfigure, and magnitudes of $\{|X[k]|\}$ – bottom subfigure

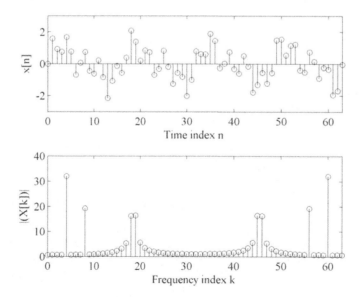

There are some important properties of the z-transform that make the z-transform representation of sequences very practical. The frequently used properties are listed in Table 1.3.

Rational z-Transforms

The most important family of z-transforms is that for which $X(z)$ can be expressed as the ratio of two polynomials in z^{-1} (or z). They are called the *rational z-transforms*. When expressed as a ratio of two polynomials in z^{-1}, the rational z-transform has the form

$$X(z) = \frac{B(z)}{A(z)} = \frac{\sum_{k=0}^{M} b_k z^{-k}}{\sum_{k=0}^{N} a_k z^{-k}}. \tag{1.36}$$

The above equation can also be represented in the product form

$$X(z) = \frac{b_0}{a_0} \cdot \frac{\prod_{k=1}^{M}\left(1 - q_k z^{-1}\right)}{\prod_{k=1}^{N}\left(1 - p_k z^{-1}\right)}. \tag{1.37}$$

Here, the roots of the numerator, $z = q_k$, are the zeros of $X(z)$, and the roots of the denominator, $z = p_k$, are the poles of $X(z)$.

The Inverse z-Transform

The sequence $\{x[n]\}$ can be derived using the inverse of $X(z)$. The *inverse z-transform* is defined by the expression

$$x[n] = \frac{1}{2\pi j} \oint_C X(z) z^{n-1} dz. \tag{1.38}$$

Table 1.3. Frequently used properties of z-transform

Property	Sequence	z-transform
	$\{x[n]\}$	$X(z)$
	$\{h[n]\}$	$H(z)$
Linearity	$a\{x[n]\} + b\{h[n]\}$	$aX(z) + bH(z)$
Time-shifting	$\{x[n-n_0]\}$	$z^{-n_0}X(z)$
Multiplication by z_0^n	$\{z_0^n x[n]\}$	$X(z/z_0)$
Convolution	$\{x[n]\} * \{h[n]\}$	$X(z)H(z)$
Modulation	$\{x[n]\}\{h[n]\}$	$\frac{1}{2\pi} \oint_C X(\vartheta)H(z/\vartheta)z^{-1}d\vartheta$

The contour of integration C is the counterclockwise contour in the region of convergence encircling the point $z = 0$.

z-Transform Representation of Discrete-Time Systems

When the sequence $\{h[n]\}$ is the impulse response of an LTI system, the z-transform,

$$H(z) = \sum_{n=-\infty}^{\infty} h[n] z^{-n} \qquad (1.39)$$

represents the *transfer function* of the LTI system. Zeros and poles of the LTI system are the zeros and poles of the transfer function $H(z)$.

An LTI system is stable when the region of convergence of the transfer function $H(z)$ includes the unit circle. The frequency response $H(e^{j\omega})$ of a stable system can be obtained by evaluating $H(z)$ on the unit circle,

$$H(e^{j\omega}) = H(z)\big|_{z=e^{j\omega}} = \sum_{n=-\infty}^{\infty} h[n] e^{-j\omega n}. \qquad (1.40)$$

The transfer function of an LTI system described by a constant-coefficient difference equation (1.13) is a rational z-transform,

$$H(z) = \frac{\sum_{k=0}^{M} b_k z^{-k}}{1 + \sum_{k=1}^{N} a_k z^{-k}}. \qquad (1.41)$$

The coefficients of $H(z)$ are those of the difference equation. The *order of the system* is defined by $\max(M, N)$.

Equation (1.41) is a general form of the rational transfer function. Being developed from the recursive difference equation (1.13) it represents the transfer function of an IIR system. In the case of an FIR system, coefficients $a_k = 0$, for $k = 1, 2, \ldots, N$, the expression (1.41) reduces to

$$H(z) = \sum_{k=0}^{M} b_k z^{-k}. \qquad (1.42)$$

Here M denotes the system order. All M poles of an FIR system are located at the origin, and therefore the FIR system is absolutely stable.

The positions of the transfer function zeros are not restricted by the stability conditions, i.e., the zeros can be placed inside or outside the unit circle, or can be placed around the unit circle. A system which includes only zeros located inside the unit circle and those located around the unit circle is called the *minimum-phase* system. At the contrary, the system which includes zeros located outside the unit circle and those located around the unit circle is called the *maximum-phase system*.

For representing the poles and zeros of the rational z-transform in the z-plane, we use the MATLAB function zplane:

zplane(B,A);

When B and A are the row vectors, the function zplane understands the numerator and denominator of the transfer function. If we write

zplane(Z,P);

where Z and P are the column vectors, the function zplane understands the zeros and poles of the transfer function, respectively.

The frequency response of the system is computed using the function freqz .

[H,f] = freqz(B,A,N,FT);

The function freqz returns the frequency response in vector H for the set of frequencies stored in vector f. Row vectors B and A contain the coefficients of the numerator and denominator of the transfer function, N is an integer that specifies the length of the vectors H and f, and FT is the sampling frequency.

The group delay of the system is computed using the function grpdelay.

[Gd,f] = grpdelay(B,A,250,1);

The function grpdelay returns the group delay in vector Gd for the set of frequencies stored in vector f. Row vectors B and A contain the coefficients of the numerator and denominator of the transfer function, N is an integer that specifies the length of the vectors Gd and f, and FT is the sampling frequency.

Next, we demonstrate the analysis of an LTI system with MATLAB using an example of the 5th order Chebyshev filter. The MATLAB program demo 1_3 computes the coefficients of the 5th order Chebyshev filter, computes and plots the frequency response and provides the pole-zero plot of the filter.

```
% Program demo_1_3
% LTI system, Chebyshev filter
% Computations of frequency response and pole-zero plot
clear all, close all
[B,A] = cheby1(5,1,0.4)    % Chebyshev filter design
[H,f] = freqz(B,A,250,2); Mag=abs(H);  % Frequency response and magnitude response
Phase = unwrap(angle(H)); [Gd,f] = grpdelay(B,A,250,2); % Phase response and group delay
figure (1)
subplot(2,2,1), zplane(B,A), subplot(2,2,2), plot(f,Mag), axis([0,1,0,1.1])
xlabel('Normalized frequency \omega/\pi'), ylabel('Magnitude')
subplot(2,2,3), plot(f,Phase), axis([0,1,-8,0])
xlabel('Normalized frequency \omega/\pi'), ylabel('Phase, rad')
subplot(2,2,4), plot(f,Gd), axis([0,1,0,15])
xlabel('Normalized frequency \omega/\pi'), ylabel('Group delay, samples')
```

Figure 1.4 displays the results. Notice that the system has all the zeros on the unit circle, at the point $z = -1$. Since there are no zeros outside the unit circle, the Chebyshev filter is a minimum phase-system.

Linear-Phase Systems

Discrete-time systems with finite impulse response can easily achieve the linear phase characteristic. This attractive property of FIR systems is particularly important in signal processing applications when the waveform of the signal has to be preserved.

Let us consider the frequency response of a noncausal LTI system whose impulse response $\{h_0[n]\}$ of a length N is defined for the time-index n in the range $\{-K, K\}$, where $K = \lfloor N/2 \rfloor$. The transfer function of the system $H_0(e^{j\omega})$ is given by,

$$H_0\left(e^{j\omega}\right) = \sum_{n=-K}^{K} h_0[n]e^{-j\omega n}.$$ (1.43)

It is easy to show that $H_0(e^{j\omega})$ is a real function of ω when the coefficients of the impulse response $\{h_0[n]\}$ are symmetric, i.e.,

$$h_0[n] = h_0[-n], \quad -K < n < K.$$ (1.44)

By substituting symmetry condition (1.44) in equation (1.43), one obtains

$$H_0\left(e^{j\omega}\right) = h_0[0] + \sum_{n=1}^{K} h_0[n]\left(e^{j\omega n} + e^{-j\omega n}\right),$$ (1.45)

and finally,

Figure 1.4. (a) Pole-zero plot: Circles are used for the zeros, while the crosses represent the poles. The filter has 5 multiple zeros at the point $z = -1$, and 5 poles inside the unit circle. (b) Magnitude response. (c) Phase response. (d) Group delay.

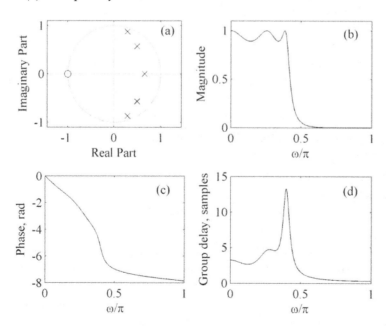

$$H_0(\omega) = h[0] + 2\sum_{n=1}^{K} h[n]\cos(\omega n). \tag{1.46}$$

Equation (1.46) shows that the frequency response for a symmetric noncausal sequence $\{h_0[n]\}$ is a real function of ω.

From the noncausal system, we obtain a causal system by simply shifting the sequence $\{h_0[n]\}$ for K samples to the right,

$$h[n] = h_0[n - K], \tag{1.47}$$

where $h[n]$ denotes the nth sample in the impulse response of the causal system. This time-shift for K samples corresponds in the frequency domain to the multiplication of the Fourier transform $H_0(e^{j\omega})$ by the exponential sequence $e^{-jK\omega}$, see Table 1.2. Thereby, the frequency response of the causal system $H(e^{j\omega})$ is given by

$$H(e^{j\omega}) = H_0(\omega)e^{-jK\omega}. \tag{1.48}$$

Since $H_0(\omega)$ is a real function, the frequency response $H(e^{j\omega})$ has a linear phase characteristic for all values of the angular frequency ω. The real *amplitude function* $H_0(\omega)$ is sometimes called the *zero-phase frequency response*.

For a causal sequence of a length N, the coefficient symmetry condition (1.47) is usually written in the form,

$$h[n] = h[N - n - 1], \qquad 0 \le n \le N - 1. \tag{1.49}$$

The linear-phase FIR system is also obtained when the impulse response satisfies the antisymmetry condition,

$$h[n] = -h[N - n - 1], \qquad 0 \le n \le N - 1. \tag{1.50}$$

In that case the phase characteristic is linear with the phase shift of $\pi/2$, i.e.,

$$H(e^{j\omega}) = H_0(\omega)e^{j(\pi/2 - K\omega)}. \tag{1.51}$$

The FIR systems satisfying symmetry condition (1.49) are used to construct the linear-phase digital filters, whereas the FIR systems with antisymmetric impulse response (1.50) are used to construct the linear-phase FIR Hilbert transformers and FIR linear-phase differentiators.

All-Pass Transfer Functions

All-pass transfer functions, called also the *all-pass filters* have a constant magnitude response for all frequencies, and a nonlinear phase characteristic. They are frequently used as building blocks to construct efficient IIR systems.

The transfer function of an Nth order all-pass filter, when expressed in the product form, is given by

$$H(z) = \prod_{k=1}^{N} \frac{z^{-1} - a_k^*}{1 - a_k z^{-1}}, \tag{1.52}$$

where a_k, $k = 1, 2, ..., N$, is a complex number representing the pole of $H(z)$. It is evident from equation (1.52) that the transfer function zeros are placed at the points $1/a_k^*$, $k = 1, 2, ..., N$, i.e. the poles and zeros are reciprocal to each other. To satisfy the stability condition, the poles should be places inside the unit circle, and consequently the module of a_k is restricted with $|a_k| < 1$. Automatically, the transfer function zeros being reciprocal to the poles should be placed outside the unit circle.

STRUCTURES FOR DISCRETE-TIME SYSTEMS

An LTI discrete system satisfying a constant-coefficient difference equation can be represented by the block diagram that interconnects basic devices: adders, scalar multipliers and delays (shifts). The block diagram, called *system structure*, describes how the arithmetic operations are performed through the system. For a given transfer function, one can develop a number of different structures, which for the given excitation produce identical outputs. Advantages and disadvantages of a particular structure depend on application, since in practice, the computations are performed in finite word-length arithmetic.

In this section, we briefly review the elementary structures for FIR and IIR systems.

Basic Implementation Structures for FIR Systems

The transfer function of an FIR filter as shown in (1.42) is the polynomial in z^{-1}, and is usually written in the form

$$H(z) = \frac{Y(z)}{X(z)} = \sum_{k=0}^{N-1} h[k] z^{-k}. \tag{1.53}$$

Here $h[0], h[1], ..., h[N-1]$ are the coefficients of the system impulse response, and $N-1$ is the filter order. The total number of coefficients, N, is usually called the *filter length*.

In time domain, an FIR system is characterized by the nonrecursive difference equation,

$$y[n] = \sum_{k=0}^{N-1} h[k] x[n-k], \tag{1.54}$$

where $x[n]$ and $y[n]$ denote samples of the input and the output sequences, respectively.

The direct realization structure depicted in Figure 1.5 is the block diagram description of difference equation (1.54). The transpose of the structure from Figure 1.5 is shown in Figure 1.6. Both structures are canonic in the respect of delays.

The number of multiplication constants in the direct realization forms of Figures 1.5 and 1.6 can be halved when implementing a linear-phase FIR filter. Figure 1.7 depicts the efficient direct realization structure, which exploits the coefficient symmetry in the impulse response of a linear-phase FIR system as given in equation (1.49). Figure 1.7 is depicted for N even where the total number of multiplication constants is $N/2$. The similar structure can be developed for N odd. In that case the number of multiplication constants reduces to $(N+1)/2$ as illustrated in Figure 1.8 on the example of the direct transpose implementation structure.

Basic Implementation Structures for IIR Systems

An infinite impulse response LTI system is characterized by the rational transfer function (1.41), or equivalently, by a linear constant-coefficient difference equation,

$$y[n] = \sum_{k=0}^{M} b_k x[n-k] - \sum_{k=1}^{N} a_k y[n-k]. \tag{1.55}$$

Here the first sum represents the nonrecursive part of the system, and the second sum represents the recursive part. Those two parts can be implemented separately and connected together. The cascade connection of the nonrecursive and recursive sections results in the realization structure called *direct form I* depicted in Figure 1.9(a), which is developed for the case $N = M$. Figure 1.9(b) shows the *direct form I_t*, which is the transpose of the direct form I. Note that the direct form I and the direct form I_t are noncanonic in respect of delays.

From the direct form I, the canonic form is obtained by simply interchanging the order of the recursive and nonrecursive sections. In the next step, it becomes obvious that the pairs of the opposite delays in the cascaded recursive and nonrecursive section store identical data thus permitting replacing each pair with a single delay. In this way, the canonic direct structure of Figure 1.10(a) is obtained. Structure depicted in Figure 1.10(a) shows the direct canonic structure called *direct form II*. The transposed direct structure shown in Figure 1.10(b) is usually called as the *direct form II_t*.

Figure 1.5. Direct implementation of FIR system

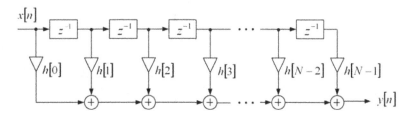

Figure 1.6. Direct transpose implementation of FIR system

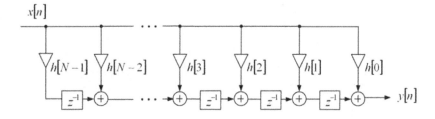

Figure 1.7. Direct implementation of a linear-phase FIR system with the reduced number of multipliers, N is even

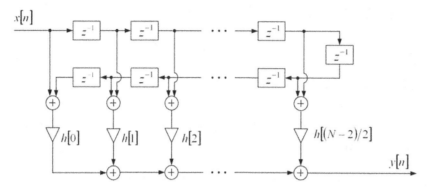

Figure 1.8. Direct transpose implementations of a linear-phase FIR system with the reduced number of multipliers, N is odd

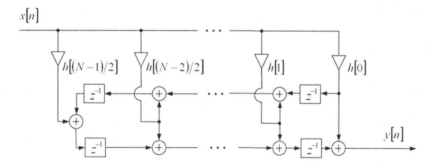

Figure 1.9. Direct Form I and Direct Form I$_t$

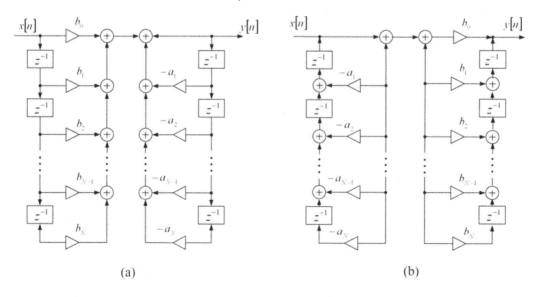

(a) (b)

Figure 1.10. Direct form II and direct form II$_t$

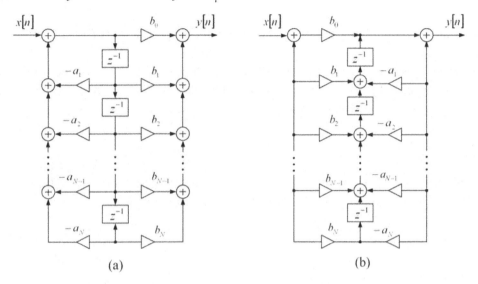

(a)   (b)

In MATLAB, the function **filter** is used to compute an output sequence $\{y[n]\}$ from an input sequence $\{x[n]\}$. The function **filter** implements the direct form II$_t$ shown in Figure 1.10(b). The following example of an IIR system illustrates the application.

Nx = 51; b = [0.5,0.7, 0.6,0.4]; a = [1,0.4,-0.3, 0.2]; n = (0:Nx-1); x = sin(2*pi*0.125*n); y = filter(b,a,x);

The following example illustrates the application of the function **filter** in the case of an FIR system.

Nx = 51; b = [0.3,0.5,0.6,0.7,0.6,0.5,0.3]; a = 1; n = (0:Nx-1); x = sin(2*pi*0.125*n); y = filter(b,a,x);

Specifying the input and output lists as

[y,zf] = filter(b,a,x),

program returns the output sequence in the vector **y**, and the vector **zf** contains the final conditions of the filter delays.

With the specifications,

[y,zf] = filter(b,a,X,zi);

program accepts initial conditions, **zi**, and returns the final conditions, **zf**, of the system delays.

SAMPLING THE CONTINUOUS-TIME SIGNAL

When the continuous-time signal $x_c(t)$ is uniformly sampled at every T seconds, the resulting sequence $\{x[n]\}$ is given by

$$x[n] = x_c(nT) \quad -\infty < n < \infty.$$

(1.56)

where T is a sampling interval or a sampling period, and its reciprocal value $F_T = 1/T$ is the sampling frequency (1.1 – 1.2).

When the continuous-time signal $x_c(t)$ is band-limited, it can be uniquely recovered from its samples if the sampling frequency is properly chosen. In frequency domain, the continuous-time signal is represented by the *continuous-time Fourier transform* (*CTFT*) given by

$$X_c(j\Omega) = \int_{-\infty}^{\infty} x_c(t) e^{-j\Omega t} dt,$$

(1.57)

where $\Omega = 2\pi F$ is frequency in radians per second. The signal $x_c(t)$ is said to be band-limited when the Fourier transform $X_c(j\Omega)$ is zero outside the prescribed frequency range

$$X_c(j\Omega) = 0, \quad \text{for} \quad |\Omega| > \Omega_N.$$

(1.58)

Here, Ω_N is the highest frequency in $X_c(j\Omega)$.

The conditions for recovering the continuous-time signal $x_c(t)$ from its samples are defined by the well-known *sampling theorem:*

If a continuous-time signal $x_c(t)$ has a band-limited Fourier transform $X_c(j\Omega)$, that is $|X_c(j\Omega)| = 0$ for $|\Omega| \geq \Omega_N = 2\pi F_N$, then $x_c(t)$ can be uniquely reconstructed without error from equally spaced samples $x_c(nT)$, $-\infty < n < +\infty$, if $F_T \geq 2F_N$, where $F_T = 1/T$ is the sampling frequency.

According to the sampling theorem, the sampling frequency $\Omega_T = 2\pi F_T$ satisfying

$$\Omega_T - \Omega_N > \Omega_N, \quad \text{or} \quad \Omega_T > 2\Omega_N,$$

(1.59)

provides that the original continuous-time signal can be reconstructed from the discrete-time signal. The frequency Ω_N is referred to as the *Nyquist frequency*; and the frequency $2\Omega_N$ is called the *Nyquist rate.*

The sampling operation is called *oversampling* if the sampling frequency is higher than the Nyquist rate, $\Omega_T > 2\Omega_N$. The term *undersampling* is used when the sampling frequency is lower than the Nyquist rate, $\Omega_T < 2\Omega_N$. Finally, the signal is *critically sampled* when the sampling frequency is exactly equal to the Nyquist rate, $\Omega_T = 2\Omega_N$.

The spectrum of the discrete-time signal $X(e^{j\omega})$ is expressible in terms of the spectrum of the continuous-time signal $X_c(j\Omega)$ by the well-known relation,

$$X(e^{j\omega}) = \frac{1}{T} \sum_{k=-\infty}^{\infty} X_c\left(j\frac{\omega}{T} - j\frac{2\pi k}{T} \right).$$

(1.60)

The spectrum of the discrete-time signal $X(e^{j\omega})$ is an infinite sum of the shifted and scaled replicas of the spectrum of the continuous-time signal $X_c(j\Omega)$. Here, the angular frequency ω in the spectrum of the discrete signal is related to the frequency Ω in the spectrum of the continuous signal by

$$\omega = \Omega T.$$

(1.61)

Equation (1.60) shows that when the sampling is performed in a sufficiently high rate, the spectrum of the discrete signal appears as a periodic repetition of the original spectrum. The original signal can be recovered by selecting with an ideal lowpass filter the baseband spectrum from the periodic spectral function $X(e^{j\omega})$. On the contrary, the undersampling causes *aliasing* in the spectrum $X(e^{j\omega})$ thus making the signal recovery impossible.

Ideally, the reconstructed signal $x_r(t)$, can be expressed in terms of the sample values $\{x[n]\}$ and the impulse response of the ideal reconstruction filter $h_r(t)$,

$$x_r(t) = \sum_{n=-\infty}^{\infty} x[n] h_r(t-nT). \tag{1.62}$$

The impulse response of the reconstruction filter $h_r(t)$ is the inverse Fourier transform of its frequency response. For an ideal low-pass filter with the cutoff frequency Ω_c, the impulse response $h_r(t)$ is given by,

$$h_r(t) = \frac{\sin(\Omega_c t)}{\Omega_c t}. \tag{1.63}$$

Usually, the cutoff frequency Ω_c is chosen as a half of the sampling frequency,

$$\Omega_c = \frac{\Omega_T}{2} = \pi F_T = \frac{\pi}{T} \tag{1.64}$$

thus giving the reconstruction formula,

$$x_r(t) = \sum_{n=-\infty}^{\infty} x[n] \frac{\sin(\pi(t-nT)/T)}{\pi(t-nT)/T}. \tag{1.65}$$

Therefore, the continuous signal $x_r(t)$ is obtained by interpolation, which is expressed in (1.65) as an infinite sum of the shifted and scaled versions of $h_r(t)$. With the ideal sampling and reconstruction of the bandlimited signal $x_c(t)$, the reconstructed signal $x_r(t)$ is equal to the original continuous signal $x_c(t)$. Unfortunately, an ideal filter is unrealizable and the reconstruction process should be implemented with some realizable approximation of $h_r(t)$.

It is of interest for the later use to review the relations between the frequencies in the spectra of the continuous-time and the discrete-time signals.

1. Continuous signals
    - Symbol F denotes frequency variable in Hz.
    - Symbol F_T denotes the sampling frequency in Hz, $F_T = 1/T$, T is the sampling interval (sampling period) in seconds.
    - Symbol Ω denotes frequency variable in radians per second, $\Omega = 2\pi F$.
    - Symbol Ω_T is used for the sampling frequency in radians per second, $\Omega_T = 2\pi F_T = 2\pi/T$.
2. Discrete signals
    - Symbol f denotes the normalized frequency in terms of the half of the sampling frequency (as in MATLAB).

- Symbol ω denotes angular frequency in radians $\omega = f\pi$, and is sometimes expressed as radians per sample.
3. Corresponding relations: discrete-to-continuous
   - Normalized frequency variable: $f = F/(F_T/2)$.
   - Angular frequency: $\omega = 2\pi F/F_T$, or $\omega = \Omega T$.
   - Sampling frequency is located at $\omega = 2\pi$ due the equalities: $2\pi F_T T = \Omega_T T = 2\pi$.

REFERENCES

Bellanger, M. (2000). *Digital processing of signals: Theory and practice.* 3rd edition. New York, NY: John Wiley.

Burrus, C.S., McClellan, J.H., Oppenheim, A.V, Parks, T.W., Schaffer, R.W., & Schussler, H.W. (1994). *Computer-based exercises for signal processing using MATLAB._Englewood Cliffs, NJ: Prentice-Hall.

Diniz, P., Netto, S., & Da Silva, E. (2002). Digital Signal Processing: System Analysis and Design. New York, NY: Cambridge University Press.

Kuc, R. (1988). *Introduction to Digital Signal Processing.* New York, NY: McGraw-Hill Book Company.

Mitra, S. K. (1999). *Digital signal processing laboratory using MATLAB.* New York, NY: The McGraw-Hill Companies, Inc.

Mitra, S. K. (2006). *Digital signal processing: A computer based approach.* 3rd edition. New York, NY: The McGraw-Hill Companies, Inc.

Oppenheim, A. V., & Schafer, R. W. (1989). *Discrete-time signal processing.* London: Prentice-Hall International.

Proakis J. G., & Manolakis D.G. (1996). *Digital signal processing: Principles, algorithms, and applications.* London: Prentice Hall.

Signal processing toolbox for use with MATLAB. User's guide. Version 6. (2006). Natick: Math-Works.

Stearns, S.D. (2002). *Digital signal processing with examples in MATLAB.* Boca Raton, Florida: CRC Press.

Chapter II
Basics of Multirate Systems

INTRODUCTION

Linear time-invariant systems operate at a single sampling rate i.e. the sampling rate is the same at the input and at the output of the system, and at all the nodes inside the system. Thus, in an LTI system, the sampling rate doesn't change in different stages of the system. Systems that use different sampling rates at different stages are called the *multirate systems*. The multirate techniques are used to convert the given sampling rate to the desired sampling rate, and to provide different sampling rates through the system without destroying the signal components of interest.

In this chapter, we consider the sampling rate alterations when changing the sampling rate by an integer factor. We describe the basic sampling rate alteration operations, and the effects of those operations on the spectrum of the signal.

TIME-DOMAIN REPRESENTATION OF DOWN-SAMPLING AND UP-SAMPLING

Converting the sampling rate means that one discrete signal is converted into another discrete signal with a different sampling rate. Two discrete signals with different sampling rates can be used to convey the same information. For example, a bandlimited continuous signal $x_c(t)$ might be represented by two different discrete signals $\{x[n]\}$ and $\{y[n]\}$ obtained by the uniform sampling of the original signal $x_c(t)$ with two different sampling frequencies F_T and $F_T{}'$

$$x[n]= x_C\left(nT\right) \text{ and } y[n]= x_C\left(nT'\right) \tag{2.1}$$

where $T= 1/F_T$ and $T'=1/F_T{}'$ are the corresponding sampling intervals. When the sampling frequencies F_T and $F_T{}'$ are chosen in such a way that each of them exceeds at least two times the highest frequency in the spectrum of $x_c(t)$, the original signal $x_c(t)$ can be reconstructed from either $\{x[n]\}$ or $\{y[n]\}$. Hence, the two signals operating at two different sampling rates are carrying the same information. By using the discrete-time operations, signal $\{x[n]\}$ can be converted to $\{y[n]\}$, or vice versa, with minimal signal distortions.

The basic operations in sampling rate alteration process are the sampling rate decrease and the sampling rate increase. Employing two operators can perform the sampling rate alteration: a *down-sampler* for the sampling rate decrease, and an *up-sampler* for the sampling rate increase. The down-sampler and the up-sampler are the *sampling rate alteration devices* since they decrease or increase the sampling rate of the input sequence.

Down-Sampling Operation

The *down-sampling* operation with a down-sampling factor M, where M is a positive integer, is implemented by discharging $M-1$ consecutive samples and retaining every Mth sample. Applying the down-sampling operation to the discrete signal $\{x[n]\}$, produces the down-sampled signal $\{y[m]\}$

$$\{y[m]\} = \{x[mM]\}. \tag{2.2}$$

The down-sampling can be imagined as a two-step operation. In the first step, the original signal $\{x[n]\}$ is multiplied with the sampling function $\{s_M[n]\}$ defined by,

$$s_M[n] = \begin{cases} 1, & n = 0, \pm M, \pm 2M, \ldots \\ 0, & \text{otherwise} \end{cases}. \tag{2.3}$$

Multiplying the sequence $\{x[n]\}$ by the sampling function $\{s_M[n]\}$ results in the intermediate signal $\{y_s[m]\}$,

$$y_s[n] = x[n]s_M[n] = \begin{cases} x[n], & n = 0, \pm M, \pm 2M, \ldots \\ 0, & \text{otherwise} \end{cases}. \tag{2.4}$$

This operation is called a *discrete sampling*. In the second step, the zero valued samples in $\{y_s[m]\}$ are omitted resulting in the down-sampled sequence $\{y[m]\}$,

$$y[m] = y_S[mM] = x[mM]. \tag{2.5}$$

Figure 2.1 illustrates the two-step description of the down-sampling operation explained above for the example down-sampling factor $M = 3$.

The down-sampling operation is sometimes called the *signal compression*, and the down-sampler is also known as a *compressor*. A block diagram representing the down-sampling operation is shown in Figure 2.2. The box with a down pointed arrow followed with the factor M is used to symbolize the down-sampling operation.

Figure 2.3 illustrates the time dimensions of down-sampling. This operation reduces the sampling frequency F_T of the original signal $\{x(nT)\}$. The sampling frequency F_T' of the signal $\{y(mT')\}$ is M times smaller than the sampling frequency of the original signal, i.e, $F_T' = F_T/M$.

Figure 2.1. Down-sampling presentation by means of discrete sampling. From top to bottom: original signal $\{x[n]\}$, sampling function $\{s_M[n]\}$, intermediate signal $\{y_s[m]\}$, and down-sampled signal $\{y[m]\}$.

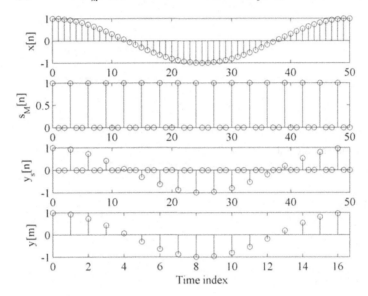

Figure 2.2. Block diagram representation of a down-sampler

$$x[n] \longrightarrow \boxed{\downarrow M} \longrightarrow y[m]$$

In Figure 2.3, the sampling periods T and T' are explicitly shown. Usually the sampling periods (or sampling frequencies) are omitted since the multirate theory can be explained without bringing T or F_T into the picture.

The MATLAB code implementing the down-sampling operation (2.2) is simply

```
y = x(1:M:N);
```

where N is the length of the original sequence x, and M (M<N) is the sampling factor. Alternatively, the MATLAB function downsample can be used to perform the down-sampling operation,

```
y = downsample(x,M);
```

In the following example, we illustrate the sampling rate reduction on the example sequence with the down-sampling factor $M=2$.

```
N = 41; % Length of the sequence
n = 0:N-1; % Time index
x = 0.6*sin(2*pi*0.0625*n)+0.3*sin(2*pi*0.2*n); % Original signal
M = 2; % Down-sampling factor
```

Figure 2.3. Converting the sampling rate with the down-sampler

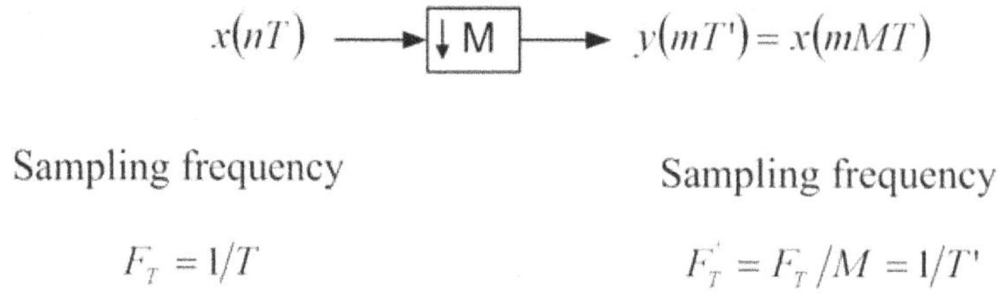

$$x(nT) \longrightarrow \boxed{\downarrow M} \longrightarrow y(mT') = x(mMT)$$

Sampling frequency

$$F_T = 1/T$$

Sampling frequency

$$F_T' = F_T/M = 1/T'$$

```
y = x(1:M:N); % Down-sampled signal
L = length(y); % Length of the down-sampled sequence
figure (1)
subplot(2,1,1), stem(0:N-1,x(1:N))
xlabel('Time index n'), ylabel('x[n]')
subplot(2,1,2)
stem(0:L-1,y(1:L))
xlabel('Time index m'), ylabel('y[m]')
```

Figure 2.4 plots the original signal {$x[n]$} and the down-sampled signal {$y[m]$}.

Up-Sampling Operation

The *up-sampling* by an integer factor L is performed by inserting L-1 zeros between two consecutive samples. Applying the up-sampling operation to the discrete signal {$x[n]$}, produces the up-sampled signal {$y[m]$} where

Figure 2.4. Down-sampling with M=2. (a) Input sequence. (b) Down-sampled (compressed) sequence

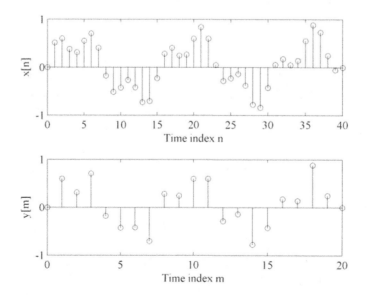

$$y[m] = \begin{cases} x[m/L], & m = 0, \pm L, \pm 2L, \cdots \\ 0, & \text{otherwise} \end{cases}. \tag{2.6}$$

A block diagram presentation of (2.6) is given in Figure 2.5. The box with an up pointed arrow followed with the factor L is used to symbolize the up-sampling operation.

The up-sampling operation increases the sampling rate F_T of the original signal $x(nT)$. The sampling frequency F_T' of the signal $y(mT')$ is L times larger than the sampling rate of the original signal, i.e, F_T' $=LF_T$. Figure 2.6 shows the time dimension of the sampling rate conversion with an up-sampler.

As in the case of a down-sampler, the sampling periods (or sampling frequencies) can be omitted in representing the up-sampling operation.

The sampling rate of the sequence $\{y[m]\}$ is L times larger than that of the sequence $\{x[n]\}$. The up-sampling is sometimes called the *sequence expansion*, and the term *expander* is sometimes used for the device. We illustrate next the process of up-sampling in MATLAB.

Using the following MATLAB code, one can implement the up-sampling operation (2.3) in two steps

```
Y = zeros(1,L*N); % N is the length of the original sequence; L is the up-sampling factor
y([1:L:length(y)]) = x;
```

where N is the length of the original sequence x, an L is the up-sampling factor. We first generate the zero valued sequence of the length N×L, and then replace each Lth zero sample with the corresponding sample of the sequence x. Alternatively, the MATLAB function upsample can be used to perform the up-sampling operation,

```
y = upsample(x,M);
```

The following MATLAB example illustrates the sampling rate increase with the up-sampling factor $L = 2$.

```
N=21;    % Length of the original sequence
n=0:N-1; % Time index
```

Figure 2.5. Block diagram representation of an up-sampler

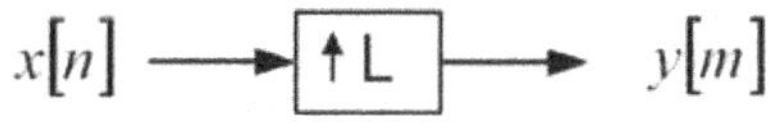

$$x[n] \longrightarrow \boxed{\uparrow L} \longrightarrow y[m]$$

Figure 2.6. Converting the sampling rate with an up-sampler

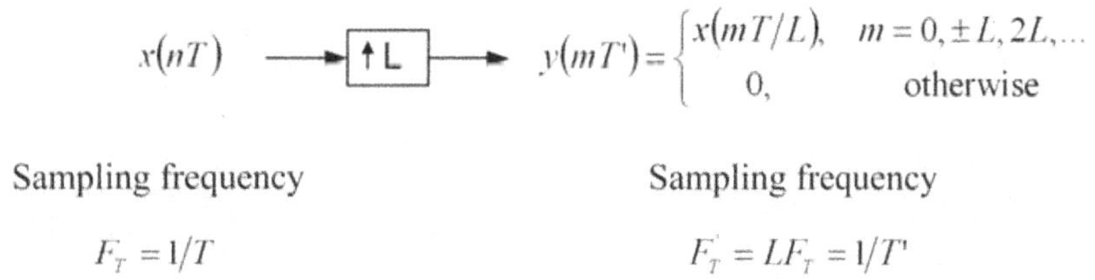

$$x(nT) \longrightarrow \boxed{\uparrow L} \longrightarrow y(mT') = \begin{cases} x(mT/L), & m = 0, \pm L, 2L, \ldots \\ 0, & \text{otherwise} \end{cases}$$

Sampling frequency   Sampling frequency

$$F_T = 1/T \qquad\qquad\qquad\qquad F_T' = LF_T = 1/T'$$

```
x=0.7*sin(2*pi*0.0625*n)+0.3*sin(2*pi*0.2*n); % Original signal
L=2; % Up-sampling factor
y=zeros(1,L*length(x));
y([1:L:length(y)])=x; % Up-sampled signal
Ny = length(y)-L+1   % Length of the up-sampled signal
figure (2)
subplot(2,1,1)
stem(0:20,x(1:21))
xlabel('Time index n'), ylabel('x[n]')
subplot(2,1,2)
stem(0:40,y(1:41))
xlabel('Time index m'), ylabel('y[m]')
```

Figure 2.7 plots the original signal and the up-sampled signal with $L=2$.

Linearity and Time-Dependence Properties

Down-sampling and up-sampling are linear time-dependent operations. The linearity property of a down-sampler and of an up-sampler is obvious since it follows directly from the definitions (2.2) and (2.6). The time-dependence property has to be examined in more details.

The time varying property of a down-sampler (up-sampler) is due to the fact that the delay of n_0 samples in the original signal does not result in the same delay of the down-sampled (up-sampled) signal.

Let us consider the responses of a down-sampler to the sequences $\{x[n]\}$, and $\{x_1[n]\} = \{x[n-n_0]\}$. The down-sampler's response $\{y[m]\}$ to the input sequence $\{x[n]\}$ is obviously

$$y[m] = x[mM],$$

Figure 2.7. Up-sampling with L = 2. (a) Input sequence. (b) Up-sampled (expanded) sequence

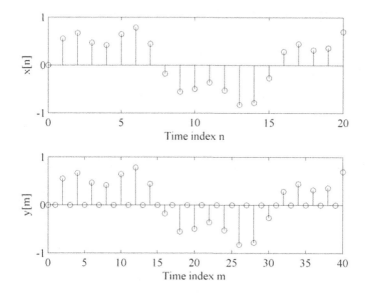

but when applying $\{x_1[n]\} = \{x[n-n_0]\}$ as the input sequence, the response $\{y_1[m]\}$ results in

$$y_1[m] = x[Mm - n_0] = y\left[\frac{mM - n_0}{M}\right] = y\left[m - \frac{n_0}{M}\right] \neq y[m - n_0], \tag{2.7}$$

where

$$y[m - n_0] = x[mM - n_0]. \tag{2.8}$$

Evidently, the delayed version of the input signal does not result in the same delay of the down-sampled signal, and according to this, the down-sampling is a time varying operation. When the ratio n_0/M is not an integer, the down-sampler produces the fractional delay of the input signal. In the case, when the delay n_0 is a multiple of the down-sampling factor M, i.e., $n_0 = kM$, the down-sampled signal $\{y_1[m]\}$ is the delayed-by-k version of the original down-sampled signal,

$$y_1[m] = x[Mm - kM] = y\left[\frac{mM - kM}{M}\right] = y[m - k]. \tag{2.9}$$

To illustrate this, let us consider the responses of a down-sampler to the example sequences $\{x[n]\}$ and $\{x[n-n_0]\}$ shown in Figure 2.8. Here, the down-sampling factor is $M = 2$, and delays $n_0 = 1$ and $n_0 = 2$ samples, respectively. Down-sampling operation produces different sequences for $\{x[n]\}$ and $\{x[n-1]\}$. Moreover, the shapes of the output sequences are different. But the down-sampled sequences for $\{x[n]\}$ and $\{x[n-2]\}$ are of the same shape. Thereby, the response of the down-sampler to $\{x[n-2]\}$ is a delayed replica of the response of the down-sampler to $\{x[n]\}$. We say that the down-sampling operation is *periodically time-invariant* with a period equal to the down-sampling factor M. The reader is recommended to examine the periodic time-invariance property of the down-sampler by solving MATLAB exercise 2.1.

The up-sampler is also a time-varying system. To show this, we observe from Eq. (2.6) that its output $y_1[m]$ for an input $x_1[n] = x[n - n_0]$ is given by

$$
\begin{aligned}
y_1[m] &= \begin{cases} x_1[m/L], & m = 0, \pm L, \pm 2L, \ldots \\ 0, & \text{otherwise} \end{cases} \\
&= \begin{cases} x[(m - Ln_0)/L], & m = 0, \pm L, \pm 2L, \ldots \\ 0, & \text{otherwise} \end{cases}
\end{aligned}
\tag{2.10}
$$

But from (2.6)

$$
y[m - n_0] = \begin{cases} x[(m - n_0)/L] & m = n_0, n_0 \pm L, n_0 \pm 2L, \ldots \\ 0 & \text{otherwise} \end{cases}
\tag{2.11}
$$

$$\neq y_1[m]$$

Figure 2.8. Illustration of the periodical time-invariance of a down-sampler with M=2

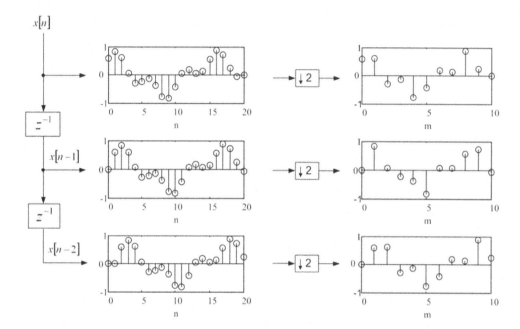

Hence, the up-sampling is a time-varying operation. Figure 2.9 illustrates the time-dependence of up-sampling operation for the factor $M = 2$, and delays $n_0 = 1$ and $n_0 = 2$ samples, respectively. It is to be noticed that the up-sampling operation does not change the shape of the original signal.

FREQUENCY-DOMAIN CHARACTERIZATION OF DOWN-SAMPLING AND UP-SAMPLING

Frequency-domain representation of down-sampling and up-sampling is used to investigate the effects of the sampling rate alterations on the spectrum of the signal. The input-output relations for the sampling rate alteration devices already defined in time domain have to be expressed in terms of z-transform and in terms of Fourier transform. This is achieved by relating the spectrum of the down-sampled (up-sampled) signal with the spectrum of the original signal.

Frequency-Domain Characterization of Down-Sampler

We consider first the z-domain representation of down-sampling. The input-output relation of a down-sampler in time domain is given in (2.2). Applying the z-transform to both sides of equation (2.2), we obtain,

$$Y(z) = \sum_{m=-\infty}^{\infty} x[Mm]z^{-m}. \tag{2.12}$$

Our goal is to express the right-hand side of equation (2.12) in terms of $X(z)$. To achieve this, we use as an intermediate sequence the sequence $\{y_s[n]\}$ obtained by the discrete sampling of the original sequence $\{x[n]\}$ as defined in (2.4).

Figure 2.9. Illustration of the time-dependence property of an up-sampler with M=2

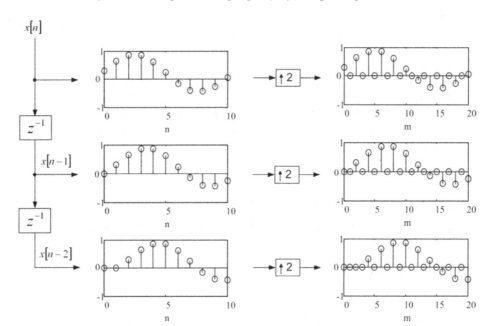

In the first step, we derive relation between $Y(z)$ and $Y_s(z)$. Since the nonzero sample values of $\{y_s[n]\}$ are the same as the sample values of $\{x[Mm]\}$, we use (2.4), (2.5) and (2.12) to obtain the desired relation.

$$Y(z) = \sum_{m=-\infty}^{\infty} x[Mm]z^{-m} = \sum_{m=-\infty}^{\infty} y_s[Mm]z^{-m} = \sum_{k=-\infty}^{\infty} y_s[k]z^{-k/M}$$
$$= Y_s\left(z^{1/M}\right) \tag{2.13}$$

In the next step, we intend to express $Y_s(z)$ in terms of $X(z)$. It is shown in equation (2.4) that the sequence $\{y_s[n]\}$ can be represented as a product of the sampling function $s_M[n]$ and the original sequence $\{x[n]\}$. Thereby, the z-transform $Y_s(z)$ is expressible by

$$Y_s(z) = \sum_{n=-\infty}^{\infty} s_M[n]x[n]z^{-n}. \tag{2.14}$$

For developing the right-hand side of (2.14) it is useful to represent $s_M[n]$ in the form

$$s_M[n] = \frac{1}{M}\sum_{k=0}^{M-1} W_M^{kn}, \tag{2.15}$$

where $W_M = e^{-j2\pi/M}$ is the factor already defined in equation (1.30). Evidently, the right-hand side of equation (2.15) equals 1 for $n = 0, \pm M, \pm 2M, \ldots$, and equals zero otherwise. This alternative representation of $s_M[n]$ can be used in equation (2.14) instead of the term $s_M[n]$. Thereby, we substitute $s_M[n]$ in (2.14) with the expression given in (2.15). With this substitution, we arrive to the relation between $Y_s(z)$ and $X(z)$,

$$Y_s\left(z\right)=\frac{1}{M}\sum_{n=-\infty}^{\infty}\left(\sum_{k=0}^{M-1}W_M^{kn}\right)x[n]z^{-n}=\frac{1}{M}\sum_{k=0}^{M-1}\left(\sum_{n=-\infty}^{\infty}x[n]W_M^{kn}z^{-n}\right)$$
$$=\frac{1}{M}\sum_{k=0}^{M-1}X\left(zW_M^{-k}\right)$$

(2.16)

The input-output relation for the down-sampler in the z-transform domain can be reached now by simply substituting $Y_s(z^{1/M})$ in (2.13) with the resulting expression from (2.16). According to this, the desired input-output relation for the down-sampler is the following

$$Y\left(z\right)=\frac{1}{M}\sum_{k=0}^{M-1}X\left(z^{1/M}W_M^{-k}\right).$$

(2.17)

From expression (2.17), it is straightforward to establish the corresponding relation between the discrete-time Fourier transform of the original signal and the discrete-time Fourier transform of the down-sampled signal. One has only to introduce the substitution $z=e^{j\omega}$ in (2.17) and obtain

$$Y\left(e^{j\omega}\right)=\frac{1}{M}\sum_{k=0}^{M-1}X\left(e^{j(\omega-2\pi k)/M}\right).$$

(2.18)

The above relation explains the implication of the down-sampling on the spectrum of the signal. Evidently, the spectrum $Y(e^{j\omega})$ is a sum of M uniformly shifted and stretched versions of $X(e^{j\omega})$ then scaled by a factor $1/M$. Equation (2.18) explicitly shows that the aliasing will occur when the bandwidth of the original signal exceeds π/M. Thus, only signals which are bandlimited to π/M can be down-sampled without distortion. For the down-sampling factor M, the highest frequency in the spectrum of $X(e^{j\omega})$ denoted by ω_H should be limited to

$$\omega_H \leq \pi / M.$$

(2.19)

The next MATLAB program illustrates the effects of down-sampling in frequency domain using MATLAB. For the original signal, we use the MATLAB function fir2 to generate a sequence whose magnitude spectrum is of a triangular shape.

```
% Program demo_2_1.m
% Spectrum of down-sampled signal
clear all, close all
% Generating the input signal 'x'
F = [0,0.1,0.46,1]; A=[0,1,0,0];        % Setting the input parameters for fir2
x = fir2(256,F,A);  % Generating the original signal 'x'
X = fft(x,1024);     % Computing the spectrum of the original signal
f = 0:1/1024:(512-1)/1024; % Normalized frequencies
figure (1)
subplot(4,1,1), plot(f,abs(X(1:512))), legend('(a)')
ylabel('¦X(e^j^\omega)¦')

M = 2;  % Down-sampling factor
y2 = x(1:M:256);      % Down-sampling
```

```
Y2 = fft(y2,1024);   % Computing the spectrum of the down-sampling signal
subplot(4,1,2), plot(f,abs(Y2(1:512))),legend('(b)')
ylabel('¦Y_2(e^j^\omega)¦')


M = 3;  % Down-sampling factor
y3 = x(1:M:256);   % Down-sampling
Y3 = fft(y3,1024);   % Computing the spectrum of the down-sampled signal
subplot(4,1,3), plot(f,abs(Y3(1:512))),legend('(c)')
ylabel('¦Y_3(e^j^\omega)¦')


M = 4;  % Down-sampling factor
y4=x(1:M:256);   % Down-sampling
Y4 = fft(y4,1024);   % Computing the spectrum of the down-sampling signal
subplot(4,1,4), plot(f,abs(Y4(1:512))),legend('(d)')
ylabel('|(Y_4(e^j^\omega)|'), xlabel('\omega/\pi')
```

The plots generated by the above program are displayed in Figure 2.10. The first plot (Figure 2.10(a)) shows the magnitude spectrum of the input signal. Evidently, the highest frequency in the spectrum satisfies (2.19) for $M = 2$. The second plot (Figure 2.10(b)) shows the magnitude spectrum of the down-sampled signal with $M = 2$. Since the spectrum from Figure 2.10(b) is only the stretched and scaled version of the spectrum shown in Figure 2.10(a), the aliasing is avoided and the original signal can be reconstructed without distortion. Effects of aliasing are visible in Figs. 2.10(c) and 2.10(d). Figure 2.10(c) displays the spectrum of down-sampled signal for $M = 3$, and Figure 2.10(c) for $M = 4$.

Frequency-Domain Characterization of Up-Sampler

The input-output relation of the up-sampler in time domain is defined in (2.4), i.e.

$$y[m] = \begin{cases} x[m/L], & m = 0, \pm L, \pm 2L, \cdots \\ 0, & \text{otherwise} \end{cases}, \tag{2.20}$$

where $\{x[n]\}$ is the original signal, and $\{y[m]\}$ is the up-sampled signal. By definition, the z-transform of the up-sampled sequence $\{y[m]\}$ is the following

$$Y(z) = \sum_{m=-\infty}^{\infty} y[m]z^{-m} \sum_{\substack{n=-\infty \\ n=mL}}^{\infty} x[n/L]z^{-n} = \sum_{m=-\infty}^{\infty} x[m]z^{-Lm} = X(z^L) \tag{2.21}$$

Therefore, the z-transform of the signal at the output of an up-sampler can be expressed as

$$Y(z) = X(z^L), \tag{2.22}$$

where $X(z)$ is the z-transform of the original signal.

When the complex variable z in (2.22) is replaced with $z = e^{j\omega}$, we obtain the input-output relation for the factor-of-L up-sampler in terms of Fourier transform

Figure 2.10. Illustration of down-sampling, representation in frequency domain: (a) original signal; (b) down-sampled signal, M = 2; (c) down-sampled signal M = 3;(d) down-sampled signal M = 4.

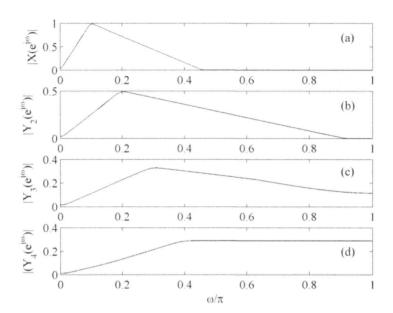

$$Y\left(e^{j\omega}\right) = X\left(e^{j\omega L}\right) \qquad (2.23)$$

Equation (2.23) shows that the factor-of-L up-sampling leads to L-fold repetition of the original spectrum $X(e^{j\omega})$ in baseband. This process is called *imaging* because we get L-1 "images" of the input spectrum. This process is illustrated on an example using the MATLAB program demo_2_2. First, we generate sequence $\{x[n]\}$ whose spectrum is of a triangular shape. Then, we apply the up-sampling with factors of $L = 2$, L=3 and $L = 4$, and compute and display the corresponding spectra.

```
% Program demo_2_2
% Spectrum of the up-sampled signal
clear all, close all
% Input signal 'x'
F = [0,0.2,0.9,1]; A = [0,1,0,0];  % Setting the input parameters for fir2
x = fir2(128,F,A); % Generating the original signal 'x'
X = fft(x,1024);   % Computing the spectrum of the original signal
f = 0:1/512:(512-1)/512;  % Normalized frequencies
figure (1)
subplot(4,1,1), plot(f,abs(X(1:512)))
ylabel('|X(e^j^\omega)|'), text(0.9,0.6,'(a)')

L = 2; % Up-sampling factor
y2 = zeros(1,L*length(x));
y2([1:L:length(y2)]) = x; % Up-sampled signal, L=2
```

```
Y2 = fft(y2,1024); % Computing the spectrum of the up-sampled signal
subplot(4,1,2), plot(f,abs(Y2(1:512)))
ylabel('|Y_2(e^j^\omega)|'), text(0.9,0.5,'(b)')


L = 3; % Up-sampling factor
y3 = zeros(1,L*length(x));
y3([1:L:length(y3)]) = x; % Up-sampled signal, L=3
Y3 = fft(y3,1024);  % Computing the spectrum of the up-sampled signal
subplot(4,1,3), plot(f,abs(Y3(1:512)))
ylabel('|Y_3(e^j^\omega)|'), text(0.9,0.6,'(c)')


L = 4; % Up-sampling factor
y4 = zeros(1,L*length(x));
y4([1:L:length(y4)]) = x; % up-sampled signal, L=4
Y4 = fft(y4,1024);  % Computing the spectrum of the up-sampled signal
subplot(4,1,4), plot(f,abs(Y4(1:512)))
ylabel('|(Y_4(e^j^\omega)|'), xlabel('\omega/\pi'), text(0.9,0.5,'(d)')
```

Figure 2.11 illustrates the effects of up-sampling in frequency domain. Figure 2.11(a) plots the spectrum of the original signal. Figs. 2.11(b-d) plot the spectra of the up-sampled signals for $L = 2$, $L = 3$ and $L = 4$, respectively. In each case, the up-sampling compresses the original spectrum by L, and causes the additional L-1 images of the original spectrum.

DECIMATION AND INTERPOLATION

Sampling rate conversion systems are used to change the sampling rate of a signal. The process of sampling rate decrease is called *decimation*, and the process of sampling rate increase is called *interpolation*. Two devices, explained above in this chapter, the down-sampler and the up-sampler, are elements that change the sampling rate of the signal. The drawback of the down-sampling is the aliasing effect, whereas the up-sampling produces the unwanted spectra in the frequency band of interest. Decimation has to be performed in such a way as to avoid the effects of aliasing, which occurs when the highest frequency in the spectrum of a down-sampled signal ω_H exceeds the value π/M, see equation (2.15) and Figure 2.10. In interpolation, the L-1 images caused by inserting L-1 zeros between the samples should be removed, see Figure 2.11.

Decimation

Decimation requires that aliasing should be prevented. Hence, prior to down-sampling with the factor of M, the original signal has to be bandlimited to π/M. This means that the factor-of-M decimation has to be implemented in two steps:

(1) Bandlimiting of the original signal to π/M
(2) Down-sampling by the factor-of-M

Figure 2.12 shows the block diagram of a decimator implemented as a cascade of the decimation filter $H(z)$, also called the antialiasing filter, and the factor-of- M down-sampler.

The role of the decimation filter $H(z)$ is to suppress aliasing to an acceptable value. Therefore, the performance of a decimator is mainly determined by the filter characteristics. Since the filter with an ideal frequency response cannot be achieved, some amount of aliasing should be tolerated. For deeper understanding of the design and implementation of decimation filters, see Chapters III, IV, V, and VI.

In MATLAB, there is a special built-in function, which performs decimation. MATLAB function decimate bandlimits the input signal to the range $[0-\pi/M]$ and down-samples the bandlimited signal by the factor of M. Hence, the function decimate implements the lowpass filtering and the down-sampling according to Figure 2.12.

The following example demonstrates the effects of down-sampling and the effects of decimation (filtering and down-sampling) on the spectrum of the signal. Decimation is performed using decimate. The input signal is composed as a superposition of two components: the sequence of a triangular-shape magnitude spectrum, and the additional sequence representing an unwanted sinusoidal signal.

Program demo_2_3 generates the input signal, computes the down-sampled signal with $M=2$ (without filtering), and computes the decimated signal with $M=2$ (filtering and down-sampling) using the function decimate. Program generates and plots the spectra of the original, down-sampled, and decimated signals.

Figure 2.11. Illustration of up-sampling, representation in frequency domain: (a) original signal, (b) up-sampled signal, L = 2, (c) up-sampled signal L = 3, up-sampled signal L = 4

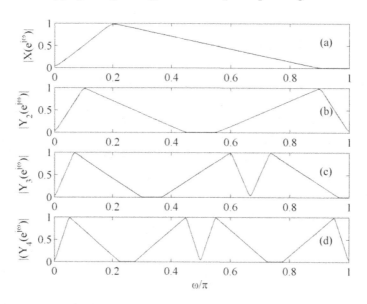

Figure 2.12. Block diagram representation of decimator

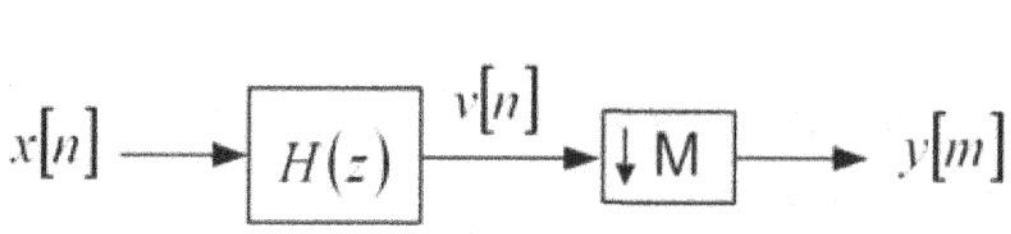

```
% Program demo_2_3
% Spectrum of decimated signal
clear all, close all
% Input signal 'x'
F =[0,0.1,0.46,1]; A = [0,1,0,0]; % Setting the input parameters for fir2
x1 = fir2(256,F,A); x2 = 0.01*cos(2*pi*0.35*(0:256)); % Generating the signal components
x = x1 + x2; % Original signal
X = fft(x,1024); % Computing the spectrum of the original signal
f = 0:1/512:(512-1)/512; % Normalized frequencies
figure (1)
subplot(3,1,1), plot(f,abs(X(1:512)))
ylabel('|X(e^j^\omega)|'), text(0.9,0.8,'(a)')

% Down-sampled signal
M = 2;  % Down-sampling factor
y = x(1:M:256); % Down-sampling
Y = fft(y,1024); % Computing the spectrum of the down-sampled signal
subplot(3,1,2), plot(f,abs(Y(1:512)))
ylabel('|Y(e^j^\omega)|'), text(0.9,0.5,'(b)')

% Decimated signal
yd = decimate(x,M);      % Decimated signal
Yd = fft(yd,1024);       % Computing the spectrum of the decimated signal
subplot(3,1,3), plot(f,abs(Yd(1:512)))
xlabel('\omega/\pi'),ylabel('|Y_d(e^j^\omega)|'), text(0.9,0.5,'(c)')
```

Figure 2.13 plots the results of the program demo_2_3. Figure 2.13(a) displays the spectrum of the original signal. Figure 2.13(b) shows the spectrum of the factor-of-2 down-sampled signal, while the spectrum of the factor-of-2 decimated signal is shown in Figure 2.13(c). It is visible from Figure 2.13(b) that the spectrum of down-sampled signal is modified by aliasing. But, when filtering is performed prior to down-sampling, the effect of aliasing is suppressed as Figure 2.13(c) illustrates.

Interpolation

Interpolation requires the removal of the images, see Figure 2.11. This means that the factor-of-L interpolation has to be implemented in two steps:

1. Up-sampling of the original signal by inserting L-1 zero-valued samples between two consecutive samples
2. Removal of the L-1 images from the spectrum of the up-sampled signal.

Figure 2.14 shows the block diagram of an interpolator implemented as a cascade of a factor-of-L up-sampler and a lowpass filter, frequently called the antiimaging filter. The cut-off frequency of the filter is π/L.

The antiimaging (interpolation) filter $H(z)$ is used to remove images from the spectrum of the up-sampled signal. Removal of images from the spectrum of the signal causes the interpolation of the sample values in time domain. The zero-valued samples in the up-sampled signal $\{x_u[m]\}$ are "filled in" with the interpolated values. As in the case of a decimator, the performance of an interpolator is mainly determined by the filter characteristics. For the design and implementation of interpolation filters, see Chapters III, IV, V, and VI.

In MATLAB, there is a special built-in function, which performs interpolation. MATLAB function interp up-samples the input signal by a factor of L, and then interpolates sample values. Hence, the function interp implements up-sampling and the lowpass filtering according to Figure 2.14.

The following example demonstrates the effects of up-sampling and the effects of interpolation (up-sampling and filtering) on the spectrum of the signal. Interpolation is performed using the function interp. The input signal is generated as a sequence whose spectrum is of a triangular shape as in the program demo_2_2.

Program demo_2_4 generates the input signal, computes the up-sampled signal with $L=2$ (without filtering), and computes the interpolated signal with $L=2$ (up-sampling and filtering) using the function interp. Program generates and plots the spectra of the original, up-sampled, and interpolated signals.

Figure 2.13. Illustration of down-sampling and decimation in frequency domain: (a) spectrum of the original signal; (b) spectrum of the down-sampled signal; (c) spectrum of the decimated signal

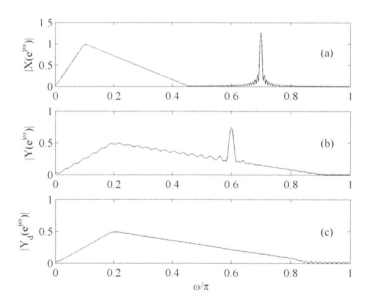

Figure 2.14. Block diagram representation of an interpolator

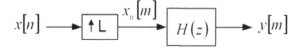

```
% Program demo_2_4.m
% Spectrum of interpolated signal
clear all, close all
% Input signal 'x'
F = [0,0.2,0.9,1]; A = [0,1,0,0]; % Setting the input parameters for fir2
x = fir2(128,F,A); % Generating the original signal 'x'
X = fft(x,1024);  % Computing the spectrum of the original signal
f = 0:1/1024:(512-1)/1024; % Normalized frequencies
figure (1)
subplot(3,1,1), plot(f,abs(X(1:512)))
ylabel('|X(e^j^\omega)|'), text(0.9,0.5,'(a))

L = 4;  % Up-sampling factor
xu = zeros(1,L*length(x));
xu([1:L:length(xu)]) = x; % Up-sampled signal
Xu = fft(xu,1024);  % Computing the spectrum of the up-sampled signal
subplot(3,1,2), plot(f,abs(Xu(1:512)))
ylabel('|X_u(e^j^\omega)|'), text(0.9,0.5,'(b)')

y = interp(x,L); % Interpolated signal
Y = fft(y,1024); % Computing the spectrum of the up-sampled signal
subplot(3,1,3), plot(f,abs(Y(1:512)))
ylabel('|(Y(e^j^\omega)|'), xlabel('\omega/(2\pi)'), text(0.9,2, '(c)')
```

Figure 2.15 displays the results of the program demo_2_4. Figure 2.15(a) plots the spectrum of the original signal. Figure 2.15(b) shows the spectrum of the factor-of-2 up-sampled signal, while the spectrum of the factor-of-2 interpolated signal is shown in Figure 2.15(c). Evidently, applying the low-pass interpolation filter the images from the spectrum of the up-sampled signal are nearly eliminated, Figure 2.15(c).

Figure 2.16 shows the portions of signals generated in the above example, program demo_2_4, in time domain. We use the following code to select and plot the segments of the signals $\{x[n]\}$, $\{x_u[n]\}$, and $\{y[n]\}$:

```
figure
subplot(3,1,1), stem(54:74,x(55:75))
ylabel('x[n]'), legend('a')
subplot(3,1,2), stem(L*55-1:L*74-1,xu(L*55-1:L*74-1))
ylabel('x_u[m]'), legend('b')
axis([L*55-1,L*74-1,-0.5,0.5])
subplot(3,1,3), stem(L*55-1:L*74-1,y(L*55-1:L*74-1))
ylabel('y[m]'), xlabel('Time index'),legend('c')
axis([L*55-1,L*74-1,-0.5,0.5])
```

Figures 2.15 and 2.16 illustrate the role of a filter in the process of interpolation. Regarding frequency domain, the interpolation filter $H(z)$ has a high attenuation in the band of images, and consequently, images are highly attenuated, nearly removed from the spectrum of the interpolated signal, compare Fig. 2.15(b) to Fig. 2.15(c). Regarding time domain, the low-pass filtering interpolates zero valued samples to some nonzero values as shown in Figs. 2.16(b) and 2.16(c).

THE SIX IDENTITIES

When constructing multirate systems, it is desirable to form an efficient implementation structure that permits the arithmetic operation to be evaluated at the lowest possible sampling rate. This goal cannot be achieved easily. Observing the structures of sampling rate converters, one concludes that filtering has to be performed at the higher sampling rate. In a decimator, filtering has to prevent aliasing and therefore has to be performed before the down-sampling. In an interpolator, the role of filtering is to remove images produced by the up-sampling operation, i.e., in an interpolator, filtering follows up-sampling. Obviously, the efficiency of the sampling rate converters may be improved if down-sampling (up-sampling) operations are incorporated into the filter structure. To achieve that, the structure of the decimator (interpolator) has to be modified. The expected properties of the new structure are twofold: the arithmetic operations are to be evaluated at the lower sampling rate, and the modification of the structure does not affect the overall performance of the decimator (interpolator).

There are six basic identities related to the multirate signal processing that when used properly, highly improve the efficiency of the system since such rules enable us to move the down-samplers and up-samplers in more desirable position in the system. In this section, we present the six identities, which sometimes are called *noble identities*. It will be shown in the next chapters how the six identities introduced in this section when properly used improve the efficiency of multirate systems.

Figure 2.15. Illustration of up-sampling and interpolation in frequency domain: (a) spectrum of the original signal; (b) spectrum of the up-sampled signal; (c) spectrum of the interpolated signal

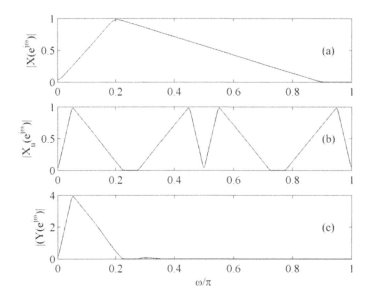

First Identity

The *first identity* is indicated in Figure 2.17. The scaling of the signals in the branches, their addition at the node, and down sampling is equivalent to down-sample the signals prior to scaling and addition. Evidently, transforming the structure from Figure 2.18 (a) to that of Figure 2.17 (b) leads to the arithmetic operations (multiplications and addition) being evaluated at the M times lower sampling rate.

Second Identity

The *second identity* states that the delay-by-M followed by a down-sampler-by-M is equivalent to the down-sampler-by-M followed by a delay-by-one. Figure 2.18 pictures the proper interchange in the cascade connection.

Third Identity

Figure 2.19 shows the *third identity*. This identity is related to the cascade connection of a linear time-invariant system $H(z)$ and a down-sampler. Filtering with $H(z^M)$ and down-sampling by M is equal to the down-sampling by M and filtering with $H(z)$. The third identity may be considered as a more general version of the second identity.

The proof of the third identity can be derived from (2.17). Detailed proofs can be found in (Jovanovic-Dolecek, 2002; Mitra, 2006; Vaidianathan, 1993). We illustrate here the third identity by means of an example developed in MATLAB. The MATLAB file demo_2_5 shows how the two different processing schemes given in Figures 2.19(a) and 2.19(b) produce identical output sequences. Program demo_2_5

Figure 2.16. Illustration of up-sampling and interpolation in time domain: (a) original signal, (b) up-sampled signal, (c) interpolated signal

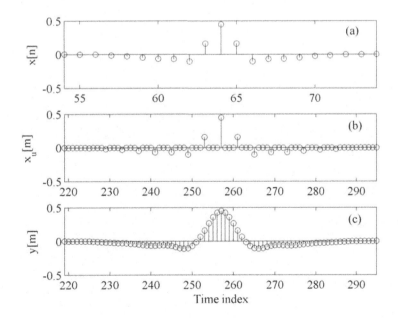

generates the first 61 samples of the signal $x = \cos(0.1\pi n)$, designs an FIR filter $H(z)$ with MATLAB function fir1 using Hamming window of the length $N = 11$, and the cut-off frequency of $f_c = 0.5$. In the sequel, program demo_2_5 forms the transfer function $H(z^4)$ by inserting 3 zero samples between the consecutive coefficients of $H(z)$. The processing is performed according to Figures 2.19(a) and 2.19(b).

```
% Program demo_2_5.m
% Illustration of the Third Identity
clear all, close all
n = 0:60;  % Time index
x = cos(2*pi*0.05*n); % Generating the original signal
h = fir1(10,0.5); % Designing the  filter transfer function H(z)
hu = upsample(h,4); % Transfer function H(z^4)
y1 = filter(hu,1,x); % Filtering
y = downsample(y1,4); % Down-sampling
m = 0:length(y)-1; % Time index
figure (1)
subplot(3,2,1), stem(n,x), ylabel('x[n]')
title('Figure 2.20(a): x[n],y_1[n],y[m]')
subplot(3,2,3), stem(n,y1), ylabel('y_1[n]')
subplot(3,2,5), stem(m,y), ylabel('y[m]')
```

Figure 2.17. First identity

Figure 2.18. Second identity

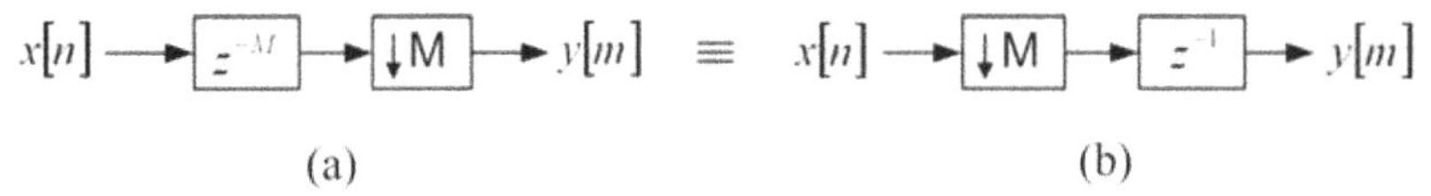

Figure 2.19. Third identity

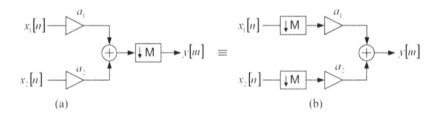

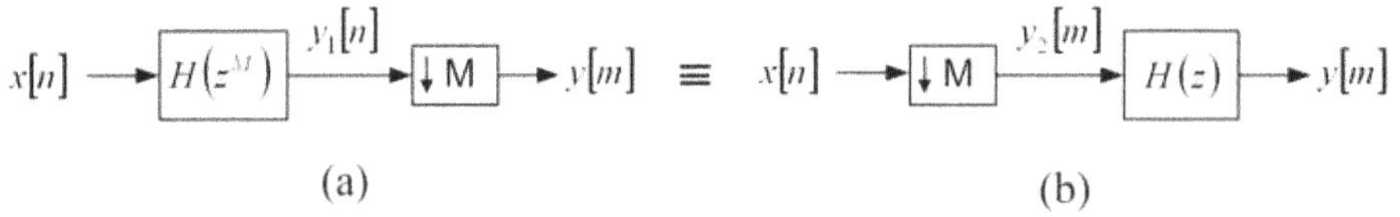

```
xlabel('Time index')
y2 = downsample(x,4); % down-sampling
y = filter(h,1,y2); % filtering
subplot(3,2,2), stem(n,x),ylabel('x[n]')
title('Figure 2.20(a): x[n],y_2[m],y[m]')
subplot(3,2,4), stem(m,y2), ylabel('y_2[m]')
subplot(3,2,6), stem(m,y),ylabel('y[m]')
xlabel('Time index')
```

The results displayed in Figure 2.20 demonstrate the equivalence of the cascade connections defined by the Third Identity

Fourth Identity

The *fourth identity* is indicated in Figure 2.21. The up-sampling prior to the branching and scaling is equivalent to branching and scaling prior to up-sampling. Evidently, transforming the structure from Figure 2.21 (a) to that of Figure 2.21 (b) leads to the arithmetic operations (multiplications and addition) being evaluated at the L times lower sampling rate.

Figure 2.20. Illustration of the Third Identity. The left-hand side shows the signals of Figure 2.19(a), and the right-hand side presents the signals of Figure 2.19(b).

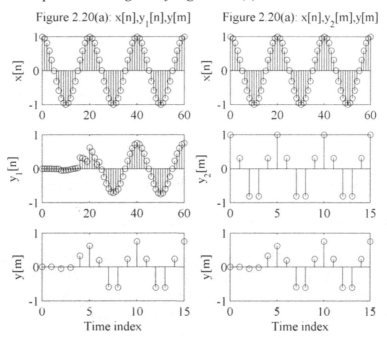

43

Figure 2.21. Fourth identity

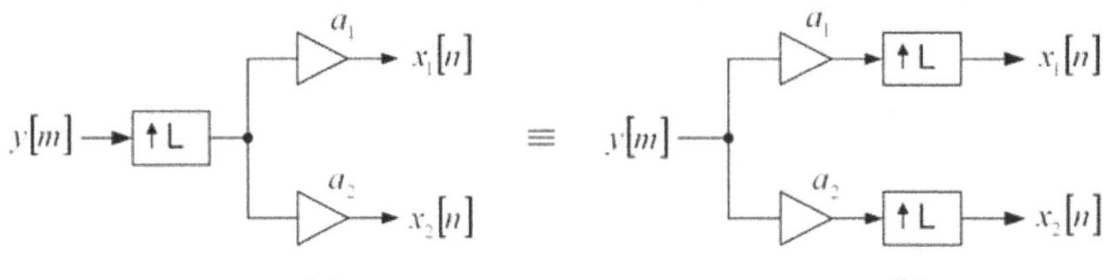

(a)   (b)

Figure 2.22. Fifth identity

(a)   (b)

Figure 2.23. Sixth identity

(a)   (b)

Fifth Identity

The *fifth identity* states that delayed-by-1 signal and up-sampled-by-L is equivalent to the signal up-sampled-by-L and delayed-by-L. Figure 2.22 indicates the proper interchange in the cascade of the delay and the up-sampler.

Sixth Identity

Figure 2.23 shows the *sixth identity*. This identity is related to the cascade connection of a linear time-invariant system and an up-sampler. Filtering with $H(z)$ and up-sampling by L is equal to the up-sampling by L and filtering with $H(z^L)$. The sixth identity may be considered as a more general version of the fifth identity.

The proof of the sixth identity follows directly from equation (2.17). Detailed proofs can be found in (Jovanovic-Dolecek, 2002; Mitra, 2006; Vaidianathan, 2003). Here, we illustrate the sixth identity by means of the MATLAB program demo_2_6. The MATLAB file demo_2_6 demonstrates how the two different processing schemes given in Figures 2.23(a) and 2.23(b) leave the input-output relations invariant. Program demo_2_6 generates the first 16 samples of the signal $x = \cos(0.2\pi n)$, designs an FIR filter $H(z)$ with MATLAB function fir1 with the 11-length Hamming window, and the cutoff frequency of f_c = 0.5. In the sequel, program demo_2_6 forms the transfer function $H(z^2)$ by inserting the zero sample between the consecutive coefficients of $H(z)$. The processing is performed according to Figures 2.23(a) and 2.23(b).

```
% Program demo_2_6.m
% Illustration of the Sixth Identity
clear all, close all
n = 0:15; % Time index
x = cos(0.2*pi*n); % Generating the original signal
h = fir1(10,0.5); % Designing the filter transfer function H(z)
hu = upsample(h,2); % Transfer function H(z^2)
y1 = filter(h,1,x); % Filtering
y = upsample(y1,2); % Up-sampling
m = 0:length(y)-1; % Time index
figure (1)
subplot(3,2,1), stem(n,x), ylabel('x[n]')
title('Figure 2.24(a): x[n],y_1[n],y[m]')
subplot(3,2,3), stem(n,y1),ylabel('y_1[n]')
subplot(3,2,5), stem(m,y), ylabel ('y[m]')
xlabel('Time index')
axis([0,30,-1,1])
y2 = upsample(x,2); % Up-sampling
y = filter(hu,1,y2); % Filtering
subplot(3,2,2), stem(n,x), ylabel('x[n]')
title('Figure 2.24(b): x[n],y_2[m],y[m]')
subplot(3,2,4), stem(m,y2), ylabel('y_2[m]')
axis([0,30,-1,1])
subplot(3,2,6), stem(m,y), ylabel('y[m]')
xlabel('Time index')
axis([0,30,-1,1])
```

The results displayed in Figure 2.24 demonstrate the equivalence of the cascade connections defined by the Sixth Identity.

CASCADING SAMPLING-RATE ALTERATION DEVICES

Two basic sampling-rate alteration devices (down-sampler and up-sampler) can be used to change the sampling rate of a signal by an integer factor only. Therefore, to change the sampling rate by a fractional factor, the cascade of up-sampler and down-sampler should be used. Figure 2.25 shows two possible combinations: (a) the input signal is first up-sampled by L and afterwards down-sampled by M, and (b) the input signal is down-sampled by M and then up-sampled by L. It is of practical interest to know whether the positions of a factor-of-M down-sampler and a factor-of-L up-sampler can be interchanged without changing the input-output relations as depicted in Figure 2.25. The answer is positive under the condition of M and L being relatively prime integers, see for example (Fliege, 1994; Jovanovic-Dolecek, 2002;

Mitra, 2006). Therefore, the two structures of Figure 2.25 are equivalent when *M* and *L* are *relatively prime*, i.e., *M* and *L* do not have a common integer factor grater than one. Cascading the sampling-rate alteration devices, we can change the sampling rate of the signal by a rational factor *L/M*.

The equivalence of Figure 2.25 can be easily demonstrated in MATLAB. Let us consider the sampling-rate conversion of the sinusoidal sequence $x=\cos(0.2\pi n)$ with $L = 2$ and $M = 3$. Program demo_2_7 illustrates the equivalences.

```
% Program demo_2_7.m
% Converting the sampling rate by the rational factor L/M=2/3
clear all, close all
n = 0:15;   % Time index
L = 2; M = 3;   % Up-sampling and down-sampling factors
x = cos(2*pi*0.1*n);   % Generating the original signal
v1 = upsample(x,L); % Up-sampling
```

Figure 2.24. Illustration of the Sixth Identity. The left-hand side shows the signals of Figure 2.23(a), and the right-hand side presents the signals of Figure 2.23(b).

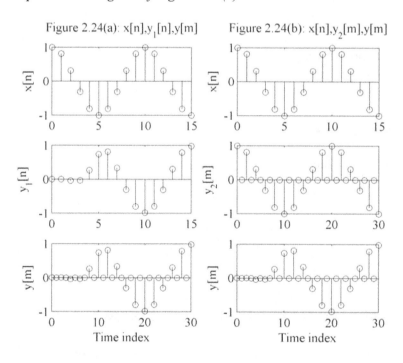

Figure 2.25. Two different realizations of a fractional sampling-rate alteration (L/M)

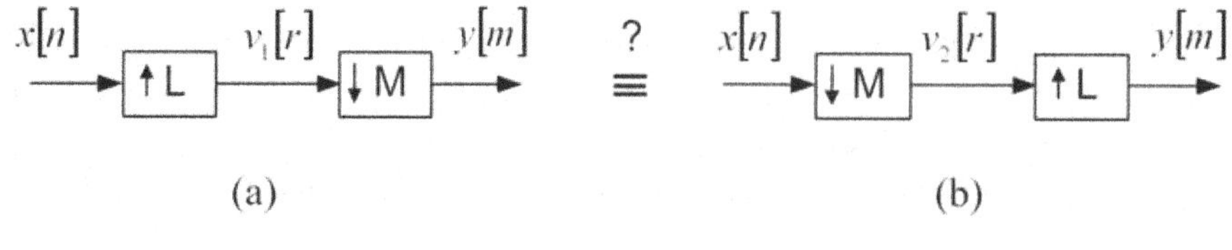

```
y = downsample(v1,M);  % Down-sampling
r = 0:length(v1)-1;  % Time index
m = 0:length(y)-1;  % Time index
figure (1)
subplot(3,2,1), stem(n,x), ylabel('x[n]')
title('Figure 2.26(a): x[n],v_1[r],y[m]')
subplot(3,2,3), stem(r,v1), ylabel('v_1[r]')
axis([0,30,-1,1])
subplot(3,2,5), stem(m,y), ylabel('y[m]')
xlabel('Time index')
v2 = downsample(x,M);  % Down-sampling
r = 0:length(v2)-1;    % Time index
y = upsample(v2,L); % Up-sampling
m = 0:length(y)-1;     % Time index
subplot(3,2,2), stem(n,x), ylabel('x[n]')
title('Figure 2.26(b): x[n],v_2[r],y[m]')
subplot(3,2,4), stem(r,v2), ylabel('v_2[r]')
axis([0,5,-1,1])
subplot(3,2,6), stem(m,y), ylabel('y[m]')
xlabel('Time index')
axis([0,10,-1,1])

axis([0,5,-1,1])
subplot(3,2,6), stem(m,y), ylabel('y[m]')
xlabel('Time Index')
axis([0,10,-1,1])
```

The results are shown in Figure 2.26. Although the intermediate sequences $v_1[r]$ and $v_2[r]$ are different, the two realizations of 2/3-sampling rate conversion produce the same output sequence $y[m]$.

The invariance of the output sequence to the interchange of the sampling rate alteration devices in 2/3-convertor described above, demonstrates the equivalence existing for the relatively prime factors L and M. The same equivalence is not true when L and M contain a common integer factor. Let us consider the effects of interchanging the up-sampler and down-sampler in the trivial case $L = M = 2$. For the example sinusoidal sequence, the results can be obtained with a slight modification in the program. demo_2_7. The resulting sequences for $L = M = 2$ are displayed in Figure 2.27. When the up-sampling operation is performed first, the down-sampling operation reconstructs the original signal, i.e. $y_1[m] = x[n]$, and evidently $m = n$. However, when the down-sampling is performed prior to up-sampling, a new different sequence $y_2[m]$ is produced, i.e., $y_2[n] \neq y_1[n]$, and therefore $y_2[n] \neq x[n]$.

SAMPLING RATE CONVERSION WITH THE PHASE OFFSET

When describing the down-sampling or up-sampling, we have assumed that the process starts at the time instant $n = 0$. However, the sampling process can include some phase offset, i.e., the starting time could be other than zero. Let us consider the down-sampling with a phase offset k, where k is an integer

Figure 2.26. Converting the sampling rate by the rational factor L/M=2/3. The left-hand side corresponds to Figure 2.25(a), and the right-hand side corresponds to Figure 2.25(b).

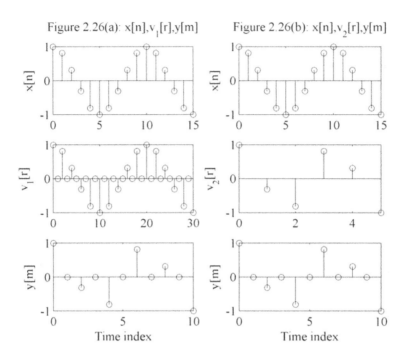

Figure 2.27. Converting the sampling rate with L = M = 2. The left-hand side corresponds to Figure 2.25(a), and the right-hand side corresponds to Figure 2.25(b).

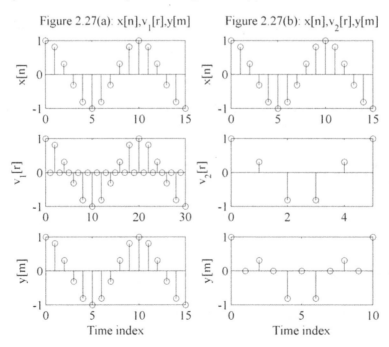

$k < M$. Including the phase-offsets into the definition of the down-sampling operation (2.2), we obtain M different signals,

$$\left\{y_k\left[m\right]\right\} = \left\{x\left[mM + k\right]\right\}, \qquad k = 0, 1, 2, \ldots, M-1. \tag{2.24}$$

For $k = 0$, the resulting signal $\{y_0[m]\}$ is the down-sampled signal without the phase offset. Each value of k defines a new signal as illustrated in Figure 2.28 for $M = 3$, and $k = 0, 1, 2$.

In MATLAB, the function downsample can be used to perform down-sampling with the given phase offset,

y = downsample(x,M,phase),

where the original sample values are stored in the vector x, M is the down-sampling factor, and phase is the offset.

Since the down-sampling is a time-dependent operation, each phase offset ($0 \le k \le M - 1$) causes a different signal. The effects of the phase offset on the spectra of the down-sampled signals can be expressed analytically, see e.g. (Fliege, 1994). It was shown that the magnitude spectrum of a down-sampled signal is independent of the phase offset when a factor-of-M down-sampler is applied to the original signal whose spectrum is bandlimited to π/M. In that case, the down-sampling with a phase offset produces a phase shift only. Considering the problem of aliasing, explained earlier in this chapter, one can arrive to the same conclusion. Namely, when the original signal is bandlimited to π/M, there is no aliasing in the signals produced by M-fold-down-sampling. However, when down-sampling the original signal whose frequency band exceeds π/M, the magnitude spectrum of a down-sampled signal should depend on the phase offset.

Let us illustrate the effects of the phase offset on down-sampled signals by means of an example. We use the MATLAB program demo_2_8 to generate two original signals $\{x[n]\}$ and $\{v[n]\}$, and to down-sample those signals with $M = 2$, and $k = 0, 1$. The spectrum of $\{x[n]\}$ is bandlimited to π/M, whereas the frequency band of $\{v[n]\}$ exceeds this limit. Program demo_2_8 computes and plots the magnitude spectra of the original signals, and the spectra of down-sampled signals for $k = 0, 1$.

```
% Program demo_2_8.m
% Down-sampling with the phase offset
% Analysis in the frequency domain
clear all, close all
% Generating the original signal 'x'
F = [0,0.1,0.46,1]; A = [0,1,0,0];  % Setting the input parameters for fir2
x = fir2(256,F,A); % Generating the original signal 'x'
X = fft(x,1024); % Computing the signal spectrum
f = 0:1/512:(512-1)/512; % Normalizad frequencies
figure (1)
subplot(3,2,1), plot(f,abs(X(1:512))), legend('(a)')
ylabel('|X(e^{j\omega})|'), axis([0,1,0,1])
```

Figure 2.28. Illustration of down-sampling with the phase offset: M = 3, and k = 0, 1, 2

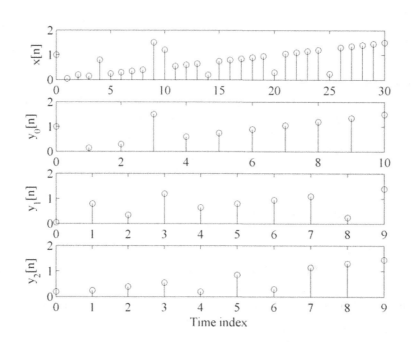

```
M = 2; % Down-sampling factor
y0 = downsample(x,M); % down-sampling without phase offset
Y0 = fft(y0,1024); % Computing the signal spectrum
subplot(3,2,3), plot(f,abs(Y0(1:512))), legend('(b)')
ylabel('|Y_0(e^{j\omega})|'), axis([0,1,0,1])

y1 = downsample(x,M,1); % Down-sampling with the phase offset
Y1 = fft(y1,1024); % Computing the signal spectrum
subplot(3,2,5), plot(f,abs(Y1(1:512))), legend('(c)')
ylabel('|Y_1(e^{j\omega})|'), xlabel('Normalized frequency \omega/\pi'), axis([0,1,0,1])

F = [0,0.1,0.7,1]; A = [0,1,0,0]; % Setting the input parameters for fir2
v = fir2(256,F,A);  % Generating the input signal 'v'
V = fft(v,1024);  % Computing the signal spectrum
f = 0:1/512:(512-1)/512;  % Normalized frequencies
figure (1)
subplot(3,2,2), plot(f,abs(V(1:512))), legend('(d)')
ylabel('|V(e^{j\omega})|'), axis([0,1,0,1])

w0=downsample(v,M); % Down-sampling without phase offset
W0=fft(w0,1024); % Computing the signal spectrum
subplot(3,2,4), plot(f,abs(W0(1:512))), legend('(e)')
ylabel('|W_0(e^{j\omega})|'), axis([0,1,0,1])
```

```
w1 = downsample(v,M,1);  % Down-sampling with the phase offset
W1=fft(w1,1024); % Computing the signal spectrum
subplot(3,2,6), plot(f,abs(W1(1:512))), legend('(f)')
ylabel('|W_1(e^{j\omega})|'), xlabel('Normalized frequency \omega/\pi'), axis([0,1,0,1])
```

The plots of Figure 2.29 demonstrate that the magnitude spectrum of the down sampled signal is independent of the phase offset when the frequency band of the original signal is properly bandlimited to π/M, see left-hand side of Figure 2.29. On the contrary, the right-hand side of Figure 2.9 shows different spectra of down-sampled signals for $k = 0$, and $k = 1$. Evidently, the frequency band of the original signal that exceeds π/M results in different magnitude spectra of the down-sampled signals.

Evidently, with the phase offset introduced into the down-sampling operation we can produce M different signals each of them operating at the M-times lower sampling rate. In the next section, we will discuss the decomposition of discrete signals to M different signal components.

A time offset can be introduced also in the up-sampling operation. The MATLAB function upsample performs the up-sampling with the time offset when the phase-shift is specified,

```
y = upsample(x,L,phase);
```

Here the original sample values are stored in the vector x, L is the up-sampling factor, and phase is the offset.

POLYPHASE DECOMPOSITION

Let us consider again the discrete sampling of the signal $\{x[n]\}$ introduced in (2.4). The reference time instant for the sampling function $\{s_M[n]\}$ in (2.4) is $n = 0$, but the reference time can be shifted from zero to some other time instant k, $0 \le k \le M$. Hence, we can define the time-shifted sampling functions $\{s_{M,k}[n]\} = \{s_M[n-k]\}$, with $k = 0, 1, 2, \ldots, M-1$,

$$s_{M,k}[n] = \begin{cases} 1, & n = k, k \pm M, k \pm 2M, \ldots, \\ 0, & \text{otherwise} \end{cases}, \qquad k = 0, 1, 2, \ldots, M-1. \tag{2.25}$$

Figure 2.30 shows the decomposition of an example signal $\{x[n]\}$ for $M = 3$. The three distinct signals $\{x_k^{(p)}[n]\}$, $k = 0, 1, 2$, are obtained by the discrete sampling of $\{x[n]\}$, i.e. by multiplying $\{x[n]\}$ with the corresponding sampling function $\{s_{3,k}[n]\}$, $k = 0, 1, 2$. Evidently, the signal $\{x[n]\}$ can be represented as the sum of three discretely sampled components, i.e.,

$$\begin{aligned} x[n] &= x_0^{(p)}[n] + x_1^{(p)}[n] + x_2^{(p)}[n] \\ &= x[n]s_{3,0}[n] + x[n]s_{3,1}[n] + x[n]s_{3,2}[n] \\ &= x[n]s_3[n] + x[n]s_3[n-1] + x[n]s_3[n-2]. \end{aligned} \tag{2.26}$$

Figure 2.29. Magnitude spectra of the down-sampled signals with the phase offset: M = 2, and k = 0, 1: (a) Spectrum of the original signal {x[n]}; (b) Spectrum of the down-sampled signal {y$_0$[n]}, k = 0; (c) Spectrum of the down-sampled signal {y$_1$[n]}, k = 1; (d) Spectrum of the original signal {v[n]}; (e) Spectrum of the down-sampled signal {w$_0$[n]}, k = 0; (f) Spectrum of the down-sampled signal {w$_1$[n]}, k = 1.

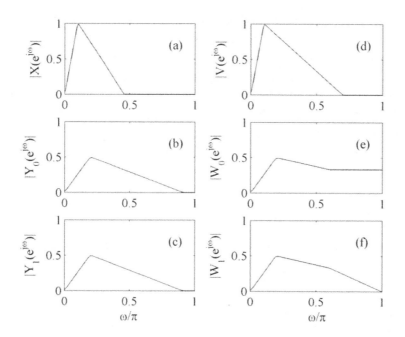

In general, for the given $\{x[n]\}$ and M, one can write

$$x[n]=\sum_{k=0}^{M-1}x_k^{(p)}[n]=\sum_{k=0}^{M-1}x[n]s_M[n-k] \tag{2.27}$$

This representation is called the polyphase representation of a discrete signal, or *polyphase decomposition*. The signal components $\{x_k^{(p)}[n]\}$, $k = 0, 1, 2, ..., M-1$, are called the *polyphase components* of the signal $\{x[n]\}$.

Equations (2.26) and (2.27) express the polyphase representation of a signal in time domain. Applying the z- transform (1.39), i.e.,

$$X(z)=\sum_{n=-\infty}^{\infty}x[n]z^{-n} \tag{2.28}$$

to (2.26) and (2.27) we arrive to the z-transform representation of polyphase decomposition. For the sake of simplicity, we first consider the z-transforms of the three polyphase components for the finite-length sequence. We take $N = 15$ for the sequence length as already chosen for the sequence $\{x[n]\}$ of Figure 2.30. Hence, the developed form of the z-transform $X(z)$ for the example sequence is the following,

$$\begin{aligned}X(z)=&x[0]+x[1]z^{-1}+x[2]z^{-2}+x[3]z^{-3}+x[4]z^{-4}+x[5]z^{-5}+x[6]z^{-6}+x[7]z^{-7}\\&+x[8]z^{-8}+x[9]z^{-9}+x[10]z^{-10}+x[11]z^{-11}+x[12]z^{-12}+x[13]z^{-13}+x[14]z^{-14}.\end{aligned} \tag{2.29}$$

The z-transform $X(z)$ can be rewritten as the sum of three polynomials,

$$X(z) = \left(x[0] + x[3]z^{-3} + x[6]z^{-6} + x[9]z^{-9} + x[12]z^{-12}\right)$$
$$+ \left(x[1]z^{-1} + x[4]z^{-4} + x[7]z^{-7} + x[10]z^{-10} + x[13]z^{-13}\right)$$
$$+ \left(x[2]z^{-2} + x[5]z^{-5} + x[8]z^{-8} + x[11]z^{-11} + x[14]z^{-14}\right). \tag{2.30}$$

Taking factors z^{-1} and z^{-2} out of the brackets, we obtain,

$$X(z) = \left(x[0] + x[3]z^{-3} + x[6]z^{-6} + x[9]z^{-9} + x[12]z^{-12}\right)$$
$$+ z^{-1}\left(x[1] + x[4]z^{-3} + x[7]z^{-6} + x[10]z^{-9} + x[13]z^{-12}\right)$$
$$+ z^{-2}\left(x[2] + x[5]z^{-3} + x[8]z^{-6} + x[11]z^{-9} + x[14]z^{-12}\right). \tag{2.31}$$

Here, the polynomials in the brackets are reduced to the polynomials in z^{-3}. Equation (2.31) is expressible in a concise form,

$$X(z) = \sum_{k=0}^{2} z^{-k} X_k^{(p)}\left(z^3\right), \tag{2.32}$$

where

$$X_k^{(p)}(z^3) = \sum_{m=0}^{4} x[3m+k]z^{-3m}, \; k = 0, 1, 2. \tag{2.33}$$

Figure 2.30. Decomposition of the signal $\{x[n]\}$ to three polyphase components: $\{x_0^{(p)}[n]\}$, $\{x_1^{(p)}[n]\}$ and $\{x_2^{(p)}[n]\}$

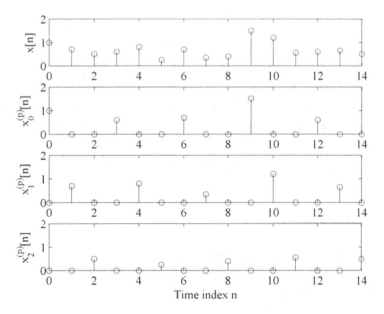

Comparing (2. 26) with (2.32) and (2.33), we conclude that the *z*-transforms for the polyphase components $\{x_k^{(p)}[n]\}$ are

$$\left\{x_k^{(p)}[n]\right\} \leftarrow z \rightarrow z^{-k} X_k^{(p)}\left(z^3\right), \qquad k = 0, 1, 2. \tag{2.34}$$

Generally, an arbitrary sequence $\{x[n]\}$ can be decomposed into *M* polyphase components. For $\{x[n]\}$ being a sequence of an infinite length, and with $n = Mm + k$, we can write,

$$X(z) = \sum_{k=0}^{M-1} \sum_{m=-\infty}^{\infty} x[mM + k] z^{-(mM+k)} = \sum_{k=0}^{M-1} z^{-k} X_k^{(p)}\left(z^M\right), \tag{2.35}$$

where

$$X_k^{(p)}(z^M) = \sum_{m=0}^{\infty} x[mM + k] z^{-mM}. \tag{2.36}$$

From (2.35) and (2.36), we observe that the *z*-transforms for a polyphase component $\{x_k^{(p)}[n]\}$ is given by

$$\left\{x_k^{(p)}[n]\right\} \leftarrow z \rightarrow z^{-k} X_k^{(p)}\left(z^M\right), k = 0, 1, 2, \ldots, M - 1. \tag{2.37}$$

Thereby, we can use (2.37) for the *z*-transform representation of the signal $\{x[n]\}$ by means of *M* polyphase components,

$$X(z) = X_0^{(p)}\left(z^M\right) + z^{-1} X_1^{(p)}\left(z^M\right) + z^{-2} X_2^{(p)}\left(z^M\right) + \ldots + z^{-(M-1)} X_{M-1}^{(p)}\left(z^M\right). \tag{2.38}$$

The polyphase decomposition is widely used in multirate signal processing. Combining the polyphase decomposition with the third and sixth identities leads to the efficient multirate implementation structures as will be shown later on. The polyphase components introduced above are obtained by the discrete sampling of the sequence $\{x[n]\}$, and consequently each polyphase component contains $M-1$ zeros between the two consecutive samples. Since only nonzero samples are needed for the further processing, the superfluous zeros can be discharged from the polyphase sequences $\{x_k^{(p)}[n]\}$, resulting in down-sampled-by-*M* polyphase components. It is obvious from the example of Figure 2.30 that for retaining the nonzero samples, down-sampling operation on the component $\{x_k^{(p)}[n]\}$ should be performed with the phase offset of *k* samples. Thereby, the polyphase component $\{x_k^{(p)}[n]\}$ has to be time-shifted first to the left by *k* samples and afterwards down-sampled by *M*. Usually the term polyphase component refers to the down-sampled polyphase component and is denoted by $\{x_k[n]\}$ (Fliege, 1994; Mitra, 2006; Hentchel, 2002). Hence, down-sampling $\{x_k^{(p)}[n+k]\}$ with the factor *M* results in the polyphase component $\{x_k[n]\}$,

$$x_k[m] = x_k^{(p)}[mM + k] \tag{2.39}$$

as shown in Figure 2.31 for the example sequence of Figure 2.30.

It is important to emphasize that the polyphase components $\{x_k[n]\}$ operate at the *M*-times lower sampling rate than the original signal $\{x[n]\}$.

Since $\{x_k[n]\}$ is the phase-shifted and down-sampled version of $\{x_k^{(p)}[n]\}$, their z-transforms are related as follows

$$\{x_k[n]\} \leftarrow z \rightarrow X_k(z) = X_k^{(p)}(z), \quad k = 0, 1, 2, \ldots, M-1. \tag{2.40}$$

Using (2.38) and (2.40), we express the z-transform of the original signal $X(z)$ in terms of z-transforms $X_k(z)$, $k = 0, 1, 2, \ldots, M-1$,

$$X(z) = \sum_{k=0}^{M-1} z^{-k} X_k(z^M). \tag{2.41}$$

In practice, we perform the polyphase decomposition by down-sampling directly the original sequence $\{x[n]\}$. This means that each of $\{x_k[n]\}$ results when down-sampling $\{x[n]\}$ with the phase offset of k samples. The purpose of the discrete-sampling based approach, as used in this section, was to introduce the concept of polyphase decomposition and to derive the relations in the domain of z-transform.

The polyphase decomposition can be regarded as an alternative representation of the sequence. Actually, the samples of $\{x[n]\}$ are distributed in a specific manner between the subsequences $\{x_k[n]\}$, $k = 0, 1, 2, \ldots, M-1$, without missing a single sample. It is straightforward to conclude that the original signal can be reconstructed from its polyphase components. The reconstruction procedure should be the following: (1) the kth component has to be up-sampled with the factor M and then shifted to the right for k samples, (2) the up-sampled and shifted components are to be added together to recompose the original signal $\{x[n]\}$. The block diagram that describes the polyphase decomposition and reconstruction is shown in Figure 2.32. The symbol z^{+1} denotes the left-shift in the process of polyphase decomposition, Figure 2.32 (a).

With the polyphase decomposition, the original signal is decomposed into M polyphase components, which operate at M-times lower sampling rate than the original signal. To reconstruct the original signal, up-sampled-by-M and phase-shifted components are added together to compose the original signal as indicated in Figure 2.32. Program demo_2_9 demonstrates in MATLAB the decomposition of the signal $\{x[n]\}$ into 4 polyphase components and the reconstruction of the original signal according to the processing scheme of Figure 2.32.

```
% Program  demo_2_9.m
% Polyphase decomposition and reconstruction
clear all, close all
% Input signal
n = 0:63;
x = zeros(size(n)); x(11:39) = 0.95.^(1:29); % Generating the sequence 'x'

% Polyphase down-sampling with the phase offset
x0 = downsample(x,4);
x1 = downsample(x,4,1);
x2 = downsample(x,4,2);
x3 = downsample(x,4,3);
```

Figure 2.31. Sequence $\{x[n]\}$ and three down-sampled polyphase components: $\{x_0[m]\}$, $\{x_1[m]\}$ and $\{x_2[m]\}$

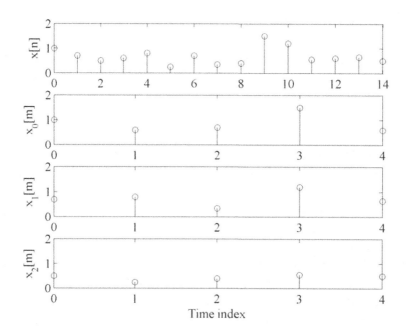

```
% Up-sampling polyphase components with the phase offset
y0 = upsample(x0,4);
y1 = upsample(x1,4,1);
y2 = upsample(x2,4,2);
y3 = upsample(x3,4,3);
figure (1)
subplot (4,1,1), stem(0:length(x0)-1,x0),ylabel('x0[m]')
subplot (4,1,2), stem(0:length(x1)-1,x1),ylabel('x1[m]')
subplot (4,1,3), stem(0:length(x2)-1,x2),ylabel('x2[m]')
subplot (4,1,4), stem(0:length(x3)-1,x3),ylabel('x3[m]')
xlabel('Time index m')

y = y0 + y1 + y2 + y3; % Adding together up-sampled polyphase components
figure (2)
subplot(2,1,1),stem(n,x), ylabel('x[n]'), axis([0,63,0,1])
subplot(2,1,2),stem(n,y), ylabel('y[n]'), axis([0,63,0,1])
xlabel('Time index n')
```

The four polyphase components are given in Figure 2.33, and Figure 2.34 presents the original signal $\{x[n]\}$ and the reconstructed signal $\{y[n]\}$. The signal $\{y[n]\}$ is the exact replica of the input signal $\{x[n]\}$, i.e., $\{y[n]\} \equiv \{x[n]\}$, as indicated in Figure 2.32.

The polyphase decomposition is widely used for efficient implementation of multirate systems as will be shown later on in this book. For the sake of simplicity, we have used the simple notations $\{x_0[n]\}$,

Figure 2.32. Block diagram presentation of polyphase decomposition and reconstruction; (a) decomposition; (b) reconstruction

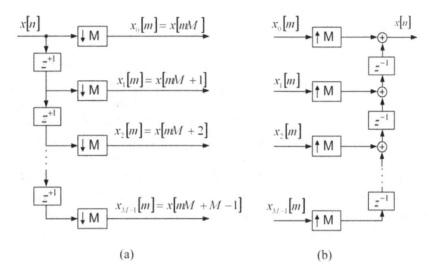

(a)   (b)

$\{x_1[n]\}, \ldots, \{x_{M-1}[n]\}$ to symbolize the polyphase components. When it is necessary, the notation of polyphase components should be given with more details. For example, the name of the original sequence, factor M, phase offset k.

MULTISTAGE SYSTEMS

The decimators and interpolators discussed so far are single-stage systems since the implementation schemes consist of a single low-pass filter and single sampling rate alteration device. When the decimation factor M can be factored into the product of integers, $M = M_1 \times M_2 \times \ldots M_K$, instead of using a single filter and factor-of -M down-sampler the overall decimator can be implemented as a cascade of K decimators. Such a cascade implementation, called a *multistage decimator*, is shown in Figure 2.35. In the same manner, the factor-of-L interpolator expressible by $L = L_1 \times L_2 \times \ldots L_K$, can be implemented as a cascade of K interpolators as depicted in Figure 2.36. The cascade implementation scheme of Figure 2.36 is called the *multistage interpolator*.

The multistage structure from Figure 2.35 replaces the single stage decimator of the factor $M = M_1 \times M_2 \times \ldots M_K$. The transfer function $H(z)$ of the equivalent single-stage decimation filter can be obtained by applying the third identity to the implementation scheme of Figure 2.35. The cascade of K decimators of Figure 2.35 gives the following equivalent transfer function $H(z)$,

$$H(z) = H_1(z) H_2\left(z^{M_1}\right) H_2\left(z^{M_1 M_2}\right) \cdots H_K\left(z^{M_1 M_2 \cdots M_{K-1}}\right). \tag{2.42}$$

Thereby, the single-stage structure indicated in Figure 2.37 is equivalent to the structure of Figure 2.35.

Similarly, the overall transfer function for the *K* stage interpolator is obtained when applying the sixth identity to the multistage implementation structure of Figure 2.36. This way, we obtain,

$$H(z)=H_1(z)H_2\left(z^{L_1}\right)H_2\left(z^{L_1L_2}\right)\cdots H_k\left(z^{L_1L_2\cdots L_{K-1}}\right).$$

(2.43)

The corresponding single-stage equivalence for the *K* stage interpolator is indicated in Figure 2.38.

The multistage structures are very useful for implementing large sampling-rate conversion factors. A single decimation/interpolation filter with a very narrow passband, usually inconvenient for the design and implementation, is replaced with the cascade of simpler filters. The specifications for those individual filters are significantly relaxed since the overall filter specification is shared between several lower-order filters. We demonstrate the effects of the multistage implementation in MATLAB on the example of a three-stage decimator with the factor *M* = 8. Since the requested factor *M* is the power of two, *M* = 8 = 2³, we use the identical filters in the cascade of Figure 2.35. Hence, we design only one filter and use this filter in each of three decimator stages. Program demo_2_10 designs a FIR filter of the length N=31 and Hamming window using MATLAB function fir1, and computes the equivalent filter frequency response for the three-stage decimator.

```
% Program demo_2_10.m
% Equivalent frequency response of the three-stage decimator
clear all, close all
h1 = fir1(30,0.5); % FIR filter H1(z)
[H1,f] = freqz(h1,1,512,2);
figure (1)
```

Figure 2.33. Polyphase components for M=4

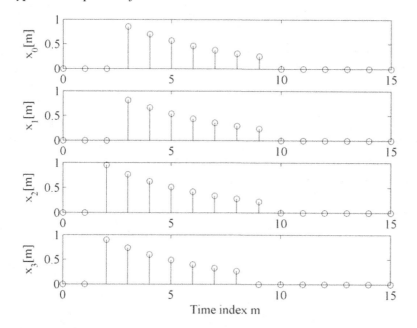

```
subplot(4,1,1), plot(f,abs(H1)), ylabel('|H_1(e^{j\omega})|')
axis([0,1,0,1.2])

M = 2;
h2 = zeros(1,M*length(h1));
h2([1:M:length(h2)]) = h1; % H2(z) = H1(z^2)
[H2,f] = freqz(h2,1,512,2);
subplot(4,1,2), plot(f,abs(H2)), ylabel('|H_1(e^{j2\omega})|')
axis([0,1,0,1.2])

M = 4;
h3 = zeros(1,M*length(h1));
h3([1:M:length(h3)]) = h1; %  H3(z) = H1(z^4)
```

Figure 2.34. Original signal {x[n]} and reconstructed signal {y[n]}

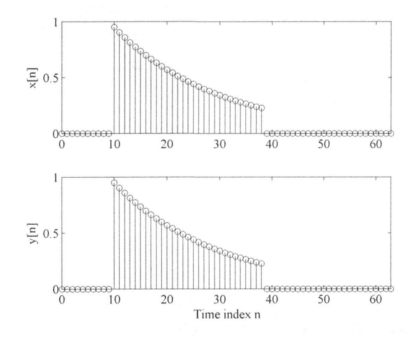

Figure 2.35. Multistage implementation of decimator

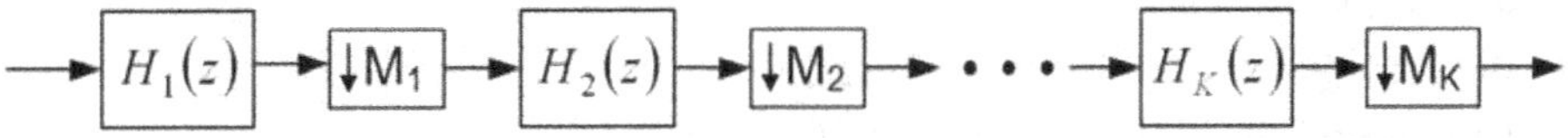

Figure 2.36. Multistage implementation of interpolator

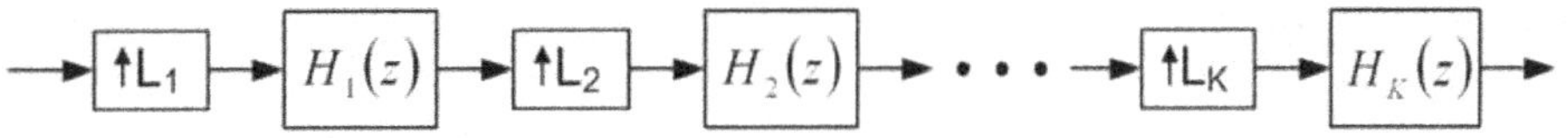

Figure 2.37. The single-stage equivalence for the multistage structure of Figure 2.35

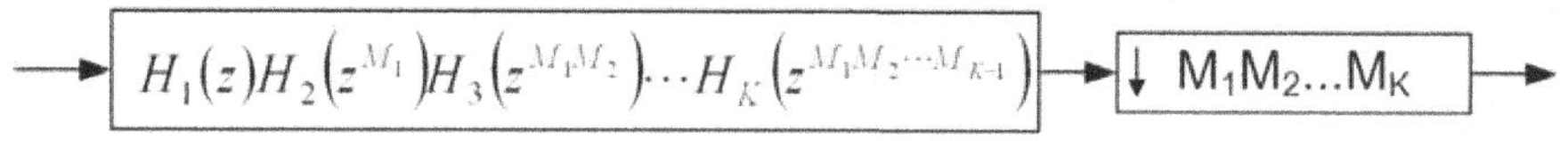

$H_1(z)H_2(z^{M_1})H_3(z^{M_1M_2})\cdots H_K(z^{M_1M_2\cdots M_{K-1}})$
↓ M1M2...MK

Figure 2.38. The single stage equivalence for the multistage structure of Figure 2.37

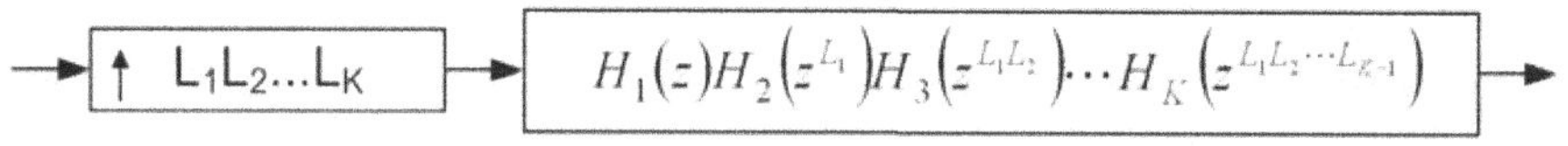

↑ L1L2...LK
$H_1(z)H_2(z^{L_1})H_3(z^{L_1L_2})\cdots H_K(z^{L_1L_2\cdots L_{K-1}})$

Figure 2.39. Three-stage decimator for $M = 8$. Plots of the magnitude responses computed in the program demo_2_10.

$|H_1(e^{j\omega})|$
$|H_1(e^{j2\omega})|$
$|H_1(e^{j4\omega})|$
$|(H(e^{j\omega})|$
ω/π

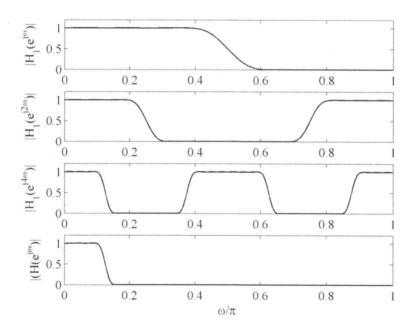

```
[H3,f] = freqz(h3,1,512,2);
subplot(4,1,3), plot(f,abs(H3)), ylabel('|H_1(e^{j4\omega})|')
axis([0,1,0,1.2])

H = H1.*H2.*H3; % Equivalent filter
subplot(4,1,4), plot(f,abs(H)),ylabel('|(H(e^{j\omega})|')
xlabel('Normalized frequency \omega/\pi'), axis([0,1,0,1.2])
```

The magnitude responses of $H_1(z)$, $H_1(z^2)$, $H_1(z^4)$ and that of the equivalent filter $H(z) = H_1(z)\,H_1(z^2)\,H_1(z^4)$ are plotted in Figure 2.39.

MATLAB EXERCISES

2.1 Generate the following sequences: (i) sinusoidal sequence of normalized frequency 0.15, (ii) sum of two sinusoidal sequences of normalized frequencies 0.1 and 0.3, (iii) product of the sinusoidal sequence of normalized frequency 0.15 and the real exponential sequence $\{0.8^n\}$. Choose the sequence lengths to be 51.

(a)   Perform the factor-of-4 down-sampling. Plot the original and down-sampled sequences.
(b)   Repeat part (a) for the factor-of-5 down-sampler.

2.2 Generate the following sequences: (i) sinusoidal sequence of normalized frequency 0.15, (ii) sum of two sinusoidal sequences of normalized frequencies 0.1 and 0.3, (iii) product of the sinusoidal sequence of normalized frequency 0.15 and the real exponential sequence $\{0.8^n\}$. Choose the sequence lengths to be 21.

(a)   Perform the factor-of-4 up-sampling. Plot the original and up-sampled sequences.
(b)   Repeat part (a) for the factor-of-5 up-sampler.

2.3 Study the time-dependence property of the down-sampling operation on the example sequence $\{x[n]\}$ composed as a sum of two sinusoidal sequences of normalized frequencies 0.0625 and 0.2. Choose $N=51$ for the sequence length. Consider the sample values outside the interval $\{0,50\}$ as zero valued. With the conversion factor $M=4$, down-sample the following sequences: $\{x[n]\}$, $\{x[n-1]\}$, $\{x[n-2]\}$, $\{x[n-3]\}$, $\{x[n-4]\}$, $\{x[n-5]\}$. Plot the down-sampled sequences. Comment on the results.

2.4 Study the spectral characteristics of the down-sampled and decimated signals by modifying Program demo_2_3. Generate 512 samples of the signal $\{x[n]\}$, composed of three sinusoidal sequences,

$$x[n] = \sin[2\pi f_1 n] + 0.9\sin[2\pi f_2 n] + 0.7\sin[2\pi f_3 n] + 0.8s[n]$$

where $\{s[n]\}$ is the additive wideband noise of normal distribution, which can be generated by using the MATLAB function randn.

Modify Program demo_2_3 to compute and plot the spectra of the (i) original signal, (ii) down-sampled-by-5 signal, and (iii) decimated-by-5 signal. Comment on the results.

2.5 Study the spectral characteristics of the up-sampled and interpolated signals by modifying the Program demo_2_4. Generate 512 samples of the signal $\{x[n]\}$, composed of two sinusoidal sequences,

$$x[n] = \sin[2\pi 0.05n] + 0.9\sin[2\pi 0.09n] + 0.8s[n]$$

where $\{s[n]\}$ is the additive wideband noise of normal distribution generated by using the MATLAB function randn.

Modify Program demo_2_4 to compute and plot the spectra of the (i) original signal, (ii) up-sampled-by-5 signal, and (iii) interpolated-by-5 signal. Comment on the results.

2.6 Modify Program demo_2_7 to study the effects of the cascade interconnection of down-sampler and up-sampler. Study the effects of interchanging the positions of the down-sampler and up-sampler when converting sampling rate with the rational factors (i) $L/M = 4/3$, and (ii) $L/M = 3/4$. Plot all sequences of interest, and comment on the results. Generate the sequence *x* on your own choice, and repeat the procedure.

2.7 Generate the triangle-shape sequence of the length $N = 100$. Perform the polyphase decomposition of the sequence for $M = 5$, and the sequence reconstruction by modifying Program demo_2_9. Plot the polyphase components, and also plot the original and the reconstructed sequences to demonstrate the exact reconstruction of the signal. Compute and plot magnitude spectra of the original and reconstructed signals and also the magnitude spectra of the polyphase components.

2.8 Design FIR filters for the two-stage decimator, which implements factor-of-15 sampling-rate conversion in two stages with $M = 3$ and $M = 5$. Compute and plot magnitude responses of two filters. Compute the frequency response of the equivalent filter for the single-stage decimator. Plot the magnitude response of the equivalent single-stage filter.

REFERENCES

Ansari,R., & Liu,B., (1993). Multirate signal processing. In Sanjit. K. Mitra and James F. Kaiser (Ed.), *Handbook for Digital Signal Processing*. Wiley-Interscience, New York, NY: 981-1084.

Bellanger, M.G., Bonnerot, G., & Coudreuse, M. (1976). Digital filtering by polyphase network: application to sample-rate alteration and filter banks. *IEEE Transactions on Acoustics, Speech, and Signal Processing*, 24 (2), 109-114.

Bellanger, M. (2000). *Digital processing of signals: Theory and practice*. 3rd edition. New York, NY: John Wiley.

Burrus, C.S., McClellan, J.H., Oppenheim, A.V, Parks, T.W., Schaffer, R.W. & Schussler, H.W. (1994). *Computer-based exercises for signal processing using MATLAB._Englewood Cliffs, NJ: Prentice-Hall.

Crochiere, R.E., & Rabiner, L.R., (1981, March). Interpolation and decimation of digital signals - A Tutorial Review. *Proceedings of the IEEE*, 69 (3), 300-331.

Diniz, P., Netto, S., & Da Silva, E. (2002). *Digital Signal Processing: System Analysis and Design* . New York, NY: Cambridge University Press.

Filter design toolbox for use with MATLAB. User's guide. Version 6. (2006). Natick: MathWorks.

Fliege, N. J. (1994). *Multirate digital signal processing*. New York, NY: John Wiley.

Harris, F. J., (2004). *Multirate signal processing for communication systems*. Upper Saddle River, NJ: Prentice Hall PTR.

Hentchel, T. (2002). *Sample rate conversion in software configurable radios*. Morwood, MA: Artech House, Inc.

Jovanović-Doleček, G.(2002). Introduction to multirate systems. In Gordana Jovanović-Doleček, (ed.), *Multirate Systems: Design & Applications*. Hershey, PA: Idea Group Publishing, 105-142.

Milić, Lj., & Lutovac, M.D. (2002). Efficient multirate filtering. In Gordana Jovanović-Doleček, (ed.), *Multirate Systems: Design & Applications*. Hershey, PA: Idea Group Publishing, 105-142.

Milić, Lj., Saramäki, T. & Bregović, R. (2006). Multirate filters: an overview. *Proc. of 2006 IEEE Asia Pacific Conference on Circuits and Systems*. Singapore, 914-917.

Mitra, S. K. (1999). *Digital signal processing laboratory using MATLAB*. New York, NY: The Mc-Graw-Hill.

Mitra, S. K. (2006). *Digital signal processing: A computer based approach*. 3rd edition. New York, NY: The McGraw-Hill.

Oppenheim, A. V., & Schafer, R. W. (1989). *Discrete-time signal processing*. 3rd edition. London: Prentice-Hall International.

Proakis J. G., & Manolakis D.G. (1996). *Digital signal processing: Principles, algorithms, and applications*. London: Prentice Hall.

Saramäki, T. *Multirate Signal Processing*. (2001). Lecture notes for a graduate course, the Institute of Signal Processing, Tampere University of Technology, Finland.

Signal processing toolbox for use with MATLAB. User's guide. Version 6. (2006). Natick: Math-Works.

Vaidyanathan, P.P., (1990). Multirate digital filters, filter banks, polyphase networks, and applications: A Tutorial. *Proceedings of the IEEE*, 78(1), 56-93.

Vaidyanathan, P.P., (1993). *Multirate systems and filter banks*. Englewood Cliffs, NJ: Prentice Hall.

Chapter III
Filters in Multirate Systems

INTRODUCTION

The role of filtering in sampling-rate conversion has been considered in Chapter II. The importance of filtering arises from the fact that the sampling theorem should be respected for all the sampling rates of the system at hand. Filters are required to bandlimit the spectrum of the signal to the prescribed bandwidth in accordance with the actual sampling rate. In sampling rate conversion systems, filters are used in decimation to suppress aliasing and in interpolation to remove imaging. Since an ideal frequency response cannot be achieved, the performance of the system for sampling rate conversion is mainly determined by filter characteristics. Obviously, an appropriate filter should enable the sampling rate conversion with minimal signal distortion. The main advantage of a multirate system is the computational efficiency, and therefore, a decimator (interpolator) that implements a high-order digital filter could not be tolerated. The specific role of a digital filter in sampling rate conversion demands high-performance filtering with the lowest possible complexity. To reach this goal one has to concentrate first on the choice of the appropriate design specifications in order to provide minimal signal distortion. Secondly, the multirate filter is to be designed in a manner to satisfy the prescribed characteristics and to provide a low-complexity implementation structure.

In this chapter, we discus first the spectral characteristics of decimators and interpolators and introduce three commonly used types of filter specifications. In the sequel, we review the MATLAB functions that are appropriate for the design of FIR and IIR filters to satisfy the specifications. An approach to computation of aliasing characteristics of decimators is given and illustrated by examples. This chapter considers also the analysis of sampling rate conversion for band-pass signals. Chapter concludes with MATLAB exercises for individual study.

SPECTRAL CHARACTERISTICS OF DECIMATORS AND INTERPOLATORS

The sampling rate reduction and sampling rate increase have been discussed in Chapter II. When reducing sampling rate, filtering should precede the sampling rate reduction to suppress aliasing. This filter, called an *antialiasing filter*, attenuates the frequency components outside the new baseband of the signal that is, bandlimits the signal spectrum to a half of the new sampling frequency. When increasing sampling rate, filtering follows the up-sampling operation. The role of the filter is to attenuate unwanted periodic spectra which appear in the new baseband. These periodic spectra are called 'images', and the filter, which is used to remove them, is called an *antiimaging filter*.

As introduced in Chapter II, the sampling rate reduction is called decimation, and the sampling rate increase is called interpolation. The performance of a decimator or an interpolator is mainly determined by filter characteristics. Therefore, the specifications for filter design should be a reasonable compromise between the performance of a decimator (interpolator) and the filter complexity.

Let us consider in more detail the unwanted spectra produced by a down-sampling operation that should be attenuated by an antialiasing filter. The structure of a factor-of-M decimator consisting of the antialiasing filter $H(z)$ and a factor-of-M down-sampler is sketched in Figure 3.1. The input signal $\{x[n]\}$ operating at the sampling frequency F_x is filtered by the antialiasing filter $H(z)$ giving the signal $\{v[n]\}$. The output signal $\{y[m]\}$ is obtained by pickking up every M^{th} sample of $\{v[n]\}$. The sampling frequency is reduced by M, $F_y = F_x/M$.

The role of the antialiasing filter $H(z)$ is to bandlimit the spectrum of the input signal to a half of the new sampling rate. The frequency response of an ideal linear-phase antialiasing filter $H_0(e^{j\omega})$ is defined as,

$$H_0\left(e^{j\omega}\right) = \begin{cases} e^{-jK\omega}, & 0 \leq \omega \leq \omega_p, \\ 0, & \omega_p \leq \omega \leq \pi, \end{cases} \tag{3.1}$$

where ω_p is the cutoff, and K is a positive constant.

If the phase characteristic is not of importance, the ideal characteristic is defined only in terms of the magnitude response,

$$\left|H_0(z)\right| = \begin{cases} 1, & 0 \leq \omega \leq \omega_p, \\ 0, & \omega_p \leq \omega \leq \pi. \end{cases} \tag{3.2}$$

The ideal characteristics defined above can be appropriately approximated with real filters. For determining the specifications for $H(z)$, we need to identify the frequency bands in which the spectrum of the input signal should be attenuated. As shown in Chapter II, the spectrum of the down-sampled signal

Figure 3.1. Structure of a factor-of-M decimator

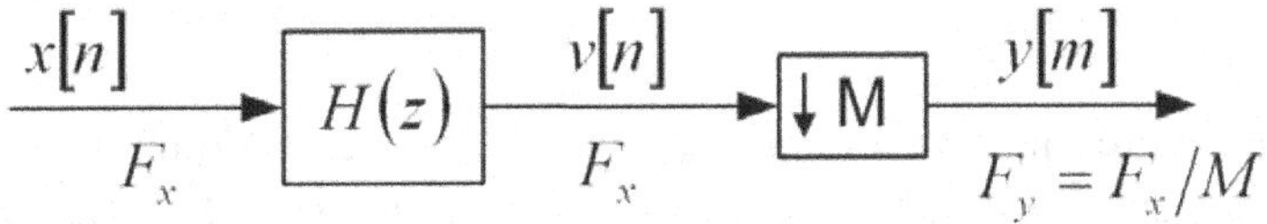

is expressible by means of the spectrum of the original signal, see equations (2.17) and (2.18). Since the output signal $\{y[m]\}$ is the down-sampled version of the signal $\{v[m]\}$, the z-transform relation is given by,

$$Y(z) = \frac{1}{M}\sum_{k=0}^{M-1} V\left(z^{1/M}W_M^{-k}\right), \quad \text{where } V(z) = X(z)H(z).$$ (3.3)

By substitution $z = e^{j\omega}$, we obtain the frequency domain relation between the original and down-sampled signals,

$$Y\left(e^{j\omega}\right) = \frac{1}{M}\sum_{k=0}^{M-1} V\left(e^{j(\omega - 2\pi k)/M}\right), \quad \text{where } V\left(e^{j\omega}\right) = X\left(e^{j\omega}\right)H\left(e^{j\omega}\right).$$ (3.4)

In expression (3.4), we observe that the spectrum of the down-sampled signal is composed of the desired unaliased component, obtained for $k = 0$, and $M-1$ unwanted aliased components, for $k = 1, 2, \ldots, M-1$. Hence, the antialiasing filter has to preserve the unaliased spectrum and to attenuate unwanted aliased spectra.

For better understanding of the role of an antialiasing filter, let us investigate decimation with $M = 2$. In that case, the sum in (3.4) has only two terms, and results in

$$Y\left(e^{j\omega}\right) = \frac{1}{2}\sum_{k=0}^{1} V\left(e^{j(\omega - 2\pi k)/2}\right) = \frac{1}{2}\left\{V\left(e^{j\omega/2}\right) + V\left(e^{j(\omega - 2\pi)/2}\right)\right\}.$$ (3.5)

Hence, the spectrum of decimated signal $Y(e^{j\omega})$ consists of two components

$$Y\left(e^{j\omega}\right) = Y_0\left(e^{j\omega}\right) + Y_1\left(e^{j\omega}\right),$$ (3.6)

where unaliased and aliased spectra are represented by $Y_0(e^{j\omega})$ and $Y_1(e^{j\omega})$, respectively, i.e.,

$$Y_0\left(e^{j\omega}\right) = \frac{1}{2}V\left(e^{j\omega/2}\right) \text{ and } Y_1\left(e^{j\omega}\right) = \frac{1}{2}V\left(e^{j(\omega - 2\pi)/2}\right).$$ (3.7)

The role of filter is to make $Y_1(e^{j\omega})$ as small as possible.

An illustration of the decimation process for $M = 2$ is shown in Figure 3.2. The input signal is composed of three sinusoids. Two of them are in the range $\{0, \pi/2\}$, and the third component is outside this range.

```
x = sin(2*pi*n*0.0625) + sin(2*pi*n*0.1250) + sin(2*pi*n*0.3200);  % Generating the sequence 'x'
```

In order to make the aliased spectral component visible in the plot, we chose a very low-order antialiasing filter. This is the 16th-order Parks-McClellan optimal FIR filter designed by using MATLAB function firgr from the *Filter Design Toolbox*. The filter is specified and designed as follows,

```
F = [0,0.45,0.55,1];  A = [1,1,0,0];  % Setting the parameters for firgr
h = firgr(16,F,A,[10,1]);  % Filter design
```

Five plots in Figure 3.2 present the decimation process in frequency domain for $M = 2$. Figure 3.2(a) depicts the decimator structure and indicates the conversion of the sampling frequency by $M = 2$. The spectrum of the original signal $\{x[n]\}$, composed of three sinusoidal components is shown in Figure 3.2(b). The antialiasing filter $H(z)$ is represented by its magnitude response in Figure 3.2(c). The magnitude spectra of the unaliased spectrum $|Y_0(e^{j\omega})|$ and the aliased spectrum $|Y_1(e^{j\omega})|$ are given in Figures 3.2(d) and 3.2(e), respectively. Finally, Figure 3.2(f) plots the magnitude spectrum of the decimated signal $|Y(e^{j\omega})|$.

From the example of Figure 3.2, it becomes evident that the aliased spectrum represented by $Y_1(e^{j\omega})$ falls in the new baseband and interferes with the unaliased spectrum $Y_0(e^{j\omega})$. The level of the unwanted component is determined by the stopband attenuation of the anti-aliasing filter. With an ideal filter, the component $Y_1(e^{j\omega})$ would be eliminated yielding the most desirable solution, $Y(e^{j\omega}) = Y_0(e^{j\omega})$. With real filters, our goal is to approximate the ideal filter in the best manner in order to provide a minimal distortion in the decimated signal, i.e. to achieve

$$Y\left(e^{j\omega}\right) \approx Y_0\left(e^{j\omega}\right). \tag{3.8}$$

Figure 3.2. Decimation with M=2, representation in frequency domain: (a) Factor-of 2-decimator. (b) Magnitude spectrum of input signal $|X(e^{j\omega})|$. (c) Magnitude response of the antialiasing filter $|H(e^{j\omega})|$. (d) Magnitude spectrum of the unaliased signal $|Y_0(e^{j\omega})|$. (e) Magnitude spectrum of the aliased signal $|Y_1(e^{j\omega})|$. (f) Magnitude spectrum of the decimated signal $|Y(e^{j\omega})|$.

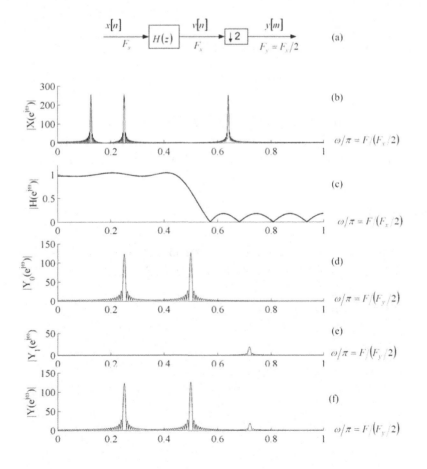

It has to be noticed that the spectrum of the down-sampled signal is multiplied by $1/M$, see equations (3.3) – (3.7) and Figure 3.2. This can be compared with sampling the continuous-time signal with two sampling periods, T and $T' = T \times M$. It is evident from equation (1.60) that the multiplication constant for two sampling rates is proportional to $1/M$.

In interpolation, the role of filtering is to remove images produced by the up-sampling operation. The structure of a factor-of-L interpolator consisting of a factor-of-M up-sampler and an antiimaging filter is sketched in Figure 3.3. The input signal $\{x[n]\}$ operating at the sampling frequency F_x is up-sampled by a factor of L by inserting $L - 1$ zeros between consecutive samples. The resulting up-sampled signal $\{v[m]\}$ operates at the sampling rate $F_y = LF_x$. The interpolated signal $\{y[m]\}$ is obtained at the output of the filter $H(z)$. The sampling frequency at the output is also $F_y = LF_x$. Considering the time domain, filtering by $H(z)$ means filling zero-valued samples with some interpolated non-zero values. In frequency domain, the role of filtering is viewed as the removal of unwanted periodic spectra called images. When interpolating the zero-valued samples, the original sample values are decreased by L. In order to keep the original sample values unchanged, the antiimaging filter has to include the multiplication by L as indicated in Figure 3.3.

The specifications for filter design are to be defined in frequency domain, and according to this, the frequency domain relations between the signals of Figure 3.3 give the appropriate insight into the problem. As shown in Chapter II, the z-transform of the up-sampled signal $V(z)$ is expressed in terms of the z-transform of the original signal $X(z)$ by

$$V(z) = X(z^L),$$
(3.9)

and accordingly, their Fourier transforms are related by

$$V(e^{j\omega}) = X(e^{j\omega L}).$$
(3.10)

The spectrum of the up-sampled signal is therefore the L-times periodically repeated spectrum of the original signal. In the new baseband, there are L spectral images. The role of the anti-imaging filter is to remove $L-1$ spectral images. Hence, with the filter $H(z)$, which follows the factor-of-L up-sampler, we have

$$Y(z) = V(z)H(z) = X(z^L)H(z).$$
(3.11)

Substituting $z = e^{j\omega}$, we obtain the spectrum of the interpolated signal $Y(e^{j\omega})$ in terms of the spectrum of the original signal $X(e^{j\omega})$ and the filter frequency response $H(e^{j\omega})$,

$$Y(e^{j\omega}) = V(e^{j\omega})H(e^{j\omega}) = X(e^{j\omega L})H(e^{j\omega}).$$
(3.12)

Figure 3.3. Structure of a factor-of-L interpolator

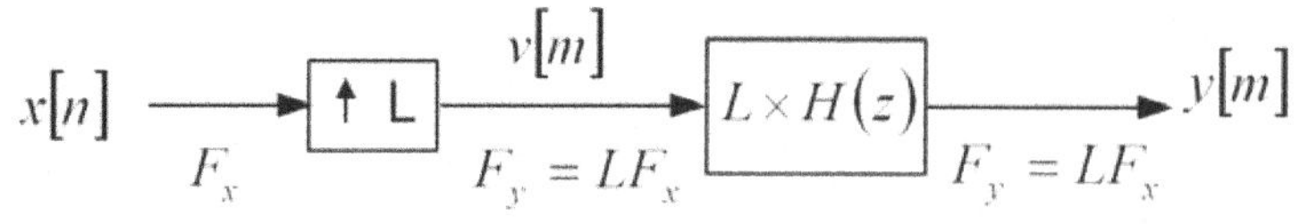

The spectrum of the up-sampled signal, the removal of images, and the spectrum of interpolated filter can be understood best by means of an example. To achieve this, we demonstrate the interpolation process for $L = 3$.

For the input signal we chose sequence $\{x[n]\}$ whose spectrum approximates a trapezoidal shape. We use MATLAB function fir2 to generate the sequence $\{x[n]\}$,

```
F = [0,0.1,0.4,0.8,0.85,1];  A = [0,0,1,1,0,0];  % Setting the parameters for fir2
x = fir2(127,F,A);   % Generating the sequence 'x'
```

The low-pass antiimaging filter $H(z)$ has to remove two images from the spectrum of the up-sampled signal. We design here the anti-imaging filter $H(z)$ as an 52-order optimal Parks-McClellan FIR filter. Using the MATLAB function firgr, we compute the filter coefficients as stated below,

```
F = [0,1/3.2,1/2.9,1];  A = [1,1,0,0];     % Setting the parameters for firgr
h = 3*firgr(52,F,A,[20,1]);     % Computing the filter coefficients
```

Notice that filter coefficients are multiplied by the interpolation factor, $L=3$.

The interpolation process is demonstrated in Figure 3.4. Figure 3.4(a) indicates the interpolator structure for $L = 3$. The spectrum of the input signal $\{x[n]\}$ is shown in Figure 3.4(b). Figure 3.4(c) plots the magnitude spectrum of the up-sampled signal $\left|V(e^{j\omega})\right|$. The magnitude response of the antiimaging filter is given in Figure 3.4(d). Finally, Figure 3.4(e) plots the magnitude spectrum of the interpolated signal $\left|Y(e^{j\omega})\right|$.

Example of Figure 3.4 shows how the images in the spectrum of an up-sampled signal are attenuated by the antiimaging filter. Obviously, with a real filter the unwanted images cannot be eliminated, but can be made negligibly small when the stopband ripple is small enough.

Filter specifications for a decimator or interpolator are to be formed in accordance with a particular application. In reality, a reasonable compromise is needed between the filter characteristics and implementation complexity.

FILTER SPECIFICATIONS FOR DECIMATORS AND INTERPOLATORS

In this section, we consider criteria for the filter magnitude response that are usually requested for decimators and interpolators. Figure 3.5 shows three typical tolerance schemes for antialiasing and antiimaging low-pass filters, which are given in the form traditionally used for finite impulse-response (FIR) filters. The boundary frequencies are expressed in terms of the angular frequency ω with the following assumptions:

- The boundary frequencies are defined in terms of the high-rate sampling frequency.
- The boundary frequencies are expressed in terms of the decimation factor M, but equivalently the interpolation factor L can be used instead of M.

The difference between the three tolerance schemes is in the stopband specifications. For all three cases, the passband is defined in the range $[0,\omega_p]$ where the magnitude response $\left|H(e^{j\omega})\right|$ has to satisfy,

Figure 3.4. Interpolation with L=3, representation in frequency domain: (a) Factor-of- 3-interpolator. (b) Magnitude spectrum of input signal $|X(e^{j\omega})|$. *(c) Magnitude spectrum of the up-sampled signal,* $|V(e^{j\omega})|$. *(d) Magnitude response of the antiimaging filter* $|H(e^{j\omega})|$. *(e) Magnitude spectrum of the interpolated signal* $|Y(e^{j\omega})|$.

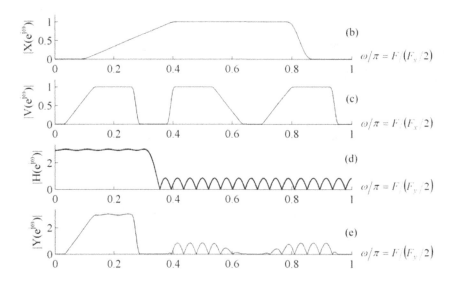

$$1-\delta_p \le \left|H\left(e^{j\omega}\right)\right| \le 1+\delta_p \qquad \text{for} \qquad 0\le\omega\le\omega_p. \tag{3.13}$$

Here, δ_p is the passband ripple, and ω_p is the passband edge frequency, also called the passband cutoff frequency.

Three different tolerance schemes are given for the stopband.

Case a specification depicted in Figure 3.5(a) is used in decimation when the aliasing in the low-rate signal should be negligible. In interpolation, this specification ensures that all images in the range $[\pi/M, \pi]$ of the high-rate signal are adequately attenuated. For the specification of Figure 3.5(a), the magnitude response has to satisfy,

$$\left|H\left(e^{j\omega}\right)\right| \le \delta_s \qquad \text{for} \qquad \pi/M \le\omega<\pi, \tag{3.14}$$

where δ_s is the stopband ripple, and the stopband edge frequency is $\omega_s = \pi/M$.

Case b specification shown in Figure 3.5(b) may be used in decimation when, in the transition band of the low-rate signal, a significant amount of aliasing can be accepted. This tolerance scheme is suitable for interpolation when the spectrum of the low-rate signal is negligible in the transition band. Hence, for the specification of Figure 3 (b), the stopband criteria are the following,

$$\left| H\left(e^{j\omega} \right) \right| \le \delta_s \qquad \text{for} \qquad 2\pi/M - \omega_p \le \omega < \pi. \tag{3.15}$$

The stopband edge frequency for the *Case b* specification is located at $\omega_s = 2\pi/M - \omega_p$.

Case c specification shown in Figure 3.5(c) contains several *don't care bands* and is suitable for interpolation for the band-limited low-rate signal. This specification is sometimes allowed for decimation if significant aliasing is acceptable in the transition band of the low-rate signal. For the *Case c* specification as given in Figure 3.5(c) the design criteria are expressible in the form,

$$\left| H\left(e^{j\omega} \right) \right| \le \delta_s \qquad \text{for} \qquad \omega \in \bigcup_{k=1}^{\lfloor M/2 \rfloor} \left(2\pi k/M - \omega_p \right), \min\left[\left(2\pi k/M + \omega_p \right), \pi \right]. \tag{3.16}$$

Case c specification defines several stopbands. The lowest stopband edge frequency is located at $\omega_s = 2\pi/M - \omega_p$.

As stated in the first paragraph of this section, the specifications shown in Figure 3.5(a-c) are presented in the form usual for finite-impulse response (FIR) filters. Traditionally, the specifications for infinite impulse response (IIR) filters are represented in a slightly different form. When specifying an IIR filter frequency response, the maximum of the magnitude is bounded to unity. Therefore, when considering the IIR filter design, we use different specification for the passband characteristic. Actually, an IIR filter passband magnitude response is limited to the range $[1 - \delta_p, 1]$. Instead of the specification given in (3.13), the passband design criterion for an IIR filter is usually defined in the form,

$$1 - \delta_p \le \left| H\left(e^{j\omega} \right) \right| \le 1 \qquad \text{for} \qquad 0 \le \omega \le \omega_p. \tag{3.17}$$

For the stopband characteristics, the specifications already defined by (3.14), (3.15), (3.16) and depicted in Figures 3.5(a-c) are also used in IIR filter design.

The tolerances for the filter magnitude response are frequently expressed in decibels. According to the tolerance scheme of Figure 5 and equation (3.13), the magnitude response in the passband oscillates between $1 + \delta_p$ and $1 - \delta_p$. Thereby, the peak passband ripple in decibels, a_p, is related to δ_p by

$$a_p = 20 \log_{10} \left(\frac{1 + \delta_p}{1 - \delta_p} \right) \tag{3.18}$$

When the passband ripple is specified by a_p, we compute δ_p by using the inverse of (3.18),

$$\delta_p = \frac{10^{a_p/20} - 1}{10^{a_p/20} + 1} \tag{3.19}$$

The minimal stopband attenuation a_s is related to the maximal stopband ripple δ_s by

$$a_s = -20 \log_{10} \left(\delta_s \right), \tag{3.20}$$

and inversely,

$$\delta_s = 10^{-a_s/20}. \tag{3.21}$$

Figure 3.5. Three types of filter specifications for sampling- rate conversion by factor M. (a) Case a tolerance scheme, (b) Case b tolerance scheme, (c) Case c tolerance scheme. Note that for IIR filter design, the passband magnitude response is limited to the range $[1 - \delta_p, 1]$.

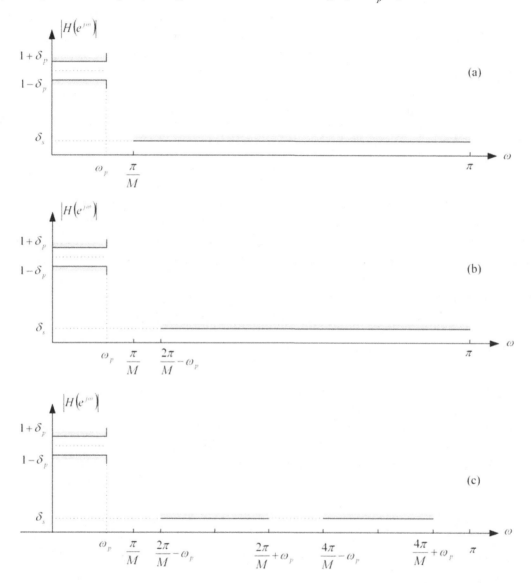

When the tolerance for the passband magnitude response is specified by (3.17), i.e. when the magnitude response in the passband oscillates within 1 and $1-\delta_p$, the peak passband ripple in decibels a_p is expressed in the form,

$$a_p = 20\log_{10}\left(\frac{1}{1-\delta_p}\right).$$

(3.22)

Inversely, when expressing δ_p in terms of a_p, we have

$$\delta_p = 1 - 10^{-a_p/20}.$$ (3.23)

In this section, we have discussed the specifications for magnitude response. In many applications, phase response, or overall delay of the system, should be considered as well. In reality, the basic design problem is to meet the given criteria with the transfer function of a reasonable complexity. In the next section, we present by means of examples the usage of MATLAB functions for designing filters according to the specifications of Figures 3.5(a-c).

MATLAB FUNCTIONS FOR FILTER DESIGN

As stated in the previous section, the choice of the appropriate tolerance scheme and the design parameters, ω_p, δ_p for the passband, and ω_s, δ_s for the stopband, defines the design criteria for the filter magnitude response. Either a finite-impulse response (FIR) or an infinite-impulse-response (IIR) transfer function can be used for constructing the sampling rate conversion system. An FIR filter easily achieves a strictly linear phase response, but requires a larger number of operations per output sample when compared with an equal magnitude response IIR filter. But multirate techniques significantly improve the efficiency of FIR filters that makes them very desirable in practice. The efficient implementation structures for FIR and IIR filters will be discussed in Chapters IV and V, respectively. In this section, we demonstrate the application of several MATLAB functions for designing FIR and IIR filters that meet the criteria given by the tolerance schemes of Figures 5(a-c).

FIR Filter Design

Finite impulse response (FIR) filters are of great importance for multirate systems. The general advantages of FIR systems, inherent system stability and phase linearity, are desirable properties for multirate signal processing. Using multirate techniques, the efficiency of a multirate FIR filter significantly increases in comparison with the single rate implementations. These are the main reasons why most of the practical multirate systems are based on FIR filtering.

We demonstrate on examples the application of MATLAB functions for computing FIR filter coefficients and for the analysis of filter transfer function. We consider the optimal linear phase and minimum phase designs based on Parks-McClellan algorithm. Two MATLAB functions are used: firpm from *Signal Processing Toolbox*, and firgr from *Filter Design Toolbox*.

Example 3.1

Design an optimal linear phase FIR filter to satisfy the *Case a* tolerance scheme of Figure 3.5(a).

Compute and plot: impulse response, magnitude response, phase response, and pole-zero locations.

Design specifications:

Sampling-rate conversion factor $M = 5$.

Passband edge at $\omega_p = 0.09\pi$. Stop-band edge at $\omega_s = \pi/5$.

Passband and stopbands ripples are given in decibels with $a_p = 0.1$ dB, and $a_s = 60$ dB.

Solution:

Step 1: Determining the input parameters for the FIR filter design using the MATLAB function firpm (firgr). The MATLAB function firpmord from *Signal Processing Toolbox* can be used to estimate the filter order N_{ord}, and to find: normalized frequency band edges, frequency band amplitudes, and weights that meet the input specifications. The MATLAB code is given below.

```
ap = 0.1; as = 60; Fp = 0.09; Fs = 1/5; % Filter specifications
dev = [(10^(ap/20)-1)/(10^(ap/20)+1),  10^(-as/20)];  % Passband and stopband ripples
F = [Fp Fs];   % Cutoff frequencies
A = [1 0];      % Desired amplitudes
[Nord,Fo,Ao,W]  = firpmord(F,A,dev) % Estimating the filter order
```

For the given input parameters, firpmord returns,

Nord = 49, Ao' = [1, 1, 0, 0], Fo' = [0, 0.0900, 0.2000,1.0000], W' = [1.0000, 5.7564]

Step 2: Filter design. Either the MATLAB function firpm from *Signal Processing Toolbox* or firgr from *Filter Design Toolbox* computes the coefficients of the optimal linear phase FIR filter for the input parameters Nord, Ao, Fo, and W.

With the value Nord= 49, as estimated in *Step 1*, the filter fails to meet the requested stopband attenuation. For achieving the pass-stopband specifications, we increase the filter order by 4,

```
Nord = Nord+4;
h = firpm(Nord,Fo,Ao,W); % Filter coefficients
```

Step 3: Display the results. The functions freqz, impz and zplane are used to compute and plot magnitude and phase responses, impulse response, and pole-zero locations, respectively.

```
figure (1)
freqz(h,1,1024) % Computes and plots the frequency response
axis([0,1,-80,5])
figure (2)
subplot(2,1,1), impz(h,1) % Plots the impulse response
subplot(2,1,2), zplane(h,1) % Pole/zero plot
```

Figures 3.6, 3.7 and 3.8 summarize the results. As shown in Figure 3.6 and 3.7, the specifications are met with an optimal linear-phase FIR filter of the length $N = N_{ord} + 1 = 54$. Figure 3.8 exposes the impulse response symmetry, and the symmetry of the *z*-plane zeros for the linear-phase filter.

Example 3.2

Design an optimal linear phase FIR filter to satisfy the *Case b* tolerance scheme of Figure 3.5(b).

Compute and plot: magnitude response, phase response, impulse response, and pole-zero locations.

Figure 3.6 Magnitude and phase responses for filter of Example 3.1

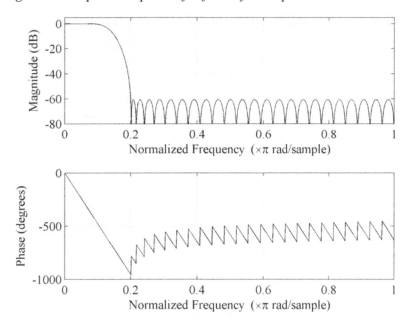

Figure 3.7. Passband details for Example 3.1

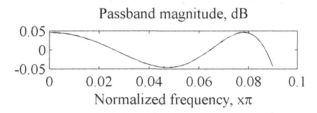

Design specifications:

Sampling-rate conversion factor $M = 5$.

Passband edge at $\omega_p = 0.09\pi$. Stop-band edge at $\omega_s = (2\pi/M - \omega_p)$.

Passband and stopbands ripples are given in decibels with $a_p = 0.1$ dB, and $a_s = 60$ dB.

Solution:

Step 1: Determining the input parameters for the FIR filter design using the MATLAB function firpm (firgr). The function firpmord estimates the filter order N_{ord}, and finds: normalized frequency band edges, frequency band amplitudes, and pass-stopband weights that meet the input specifications. The MATLAB code is a slightly modified code of *Example 3.1.*

```
M = 5; ap = 0.1; as = 60; Fp = 0.09;   Fs = 2/M-Fp; % Filter specifications
dev  = [(10^(ap/20)-1)/(10^(ap/20)+1),  10^(-as/20)] % Passband and stopband ripples
F = [Fp Fs];   % Cutoff frequencies
A = [1 0];     % Desired amplitudes
```

Figure 3.8. Impulse response, and the pole-zero plot for the filter of Example 3.1

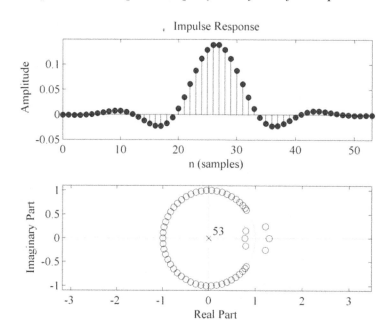

[Nord,Fo,Ao,W] = firpmord(F,A,dev); % Estimating the filter order

For the given input, firpmord returns,

Nord = 24, Ao' = [1, 1, 0, 0], Fo'= [0, 0.0900, 0.3100,1.0000], W' = [1.0000, 5.7564]

Step 2: Filter design. We use MATLAB function firpm from *Signal Processing Toolbox* to compute the coefficients of the optimal linear phase FIR filter for the input parameters Nord, Ao, Fo, and W computed above.

With the value Nord= 24 obtained in *Step 1*, the filter fails to meet the specified stopband attenuation of a_s = 60 dB. For achieving the pass-stopband specifications, we increase the filter order by 3,

```
Nord = Nord + 3;
h = firpm(Nord,Fo,Ao,W);         % Computing the filter coefficients
```

Step 3: Display the results. The functions freqz, impz and zplane are used to compute and plot magnitude and phase responses, impulse response, and pole-zero locations, respectively.

```
figure (1)
freqz(h,1,1024)                  % Computes and plots the frequency response
axis([0,1,-80,5])
figure (2)
subplot(2,1,1), impz(h,1)        % Plots the impulse response
subplot(2,1,2), zplane(h,1)       % Pole/zero plot
```

Results are given in Figures 3.9, 3.10 and 3.11. The magnitude responses displayed in Figures 3.9 and 3.10 show that the optimal linear-phase FIR filter of the length $N = N_{ord} + 1 = 28$ satisfies the specifications. Figure 3.11 exposes the impulse response symmetry, and the symmetry of the z-plane zeros for the linear-phase filter.

This example demonstrates the efficiency of the *Case b* tolerance scheme of Figure 3.5(b). The specifications are met with the linear-phase FIR filter of the length $N = 28$. For the *Case a* tolerance scheme of Figure 3.5(a) the filter length should be $N = 54$ as shown in *Example 3.1* and in Figures 3.6 and 3.7. Comparing the transition bands of two filters, we observe that when the transition bandwidth is doubled, the filter complexity can be nearly halved.

Example 3.3

Design an optimal linear phase FIR filter to satisfy the *Case c* tolerance scheme of Figure 3.5(c).

Compute and plot: magnitude response, phase response, impulse response, and pole-zero locations.

Design specifications:

Sampling-rate conversion factor $M = 5$.

Passband edge at $\omega_p = 0.09\pi$.

Stop-bands: 1st stopband: $[(2\pi/M - \omega_p), (2\pi/M + \omega_p)]$, 2nd stopband $[(4\pi/M - \omega_p), (4\pi/M + \omega_p)]$.

Passband and stopbands ripples are given in decibels with $a_p = 0.1$ dB, and $a_s = 60$ dB.

Solution:

Step 1: Determining the input parameters for the FIR filter design using the MATLAB function firpm (firgr).

- For the specified passband and two stopbands, we compute the boundary frequencies and determine vector of normalized frequency points, fo. The length of fo should be even.

Figure 3.9. Magnitude and phase responses for the filter of Example 3.2

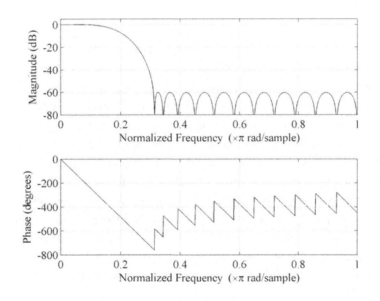

Figure 3.10. Passband details for Example 3.2

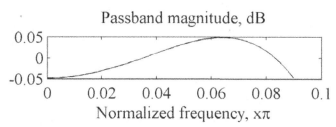

Figure 3.11. Impulse response and the pole-zero plot for the filter of Example 3.2

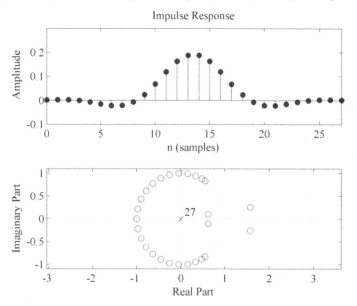

```
fop = [0, fp];                  % Passband boundary frequencies
fos1 = [2/M-fp,2/M+fp];         % Stopband 1 boundary frequencies
fos2 = [4/M-fp,4/M+fp];         % Stopband 2 boundary frequencies
fo= [fop, fos1, fos2];          % Vector of normalized frequency points
```

- For the specified pass and stopbands, we determine the vector of desired amplitudes Ao, Ao = [1,1,0,0,0,0,0,0]; That is that the desired magnitude response in the passband is 1, and 0 in the two stopbands. Notice that vectors Ao and Fo should be of the same length.

- The vector of passband/stopband weights W specifies different weights per bands. The relative weight between bands is given as the ratio of the ripples assigned per band. Hence, we have to compute passband ripple δ_p for the given $a_p = 0.1$ dB, and the stopband ripple δ_s for $a_s = 60$ dB. From (3.19) and (3.21), we obtain $\delta_p = 0.00575639914962$, and $\delta_s = 0.001$, respectively. If we take weight 1 for the passband, the stopband weight is determined by the ratio $\delta_p/\delta_s = 5.75639914962$. Vector W is of the half length of Fo (Ao) and contains one weight per band. For the specifications of this example, the weights for one passband and two stopbands are given by W = [1.0000, 5.7564, 5.7564];

- After a few attempts, we find that the specifications can be met with the filter order Nord = 26.

Step 2: Filter design. The MATLAB function firpm from *Signal Processing Toolbox* is used to compute the coefficients of the optimal linear phase FIR filter.

h = firpm(Nord,fo,Ao,W); % Computing the filter coefficients

Step 3: Display the results. The functions freqz, impz and zplane are used to compute and plot magnitude and phase responses, impulse response, and pole-zero locations, respectively.

figure (1)
freqz(h,1,1024) % Computes and plots the frequency response
axis([0,1,-80,5])
figure (2)
subplot(2,1,1), impz(h,1) % Plots the impulse response
subplot(2,1,2), zplane(h,1) % Pole/zero plot

Results are given in Figures 3.12, 3.13 and 3.14. The magnitude responses displayed in Figures 3.12 and 3.13 show that the optimal linear-phase FIR filter of the length $N = N_{\text{ord}} + 1 = 27$ satisfies the specifications. Figure 3.14 exposes the impulse response symmetry, and the symmetry of the z-plane zeros for the linear-phase filter.

Comparing the results of *Example 3.3* with those of *Example 3.2*, we notice that the filter length is only slightly decreased. The specifications are satisfied with $N = 28$ in *Example 3.2*, whereas $N = 27$ was needed for the specifications of *Example 3.3* regardless of the existing don't care bands in the stopband region. Savings could be expected in the cases with a wider space between the stopbands.

Example 3.4

Design an optimal minimum-phase FIR filter to satisfy the *Case a* tolerance scheme of Figure 3.5(a). Compute and plot: impulse response, magnitude response, phase response, and pole-zero locations.
Design specifications from Example 3.1:
   Sampling-rate conversion factor $M = 5$.
   Passband edge at $\omega_p = 0.09\pi$. Stopband edge at $\omega_s = \pi/5$.
   Passband and stopbands ripples are given in decibels with $a_p = 0.1$ dB, and $a_s = 60$ dB.
Solution:
Step 1: Determining the input parameters for the FIR filter design using the MATLAB function firgr from *Filter Design Toolbox*.

We take the vector Fo for frequency bands, and also the vector Ao for amplitudes from *Example 1*, since the passband/stopband specifications for this example are those of *Example 3.1*. The filter order Nord, and the passband/stopband weights W are to be chosen for the minimum-phase filter design. After several attempts, a reasonable compromise is found between Nord and W. Finally, the input parameters for firgr are the following,

Figure 3.12. Magnitude and phase responses for the filter of Example 3.3

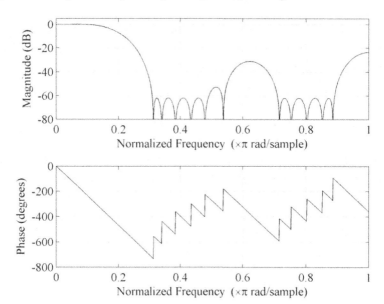

Figure 3.13. Passband details for Example 3.3

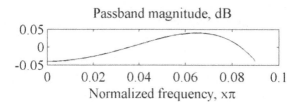

Figure 3.14. Impulse response and the pole-zero plot for the filter of Example 3.3

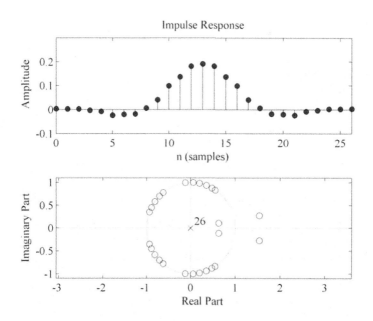

```
fp = 0.09;     ; fs = 1/5;
Nord = 38; fo = [ 0, fp, fs, 1.0000]; Ao = [1, 1, 0, 0]; W =  [1.0000, 200];
```

Step 2: Filter design. In MATLAB, a minimum-phase FIR filter can be designed by using function firgr from *Filter Design Toolbox*, which computes the coefficients of the optimal linear-phase FIR filter for the input parameters Nord, Ao, fo, and W determined in *Step 1*.

```
h = firgr (Nord,fo,Ao,W,'minphase'); % Filter coefficients
```

Step 3: Display the results. The functions freqz, impz and zplane are used to compute and plot magnitude and phase responses, impulse response, and pole-zero locations, respectively.

```
figure (1)
freqz(h,1,1024) % Computes and plots the frequency response
axis([0,1,-80,5])
figure (2)
subplot(2,1,1), impz(h,1) % Plots the impulse response
subplot(2,1,2), zplane(h,1) % Pole/zero plot
```

Figures 3.15, 3.16 and 3.17 plot the results. As shown in Figure 3.15 and 3.16, the specifications are met with an optimal minimum-phase FIR filter of the length $N = N_{ord} + 1 = 39$. Apparently, the phase characteristic is nonlinear. The impulse response of a minimum-phase filter is non-symmetric as demonstrated in Figure 3.17. Figure 3.17 exposes also the *z*-plane zeros, which are located around the unit circle and inside the unit circle. Since the filter is minimum-phase, there are no zeros outside the unit circle.

In *Example 3.1*, the requested magnitude response is met with the linear-phase FIR filter of the length $N = 54$, whereas the filter length of only $N = 39$ was shown to be sufficient for satisfying the requirements for the minimum-phase design of *Example 3.4*.

In this subsection, we have considered the design of FIR filters based on the optimal equiripple approach, which provides desired pass- stopband characteristics with a minimal filter order. However, one can use other methods for computing the coefficients of the filter transfer functions. FIR filter design based on the window functions is frequently used in practice. The MATLAB function fir1 from *Signal Processing Toolbox* implements the classical method of windowed linear-phase FIR filter design. Readers can use the window method for solving MATLAB Exercises given at the end of this chapter.

Interpolated FIR (IFIR) Filter Design

The interpolated FIR (IFIR) filters based on the cascade of two filters were first suggested in (Neuvo, Dong, & Mitra 1984). The realization structure of an IFIR filter is indicated in Figure 3.18. The first filter in cascade is the periodic filter $H(z^L)$, and the second filter $G(z)$ is called the image suppression filter.

The overall transfer function for the structure of Figure 3.18, $H_{IFIR}(z)$ is evidently given by

$$H_{IFIR}(z) = H(z^L)G(z)$$ (3.24)

81

Figure 3.15. Magnitude and phase responses for the minimum-phase filter of Example 3.4

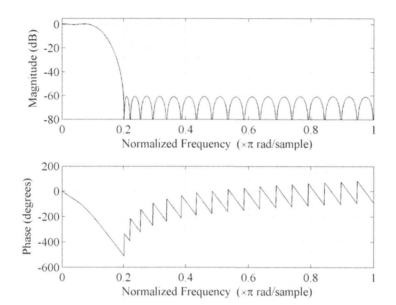

Figure 3.16. Passband details for Example 3.4

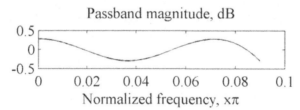

The impulse response of the periodic filter $H(z^L)$ is the up-sampled-by-L impulse response of the filter $H(z)$. The role of the filter $G(z)$ is to interpolate the zero valued samples. In the frequency domain, the role of $G(z)$ is viewed as the suppression of unwanted images.

For low-pass filter design, the IFIR based approach is of interest when the cutoff frequency is low in comparison with the sampling frequency. The advantage of IFIR filter design is due to the fact that the overall frequency response is shared between two low-order filters. In *Example 3.5*, we demonstrate the efficiency of an IFIR filter on the specifications given in *Example 3.1* for the *Case a* tolerance scheme of Figure 3.5(a).

Example 3.5

Design an IFIR filter to satisfy the *Case a* tolerance scheme of Figure 3.5(a). Compute and plot: impulse response, magnitude response, phase response, and pole-zero locations.

Figure 3.17. Impulse response and the pole-zero plot for the filter of Example 3.4

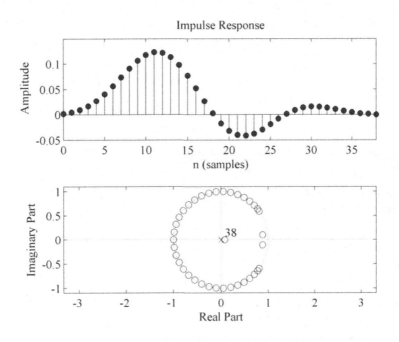

Figure 3.18. Realization structure for an IFIR filter

Design specifications:

Sampling-rate conversion factor $L = 5$.

Passband edge at $\omega_p = 0.09\pi$. Stop-band edge at $\omega_s = \pi/5$.

Passband and stopbands ripples are given in decibels with $a_p = 0.1$ dB, and $a_s = 60$ dB.

Solution:

Step 1: Determining the input parameters for the IFIR filter design using the MATLAB function ifir from the *Filter Design Toolbox*. The input parameters for ifir are: the interpolation parameter L, the filter type: low-pass, high-pass, the two-element vector Fo containing the cutoff frequencies, and the two-element vector dev containing the peak ripple values in the passband and in the stopband. For the given specifications, we chose the following values:

L = 3; Fo = [.09, 0.2]; dev = [0.00575639914962, 0.001]);

Step 2: Filter design. We use the MATLAB function ifir from *Filter Design Toolbox* and the input parameters L, Fo, and dev determined in *Step 1*, to compute the coefficients of the periodic filter $H(z^L)$, image-suppression filter $G(z)$, and of the overall filter $H_{IFIR}(z)$. We also compute the magnitude responses of the three filters. The MATLAB code is the following,

```
[h,g] = ifir(3,'low',Fo,dev); % Design of H(z^L) and G(z)
[H,w] = freqz(h,1,1024); G=freqz(g,1,1024); % Computing the frequency responses of H(z) and G(z)
hIFIR = conv(h,g); % Computing the overall impulse response of the interpolated filter
HIFIR = freqz(hIFIR,1,1024); % Computing the frequency response of the interpolated filter
```

Step 3: Display the results. In this step, we plot magnitude responses for $H(z^L)$, $G(z)$, and $H_{IFIR}(z)$, their impulse responses, and pole-zero locations for $H_{IFIR}(z)$.

```
figure (1)
subplot(2,1,1), freqzplot([H,G],w,'mag'); axis([0,1,-80,5])
legend('Periodic Filter','Image Suppressor Filter');
subplot(2,1,2), freqzplot(HIFIR,w,'mag');
axis([0,1,-150,5]), legend('Overall Filter');
figure (2)
subplot(2,2,1), stem(0:length(h)-1,h)
ylabel('h[n]'),xlabel('n'), axis([0,51,-0.1,0.4])
subplot(2,2,2), stem(0:length(g)-1,g)
ylabel('g[n]'),xlabel('n'), axis([0,14,-0.1,0.3])
subplot(2,2,3), stem(0:length(hIFIR)-1,hIFIR)
ylabel('h_I_F_I_R[n]'),xlabel('n'), axis([0,length(hIFIR)-1,-0.05,0.15])
subplot(2,2,4), zplane(hIFIR,1)
```

Figures 3.19, 3.20 and 3.21 display the results. Figure 3.19 illustrates the concept of IFIR approach. The periodic filter is "responsible" for achieving the desired transition band, whereas the image suppressor filter provides a desired stopband attenuation in unwanted periodic passbands of the first filter. The good passband behavior of the overall filter is exposed in Figure 3.20. Figure 3.21 shows the three impulse responses, and also the z-plane locations for the overall filter. The phase responses are not displayed for this example. Evidently, the phase responses of $H(z^L)$, $G(z)$, and $H_{IFIR}(z)$ are linear due to the symmetric impulse responses indicated in Figure 3.21.

The requested specifications are met with 18 nonzero coefficients in the periodic filter $H(z^L)$, and 15 coefficients in the image-suppressor filter $G(z)$. Thereby, the IFIR filter design meets the specifications with only 33 nonzero coefficients. It was shown in *Example 3.1*, that with the single-filter optimal FIR design the specifications are met with 54 nonzero coefficients. In the following chapters, especially in Chapter IV, we will consider the efficient multirate implementation of FIR digital filters. The IFIR approach is an efficient single-rate implementation for a digital filter of high complexity, and according to this the IFIR filter offers an alternative solution to the multirate filtering.

IIR Filter Design

Infinite impulse response (IIR) filters are used in multirate systems when the high computational efficiency or low signal delay are of highest priority. Compared with an FIR filter, the IIR filter attains the same pass- stopband requirements with the transfer function of a significantly lower order. The disadvantages of IIR filters are the nonlinear phase response and the sensitivity of filter characteristics to the quantization error arising in arithmetic operations.

Figure 3.19. Magnitude responses of H(z^L), G(z), H_{IFIR}(z) for the IFIR filter of Example 3.5

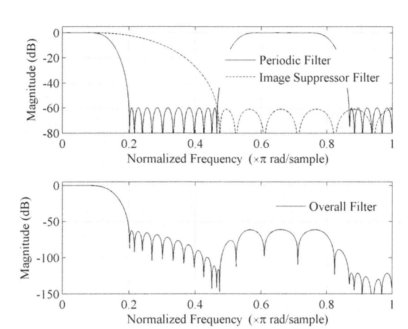

Figure 3.20. Passband details for the overall filter H_{IFIR}(z) of Example 3.5

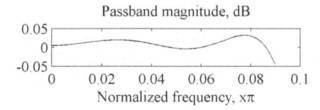

Here, we demonstrate the application of MATLAB functions in the design and analysis of IIR filters. We illustrate IIR filter design on example elliptic (Cauer) and Chebyshev filters. We are concentrated on *Case a* and *Case b* tolerance schemes of Figure 3.5.

Example 3.6

Design elliptic IIR filters to satisfy: (a) *Case a* tolerance scheme from Figure 3.5(a), and (b) *Case b* tolerance scheme of Figure 3.5(b). Compute and plot the magnitude and phase responses, impulse response and pole/zero plot.

Design specifications:

Sampling-rate conversion factor $M = 5$.

Passband edge at $\omega_p = 0.09\pi$.

Figure 3.21. Impulse responses $\{h[n]\}$, $\{g[n]\}$, $\{h_{IFIR}[n]\}$, and the pole-zero plot for $H_{IFIR}(z)$

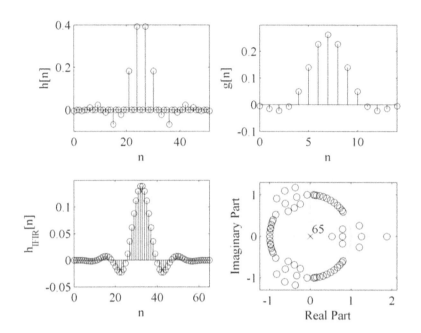

Stop-band edge at: *Case a:* $\omega_s = \pi/M$; *Case b:* $\omega_s = (2/M-0.09)\pi$.

Passband and stopbands ripples are given in decibels with $a_p = 0.1$ dB, and $a_s = 60$ dB.

Solution:

Step 1: Determining the input parameters for the elliptic filter design using the MATLAB function ellip. The MATLAB function ellipord from *Signal Processing Toolbox* calculates the minimum order N and the cutoff frequency Fn of an elliptic filter that meets the set of filter design specifications defined above.

```
[N,Fn] = ellipord(Fp,Fs,ap,as)
```

Input arguments for *Case a:* Fp=0.09; Fs=0.20; ap=0.1; as=60; returns: N = 5; Fn = 0.09.

Input arguments for *Case b:* Fp=0.09; Fs=0.31; ap=0.1; as=60; returns: N = 4; Fn = 0.09.

Step 2: Filter design. The MATLAB function ellip from *Signal Processing Toolbox* computes the coefficients (or poles and zeros) of an elliptic filter transfer function. The input arguments are the filter order N, the peak attenuation in the passband ap (dB), the minimal stopband attenuation as (dB), and normalized passband edge frequency Fn. When specified with two output arguments, ellip computes the coefficients of the transfer function,

```
[B,A] = ellip(N,ap,as,Fn);  % Elliptic filter design
```

where the numerator and denominator coefficients in descending powers of z are stored in row vectors B and A, respectively. When specified with three output arguments, ellip returns the poles and zeros, and the gain factor, [z,p,k] = ellip(N,ap,as,Fn); % Elliptic filter design

Here the column vectors z and p store zeros and poles, and the scalar k is the gain factor.

Step 3: Display the results. The magnitude and phase responses, impulse responses (first 61 samples), and pole/zero plots for the 5th order and 4th order elliptic filters are computed by using freqz, impz and zplane, respectively. The magnitude responses are plotted in Figures 3.22 and 3.23. Figure 3.24 displays the impulse responses and the pole/zero locations for both designs.

This example demonstrates the efficiency of IIR filters and elliptic transfer functions. The specifications of the *Case a* tolerance scheme are met with the 5th order elliptic filter, and the 4th order filter succeeds in meeting the requirements of the *Case b* tolerance scheme. The disadvantage of those efficient designs is the nonlinearity of phase responses. Furthermore, the filter poles are close to the unit circle, particularly those of the 5th order filter, see Figure 3.24. This may cause the unwanted finite wordlength effects in implementation.

Example 3.7

Design Chebyshev IIR filters to satisfy: (a) *Case a* tolerance scheme, Figure 3.5(a), and (b) *Case b* tolerance scheme of Figure 3.5(b). Compute and plot the magnitude and phase responses, impulse response and pole/zero plot.

Design specifications:

Sampling-rate conversion factor $M = 5$.

Passband edge at $\omega_p = 0.09\pi$.

Stop-band edge at: *Case a:* $\omega_s = \pi /M$, *Case b:* $\omega_s = (2/M - 0.09)\pi$.

Passband and stopbands ripples are given in decibels with $a_p = 0.1$ dB, and $a_s = 60$ dB.

Solution:

Step 1: Determining the input parameters for the Chebyshev filter design using the MATLAB function cheby1. The MATLAB function cheb1ord from *Signal Processing Toolbox* can be used to calculate the minimum order N and the normalized cutoff frequency Fn of a Chebyshev filter that meets the given set of design specifications.

[N,Fn] = cheb1ord(Fp,Fs,ap,as) % Estimating the filter order

Input arguments for *Case a:* Fp = 0.09; Fs = 0.20; ap = 0.1; as = 60; returns: N = 7; Fn = 0.09.
Input arguments for *Case b:* Fp = 0.09; Fs = 0.31; ap = 0.1; as = 60; returns: N = 5; Fn = 0.09.

Step 2: Filter design. The MATLAB function cheby1 from *Signal Processing Toolbox* computes the coefficients (or poles and zeros) of the Chebyshev filter transfer function. The input arguments are the filter order N, the peak attenuation in the passband ap (dB) and normalized passband edge frequency Fn. When specified with two output arguments, cheby1 computes the coefficients of the transfer function,

[B,A] = cheby1(N,ap, Fn); % Chebyshev filter design

Figure 3.22. Magnitude and phase responses for elliptic filters of Example 3.6. Solid line is for Case a design, N=5, and dotted line for Case b design, N=4.

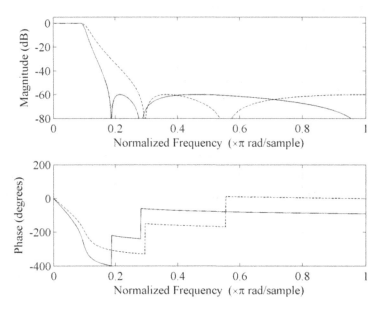

Figure 3.23. Passband details for elliptic filters of Example 3.6. Solid line is for Case a design, N=5, and dotted line for Case b design, N=4.

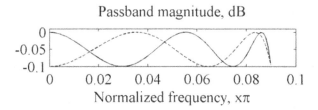

where the numerator and denominator coefficients in descending powers of z are stored in row vectors B and A, respectively. When specified with three output arguments, cheby1 returns the poles and zeros, and the gain factor,

[z,p,k] = cheby1(N,ap,Fn); % Chebyshev filter design

Here, the zeros and poles are stored in the column vectors z and p, and the scalar k is the gain factor.

Step 3: Display the results. The magnitude and phase responses, impulse responses, and pole/zero plots for the 7th order and 5th order Chebyshev filters are computed by using freqz, impz and zplane, respectively. The magnitude responses are plotted in Figures 3.25 and 3.26. Figure 3.27 depicts the impulse responses and the pole/zero locations for both designs.

Figure 3.24. Impulse responses and pole/zero plots for elliptic filters of Example 6

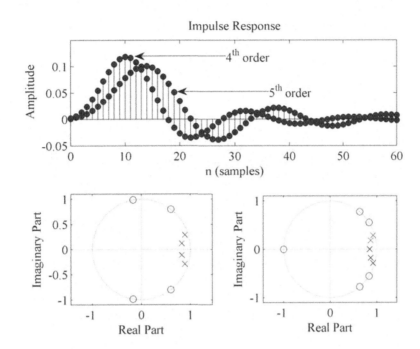

The specifications of the *Case a* tolerance scheme are met with the 7th order Chebyshev filter, and the 5th order filter meets the requirements of the *Case b* tolerance scheme. The phase characteristics are nonlinear as shown in Figure 3.25. As in the case of elliptic filter design of *Example 3.6*, the filter poles are close to the unit circle, and unwanted finite wordlength effects can be expected in implementation.

COMPUTATION OF ALIASING CHARACTERISTICS

Decimation inevitably causes aliased spectra in the baseband of the low-rate signal. Consequently, aliasing introduces the unwanted interference in the decimated signal, and decreases the signal-to-noise ratio in the baseband. The role of the decimation filter is to suppress the aliased spectra to the level which is acceptable for the application at hand. This can be achieved with an appropriate attenuation of the frequency bands in the high-rate signal that overlap in the baseband of the low-rate signal.

In this section, we give an insight into the filter contribution to the suppression of aliasing, and demonstrate on examples how the *aliasing characteristics* of the decimation filter can be computed.

The role of decimation (antialiasing) filter $H(z)$ is to attenuate the spectral components of the high-rate signal above a half of sampling-frequency of the low-rate signal. As already given in (3.4), the spectrum of the low-rate signal $Y(e^{j\omega})$ is expressible in terms of the spectrum of the high-rate signal $X(e^{j\omega})$ and the frequency response of the antialiasing filter $H(e^{j\omega})$,

$$Y\left(e^{j\omega}\right) = \frac{1}{M}\sum_{k=0}^{M-1} X\left(e^{j(\omega-2\pi k)/M}\right) H\left(e^{j(\omega-2\pi k)/M}\right). \tag{3.25}$$

We can write (3.25) also in the form,

Figure 3.25. Magnitude and phase responses for Chebyshev filters of Example 3.7. Solid line is for Case a design, N=7, and dotted line for Case b design, N=5.

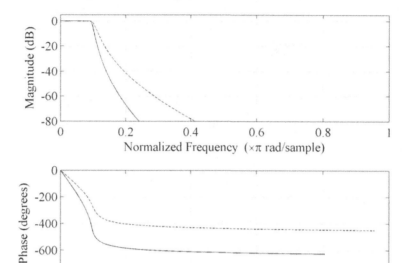

Figure 3.26. Passband details for Chebyshev filters of Example 3.7. Solid line is for Case a design, N=7, and dotted line for Case b design, N=5.

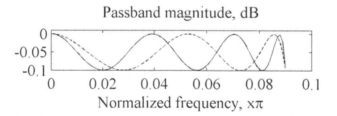

$$Y\left(e^{j\omega}\right)=\frac{1}{M}\sum_{k=0}^{M-1}X\left(e^{j(\omega-2\pi k)/M}\right)H_{al}^{k}\left(e^{j\omega}\right),\qquad(3.26)$$

where

$$H_{al}^{k}\left(e^{j\omega}\right)=H\left(e^{j(\omega-2\pi k)/M}\right),\quad k=0,1,\ldots,M-1.\qquad(3.27)$$

Here, for $k = 0$, function $H_{al}^{k}(e^{j\omega})$ represents the unaliased characteristic, whereas $H_{al}^{k}(e^{j\omega})$ for $k = 1, \ldots,$ $M-1$, represent $M-1$ aliasing characteristics for a factor-of-M decimator. To determine the suppression of the aliased spectra, we compute the absolute values of $H_{al}^{k}(e^{j\omega})$, and usually express in decibels,

$$a_{al}^{k}\left(\omega\right)=20\log_{10}\left|H_{al}^{k}\left(e^{j\omega}\right)\right|,\quad k=0,1,\ldots,M-1.\qquad(3.28)$$

Figure 3.27. *Impulse responses and pole/zero plots for Chebyshev filters of Example 7*

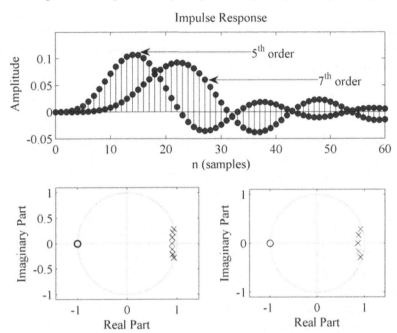

The aliasing characteristics of a decimator are easier to analyze in terms of "real" frequencies expressed in Hz. Let us denote by F_x and $F_y = F_x/M$ the input and output frequencies of a factor-of-M decimator, respectively. The angular frequency ω is to be normalized to the output sampling frequency F_y, since we compute the frequency components in the baseband of the lower-rate signal, i.e.,

$$\omega = 2\pi \, F/F_y \, . \tag{3.29}$$

Therefore, the alternative expression for (3.27) is the following,

$$H_{al}^k \left(e^{j2\pi F/F_y} \right) = H \left(e^{j2\pi (F - kF_y)/F_x} \right) \quad k = 0,1,\ldots,M-1. \tag{3.30}$$

We also express (3.28) in the form

$$a_{al}^k \left(F \right) = 20\log_{10} \left| H_{al}^k \left(e^{j2\pi F/F_y} \right) \right|, \quad k = 0,1,\ldots,M-1. \tag{3.31}$$

Using equations (3.30) and (3.31), we can compute the aliasing characteristics of the filter $H(z)$ in the frequency band of the low rate signal $[0, F_y/2]$. In the following example, we demonstrate in MATLAB the computation of aliasing characteristics for FIR filters of *Examples 3.1, 3.2, 3.3* from the preceding section.

Example 3.8

Program demo_3_1 can be employed to compute the aliasing characteristics of an FIR antialiasing filter, which is used in the factor-of-5 decimator. The input sampling frequency of the decimator is 10000Hz.

Program calls the input data vector h with FIR filter coefficients. Program computes and plots:

- Magnitude response of the antialiasing filter $H(z)$ in the range of the high-rate signal [0, 5000 Hz].
- Unaliased characteristic, according to (3.30) with $k = 0$, in the range of the low-rate signal [0, 1000 Hz].
- Four aliased characteristics, according to (3.30) with $k = 1, 2, 3, 4$, in the range of the low-rate signal [0, 1000 Hz].

```
% Program demo_3_1
% Computes and plots the attenuation of aliasing components in FIR decimator.
% FIR filter: program requests the FIR filter coefficeints.
% Sampling frequencies: input Fx=10000 Hz, output Fy=2000 Hz.
% Decimation factor M=5.
close all
h = input('Insert the FIR filter coefficients ')
n = 0:length(h)-1; % Time index
Fx = 10000; % Input sampling frequency (Hz)
M = 5; % Decimation factor
Fy = Fx/M;  % Output sampling frequency (Hz)
% Computing the filter magnitude response
for r=1:1001
    F(r) = (r-1)*Fx/2000;
omega = 2*pi*F(r)/Fx;
W = exp(-i*n*omega);
H(r) = sum(h.*W);
end

figure (1)
subplot(2,1,1)
plot(F,20*log10(abs(H)))
xlabel('Frequency, Hz'),ylabel('Magnitude, dB')
axis([0,5000,-80,5]), grid

subplot(2,1,2)
% Computing the unaliased characteristic and 4 aliased characteristics according to (3.30)
k = 0;
for l=1:5  %
for r=1:201
    FF(r) = (r-1)*Fy/400;
omega = 2*pi*FF(r)/Fy;
W = exp(-i*n*(omega-2*k*pi)/M);
Hal(r) = sum(h.*W);
end
k = k + 1;
```

```
plot(FF,20*log10(abs(Hal)))
hold on
end
xlabel('Frequency, Hz'),ylabel('Magnitude, dB')
axis([0,1000,-80,5]), grid on
hold off
```

Using program demo_3_1, the aliasing characteristics are computed for FIR filters of *Examples 3.1, 3.2* and *3.3*. Figure 3.28 presents the results obtained for the filter of *Example 3.1*, which was designed to satisfy *Case a* specification of Figure 3.5(a). Figure 3.28 presents: the filter magnitude response in the baseband of the high-rate signal [0, 5000 Hz], and the plots of the unaliasing characteristic and four aliasing characteristics in the baseband of the low-rate signal [0, 1000 Hz]. The plots of Figure 3.28 demonstrate that in the whole baseband of the low-rate signal aliasing is suppressed by 60 dB. This is ensured by the minimal attenuation of 60 dB in the whole range [1000 Hz, 5000 Hz] at the high-rate side. Hence, with the *Case a* specification, aliasing in the transition band is negligable.

Figure 3.29 presents the results obtained for the filter of *Example 3.2* designed according to the *Case b* specification of Figure 3.5(b). Figure 3.29 presents: the filter magnitude response in the baseband of the high-rate signal [0, 5000 Hz], and the plots of the unaliasing characteristic and four aliasing characteristics in the baseband of the low-rate signal [0, 1000 Hz]. The plots of Figure 3.29 show the significant amount of aliasing from the adjacent band, whereas the aliasing from the remaining three bands is suppressed by 60 dB. As indicated in Figure 3.29, aliasing from the adjacent band appears in the transition band of the low-rate signal. Some amount of aliasing in the transition band is acceptable in many applications.

Figure 3.30 presents the results obtained for the filter of *Example 3.3*, which was designed to satisfy *Case c* specification of Figure 3.5(c). Figure 3.30 shows the filter magnitude response in the baseband of the high-rate signal [0, 5000 Hz]. In the baseband of the low-rate signal [0, 1000 Hz], Figure 3.30 plots the unaliasing characteristic and four aliasing characteristics. The curves of Figure 3.30 show a significant amount of aliasing in the transition band of the low-rate signal that comes from the adjacent band and also from the don't-care bands. It can be observed that aliasing is suppressed by 60 dB in the passband of the low-rate signal. This is ensured by the minimal attenuation of 60 dB in the subbands, see the filter magnitude response in Figure 3.30. Thereby, *Case c* specification permits a high amount of aliasing in the transition band of the low-rate signal.

In this example, we have demonstrated the aliasing characteristics on three different FIR filters designed to meet *Case a*, *Case b*, and *Case c* tolerance schemes. It is left to the reader to modify program demo_3_1 for computing the aliasing characteristics for decimators with IIR filters.

SAMPLING RATE ALTERATION OF BANDPASS SIGNALS

In preceding sections, we have assumed that the spectrum of the signal whose sampling frequency is to be altered is concentrated in the frequency range $[0, \omega_p]$ with $\omega_p < \pi$, i.e., we have considered the sampling rate alteration of lowpass signals. However, the need for sampling rate alteration of bandpass signals frequently occurs in practice. The spectrum of a bandpass signal is usually concentrated in the narrow band of frequencies above the zero frequency. The central frequency ω_c of such a signal is generally

Figure 3.28. Magnitude response of the antialiasing filter. Magnitude responses of the unaliased characteristic (thick line) and the magnitude response of the four aliasing characteristics (thin lines). Case a FIR filter design.

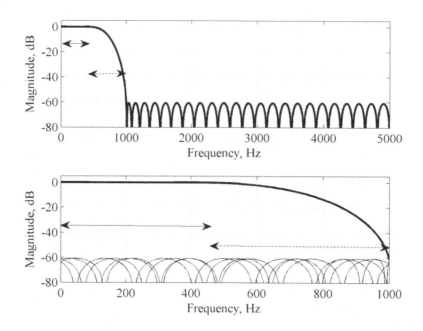

Figure 3.29. Magnitude response of the antialiasing filter. Magnitude responses of the unaliased characteristic (thick line) and the magnitude response of the four aliasing characteristics (thin lines). Case b FIR filter design.

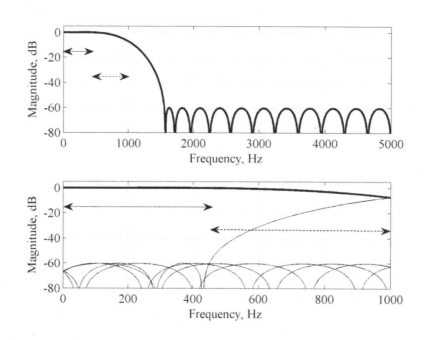

much larger than the signal bandwidth Δ_B, $\omega_c \gg \Delta_B$. Generally, the spectrum of the bandpass signal can occupy any band in the range $[0, \pi]$. Using modulation techniques, the bandpass signal can be translated into the lowpass range $[0, \Delta_B]$, and then decimated or interpolated as a lowpass signal.

The most popular technique for the sampling rate alteration of bandpass signals is based on the *integer-band decimation and interpolation*. This technique can be applied when the signal spectrum is confined to the frequency range

$$\frac{k\pi}{M} \leq \omega \leq \frac{(k+1)\pi}{M}, \tag{3.32}$$

where k is a positive integer. The bandwith of the signal is restricted to $\Delta_B \leq \pi/M$, and the boundary frequencies of the spectrum are the integer multiples of π/M as indicated in Figure 3.31.

In decimation, prior to down-sampling, the signal components outside the desired frequency range have to be eliminated. Thereby, the signal is filtered first with an antialiasing bandpass filter $H_{BP}(z)$ whose ideal magnitude response for the integer-band decimation is defined by

$$\left| H_0^{BP}\left(e^{j\omega}\right) \right| = \begin{cases} 1, & k\pi/M \leq |\omega| \leq (k+1)\pi/M \\ 0, & \text{otherwise} \end{cases} \tag{3.33}$$

In practice, the ideal magnitude response is approximated with an appropriate transfer function. Filtered signal is bandlimited to the selected frequency range defined by (3.32) and can be directly down-sampled by M.

Figure 3.32 pictures the decimation process in terms of 'real' frequencies. The desired frequency band is selected by the bandpass filter $H_{BP}(z)$. The direct factor-of-M downsampling of the filtered bandpass

Figure 3.30. Magnitude response of the antialiasing filter. Magnitude responses of the unaliased characteristic (thick line) and the magnitude response of the four aliasing characteristics (thin lines). Case c FIR filter design.

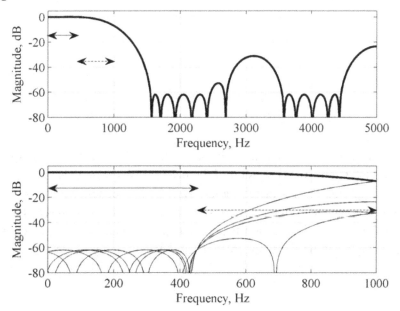

Figure 3.31. Spectrum of the bandpass signal

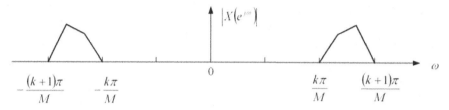

Figure 3.32. Integer-band decimation. (a) Structure of the bandpass decimator. (b) Spectrum of the bandpass signal. (c) Spectrum of the decimated signal for k even. (d) Spectrum of the decimated signal for k odd.

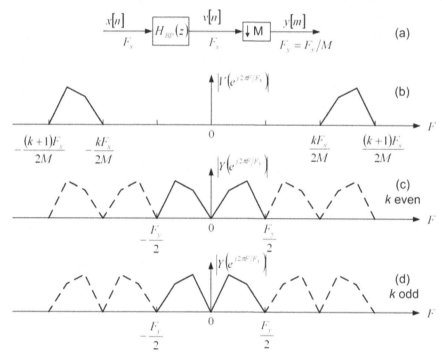

signal $\{v[n]\}$ results in the spectrum shown in Figure 3.32(c) for k even, and in the spectrum shown in Figure 3.32(d) for k odd. There is an inversion of the original spectrum for k odd. This inversion can be turned back by multiplying the decimated signal $\{y[n]\}$ by the sequence $(-1)^n$, $n=0, 1,2,\dots$ Note that the passband boundary frequencies of the antialiasing bandpass filter are to be chosen according to the constraint given by (3.32). Otherwise, a significant amount of aliasing can be expected.

The following example demonstrates the integer-band decimation process.

Example 3.9

The decimation process of the bandpass signal is demonstrated in the program demo_3_2. The program generates 512 samples of the input signal $\{x[n]\}$ using the MATLAB function fir2, designs the 68th order linear-phase FIR bandpass filter $H_{BP}(z)$, performs bandpass filtering, and performs the down-sampling of the filtered signal. The sampling rate at the input is $F_x = 10000$ Hz, and the down-sampling factor is $M = 5$. The program plots the spectrum of the input signal, filter magnitude response, spectrum of the filtered signal and spectrum of the decimated signal.

```
% Program demo_3_2
% Integer-band decimation
clear all, close all
Fx = 10000; M = 5; Fy = Fx/M;
F = [0,2100/5000,2800/5000,2900/5000,3100/5000,3400/5000,3800/5000,4000/5000,4500/5000,1];
A = [0,0.05,1,0,0,0.7,0.8,0,0.6,0];
x = fir2(511,F,A);       % Generating the input signal signal 'x'
[X,f] = freqz(x,1,1024,10000);    % Computing the spectrum of the signal 'x'
figure (1)
subplot(4,1,1)
plot(f,abs(X)), ylabel('|X(e^j^\omega)|')
% Bandpass filter design
hBP = firgr(68,[0,1800/5000,2200/5000,2800/5000,3200/5000,1],[0,0,1,1,0,0]);
[HBP,f] = freqz(hBP,1,1024,10000);    % Computing the bandpass filter frequency response
subplot(4,1,2), plot(f,20*log10(abs(HBP))), ylabel('|H_B_P(e^j^\omega)|')
axis([0,5000,-60,5])
v = filter(hBP,1,x); % Bandpass filtering
[V,f] = freqz(v,1,1024,10000);  % Computing the spectrum of the bandpass signal
subplot(4,1,3), plot(f,abs(V)), ylabel('|V(e^j^\omega)|')
y = downsample(v,M);   % Down-sampling of the bandpass signal
[Y,ff] = freqz(y,1,1024,2000); % Computing the spectrum of the decimated signal
subplot(4,1,4); plot(ff,abs(Y)), ylabel('|Y(e^j^\omega)|')
xlabel('Frequency, Hz')
```

The results obtained for $k = 2$ are displayed in Figure 3.33. Filter selects the frequency range 2000 – 3000 Hz and forms the bandpass signal $v[n]$, whose spectrum is nonzero in the band 2000 – 3000 Hz and approximately zero outside this range. The decimated signal $y[n]$ is obtained by direct down-sampling of bandpass signal $v[n]$ by $M = 5$. Figure 3.33 plots the spectrum of the original signal, filter magnitude response, spectrum of the bandpass signal, and the spectrum of the decimated signal. Notice that for the selected k being an even number, the spectrum of the decimated signal is not inverted.

The integer-band interpolation actually means extracting the kth image from L spectral images of an up-sampled-by-L signal. Hence, the integer bandpass interpolator is the cascade connection of an up-sampler and the antiimaging bandpass filter as indicated in Figure 3.34.

For selecting the desired k^{th} image from the spectrum of the up-sampled signal, the cutoff frequencies of the antiimaging bandpass filter should obey constraints given in (3.32). According to (3.32), the ideal antiimaging filter magnitude response $|H_0^{BP}(e^{j\omega})|$ for

A factor-of-L interpolator is defined by

$$\left|H_0^{BP}\left(e^{j\omega}\right)\right| = \begin{cases} L, & k\pi/L \le |\omega| \le (k+1)\pi/L \\ 0, & \text{otherwise} \end{cases}. \tag{3.34}$$

This ideal magnitude response is approximated with the filter transfer function that selects the desired frequency band and attenuates sufficiently the unwanted images. The process of integer-band interpolation is illustrated in the following example.

Example 3.10

Program demo_3_3 demonstrates the bandpass interpolation process for $k = 3$. First, program generates the 103 samples of the lowpass input signal x which has to be interpolated. The program performs up-sampling of the input signal, designs the 68th order linear-phase FIR bandpass filter $H_{BP}(z)$, and performs the bandpass filtering of the up-sampled signal. The sampling rate at the input is $F_x = 2000$ Hz, and the up-sampling factor is $M = 5$. The program plots the spectrum of the input signal, spectrum of the up-sampled signal, filter magnitude response, and spectrum of the filtered (interpolated) signal.

```
% Program demo_3_3
% Integer-band interpolation
clear all,close all
Fx = 2000; M = 5; Fy = 10000;
F = [0,100/1000,800/1000,900/1000,1]; A=[0,0,1,0,0];
```

Figure 3.33. Illustration of the integer-factor decimation of the bandpass signal for k = 2, program demo_3_2

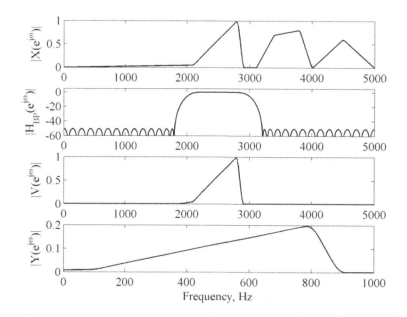

Figure 3.34. Structure of the bandpass interpolator

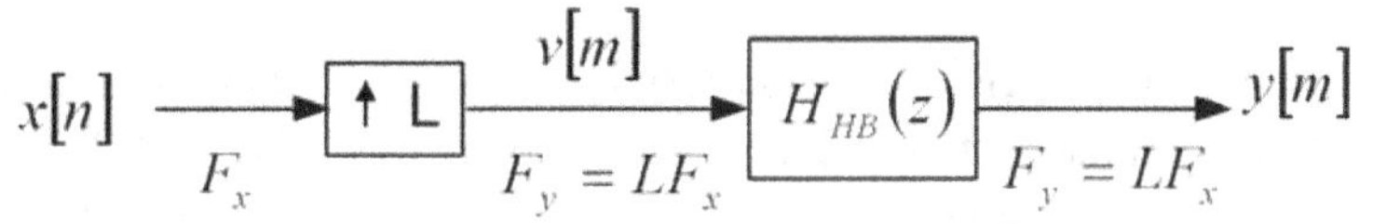

```
x = fir2(102,F,A);     % Generating the input signal signal 'x'
figure (2)
[X,ff] = freqz(x,1,1024,2000);  % Computing the spectrum of the signal 'x'
subplot(4,1,1); plot(ff,abs(X)), ylabel('|Y(e^j^\omega)|')
xlabel('Frequency, Hz')
% Bandpass filter design
hBP=5*firgr(64,[0,2800/5000,3200/5000,3800/5000,4200/5000,1],[0,0,1,1,0,0]);
v = upsample(x,5);     % Up-sampling
[V,f] = freqz(v,1,1024,10000); % Spectrum of the up-sampled signal
subplot(4,1,2)
plot(f,abs(V)), ylabel('|V(e^j^\omega)|')
[HBP,f] = freqz(hBP,1,1024,10000);  % Computing the bandpass filter frequency response
subplot(4,1,3)
plot(f,20*log10(abs(HBP))), ylabel('|H_B_P(e^j^\omega)|')
axis([0,5000,-60,5])
y = filter(hBP,1,v);   % Filtering
[Y,f] = freqz(y,1,1024,10000); % Computing the spectrum of the bandpass interpolated signal
subplot(4,1,4)
plot(f,abs(Y)), ylabel('|Y(e^j^\omega)|')
xlabel('Frequency, Hz')
```

The results of the integer-band interpolation obtained for $k = 3$ are displayed in Figure 3.35. The low-pass signal whose spectrum is of a triangular-shape is shown on the top. The next plot displays the spectrum of the up-sampled signal for $L = 5$. The bandpass filter $H_{BP}(e^{j\omega})$ selects the frequency band 3000-4000 Hz. The interpolation is accomplished by bandpass filtering of the up-sampled signal. Finally, the spectrum of the interpolated bandpass signal is shown.

The plots of Figure 3.35 illustrate the inversion of the signal spectrum in the case when k is an odd number. The spectrum can be re-inverted by multiplying the interpolated signal $\{y[n]\}$ by the sequence $(-1)^n$, $n=0, 1,2,\dots$.

The integer-band decimation and interpolation discussed in this section provide also a modulation-free method for frequency translations. When the bandpass signal is decimated by M and then up-sampled by the same conversion factor, $L = M$, the interpolation band-pass filter can be designed to select a desired spectral image and suppress others. This is illustraded in Figures 3.33 and 33.35. The frequency band 2000-3000 Hz from Figure 3.33, after decimation-by-5 with $k = 2$, and interpolation-by-5 with $k = 3$ is translated into the frequency band 3000-4000 Hz, see Figure 3.35.

MATLAB EXERCISES

3.1 Design an optimal linear-phase FIR filter to satisfy the *Case a* tolerance scheme of Figure 3.5 (a), for the sampling-rate conversion factor $M = 4$. The peak passband ripple is $a_p = 0.1$ dB, and the minimal stopband attenuation is $a_s = 50$ dB. Choose the passband edge frequency $\omega_p < \pi/M$. Compute and plot: impulse response, magnitude response, phase response, and pole-zero locations.

3.2 Compute and plot the aliasing characteristics of the 4-fold-decimator with antialiasing filter from Exercise 3.1

3.3 Design an optimal linear-phase FIR filter to satisfy the *Case b* tolerance scheme of Figure 3.5 (b), for the sampling-rate conversion factor $M = 4$. The peak passband ripple is $a_p = 0.1$ dB, and the minimal stopband attenuation is $a_s = 50$ dB. Choose the passband edge frequency $\omega_p < \pi/M$. Compute and plot: impulse response, magnitude response, phase response, and pole-zero locations.

3.4 Compute and plot the aliasing characteristics of the 4-fold-decimator with antialiasing filter from Exercise 3.3.

3.5 Design an optimal linear-phase FIR filter to satisfy the *Case c* tolerance scheme of Figure 3.5 (c), for the sampling-rate conversion factor $M = 4$. The peak passband ripple is $a_p = 0.1$ dB, and the minimal stopband attenuation is $a_s = 50$ dB. Choose the passband edge frequency $\omega_p < \pi/M$. Compute and plot: impulse response, magnitude response, phase response, and pole-zero locations.

Figure 3.35. Illustration of the integer-factor interpolation for k = 2, program demo_3_3

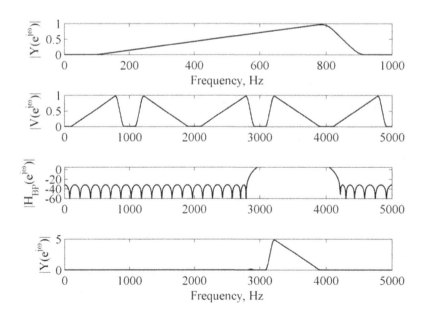

3.6 Compute and plot the aliasing characteristics of the 4-fold-decimator with antialiasing filter from Exercise 3.5.

3.7 Design an elliptic IIR filter to satisfy the *Case b* tolerance scheme of Figure 3.5 (b), for the sampling-rate conversion factor $M = 4$. The peak passband ripple is $a_p = 0.1$ dB, and the minimal stopband attenuation is $a_s = 50$ dB. Choose the passband edge frequency $\omega_p < \pi/M$. Compute and plot: impulse response, magnitude response, phase response, and pole-zero locations.

3.8 Compute and plot the aliasing characteristics of the 4-fold-decimator with IIR anti-aliasing filter from Exercise 3.7.

3.9 Design a Chebyshev IIR filter to satisfy the *Case b* tolerance scheme of Figure 3.5 (b), for the sampling-rate conversion factor $M = 4$. The peak passband ripple is $a_p = 0.1$ dB, and the minimal stopband attenuation is $a_s = 50$ dB. Choose the passband edge frequency $\omega_p < \pi/M$. Compute and plot: impulse response, magnitude response, phase response, and pole-zero locations.

3.10 Compute and plot the aliasing characteristics of the 4-fold-decimator with IIR antialiasing filter from Exercise 3.9.

3.11 Modify program `demo_3_2` to study the integer-band decimation for the factor $M = 6$, and the input sampling frequency of 120 kHz.

3.12 Using the integer-band decimation and the integer-band interpolation, develop the MATLAB program which translates the selected band of frequencies into a desired position.

REFERENCES

Ansari,R., & Liu,B., (1993). Multirate signal processing. In Sanjit. K. Mitra and James F. Kaiser (ed.), *Handbook for Digital Signal Processing*. New York, NY: John Wiley-Interscience, 981-1084.

Bellanger, M.G., Bonnerot, G., & Coudreuse, M. (1976). Digital filtering by polyphase network: application to sample-rate alteration and filter banks. *IEEE Transactions on Acoustics, Speech, and Signal Processing, 24*(2), 109-114.

Bellanger, M. (2000). *Digital processing of signals: Theory and practice.* 3rd edition. New York, NY: John Wiley.

Burrus, C.S., McClellan, J.H., Oppenheim, A.V, Parks, T.W., Schaffer, R.W. & Schussler, H.W. (1994). *Computer-based exercises for signal processing using MATLAB._*Englewood Cliffs, NJ: Prentice-Hall.

Crochiere, R.E., & Rabiner, L.R., (1981). Interpolation and decimation of digital signals - A Tutorial Review. *Proceedings of the IEEE, 69*(3), 300-331.

Crochiere, R.E., & Rabiner, L.R. (1983). *Multirate digital signal processing.* Englewood Cliffs, NJ: Prentice-Hall.

Diniz, P., Netto, S., & Da Silva, E. (2002). *Digital Signal Processing: System Analysis and Design* . New York, NY: Cambridge University Press.

Filter design toolbox for use with MATLAB. User's guide. Version 6. (2006). Natick: MathWorks.

Fliege, N. J. (1994). *Multirate digital signal processing*. New York, NY: John Wiley.

Harris, F. J., (2004). *Multirate signal processing for communication systems*. Upper Saddle River, NJ: Prentice Hall PTR.

Hentchel, T. (2002). *Sample rate conversion in software configurable radious*. Morwood, MA: Artech House.

Lutovac, M. D. Tošić, D. V., & Evans, B. L. (2001). *Filter Design for Signal Processing Using MATLAB and Mathematica*, Upper Saddle River, NJ, Prentice Hall.

McClellan, J. H., & Parks, T. W., & Rabiner, L. R. (1973). A computer program for designing optimum FIR linear-phase digital filters. *IEEE Transactions on Audio Electroacoustics*, *21*(6), 506-526.

Milić, Lj., & Lutovac, M.D. (2002). Efficient multirate filtering. In Gordana Jovanović-Doleček, (ed.), *Multirate Systems: Design & Applications*. Hershey, PA: Idea Group Publishing, 105-142.

Milić, Lj., Saramäki, T., & Bregović, R. (2006). Multirate filters: an overview. *Proc. of 2006 IEEE Asia Pacific Conference on Circuits and Systems*. Singapore, 914-917.

Mitra, S. K. (2006). *Digital signal processing: A computer based approach*. 3rd edition. New York: The McGraw-Hill Companies, Inc.

Neuvo, Y., Dong, C.-Y., & Mitra S. K. (1984). Interpolated finite impulse response filters. *IEEE Trans. on Acoustics, Speech and Signal Processing*, ASSP-*32*(3), 563-570.

Oppenheim, A. V., & Schafer, R. W. (1989). *Discrete-time signal processing*. London: Prentice-Hall.

Proakis J. G., & Manolakis D.G. (1996). *Digital signal processing: Principles, algorithms, and applications*. London: Prentice Hall.

Saramäki, T. *Multirate Signal Processing*. (2001). Lecture notes for a graduate course, the Institute of Signal Processing, Tampere University of Technology, Finland.

Signal processing toolbox for use with MATLAB. User's guide. Version 6. (2006). Natick: Math-Works.

Shpak, D.J., & Antoniou, A. (1990). A generalized Remez method for the design of FIR digital filters. *IEEE Transactions on Circuits and Systems*, *37*(2), 161-174.

Vaidyanathan, P.P., (1990). Multirate digital filters, filter banks, polyphase networks, and applications: A Tutorial. *Proceedings of the IEEE*, *78*(1), 56-93.

Vaidyanathan, P.P., (1993). *Multirate systems and filter banks*. Englewood Cliffs, NJ: Prentice Hall.

Chapter IV
FIR Filters for Sampling Rate Conversion

INTRODUCTION

The role of filters in sampling-rate conversion process has been discussed in Chapters II and III. Filters are used to suppress aliasing in decimators and to remove images in interpolators. The overall performance of a decimator or of an interpolator mainly depends on the characteristics of antialiasing and antiimaging filters. In Chapter III, we have considered the typical filter specifications and several methods for designing filter transfer functions that can meet the specifications. In this chapter, we are dealing with the implementation aspects of decimators and interpolators. The implementation problem arises from the unfavorable facts that filtering has to be performed on the side of the high-rate signal: in decimation filtering precedes the down-sampling, and in interpolation up-sampling precedes filtering. The goal is to construct a multirate implementation structure providing the arithmetic operations to be performed at the lower sampling rate. In this way, the overall workload in the sampling-rate conversion system can be decreased by the conversion factor $M(L)$.

The multirate filter implementation means that down-sampling or up-sampling operations are embedded into the filter structure. In this chapter, we are focused on the structures developed for finite impulse response (FIR) filters. The nonrecursive nature of FIR filters offers the opportunity to create implementation schemes that significantly improve the overall efficiency of FIR decimators and interpolators.

This chapter concentrates on the direct implementation forms for decimators and interpolators and the implementation forms based on the polyphase decompositions. Memory saving solutions for polyphase decimators and interpolators are also presented. Finally, the efficiency of FIR polyphase decimators and interpolators is discussed. The chapter concludes with MATLAB exercises for the individual study.

DIRECT IMPLEMENTATION STRUCTURES FOR FIR DECIMATORS AND INTERPOLATORS

The simplest implementation of an FIR filter is based on the direct transversal forms presented in Chapter I, Figures 1.14 and 1.15. When used as an antialiasing filter in decimators or as an antiimaging filter in interpolators, the direct implementation filter structures can be modified to the computationally efficient implementation forms. In this section, we present the efficient decimators and interpolators that provide the arithmetic operations to be evaluated at the sampling rate of the low-rate signal.

Direct Implementation Structures for FIR Decimators

Let us consider the factor-of-M decimator of Figure 4.1, which uses an FIR antialiasing filter with the impulse response $\{h[n]\}$. The time-domain input-output relations for the filter are expressed by the convolution,

$$v[n] = \sum_{k=0}^{N-1} h[k] x[n-k].$$ (4.1)

The decimated signal $\{y[m]\}$ is obtained after applying down-sampling operation to the filtered signal $\{v[n]\}$,

$$y[m] = v[mM].$$ (4.2)

For evaluating the decimated signal $\{y[m]\}$ we use only every M^{th} sample of the filtered signal. Accordingly, it is sufficient to compute only every M^{th} sample of $\{v[n]\}$, i.e.,

$$v[mM] = \sum_{k=0}^{N-1} h[k] x[mM-k].$$ (4.3)

From equations (4.1), (4.2) and (4.3) it follows that the computation of decimated signal $\{y[n]\}$ can be performed directly by computing every Mth sample of the convolution sum,

$$y[m] = \sum_{k=0}^{N-1} h[k] x[mM-k].$$ (4.4)

Two direct implementation structures for a factor-of-M decimator are shown in Figures 4.2(a) and 4.2(b). Figure 4.2(a) presents the cascade of an FIR filter and down-sampler as defined by equations (4.1) and (4.2) and indicated in Figure 4.1. The efficient structure that implements equation (4.4) is depicted in Figure 4.2(b). The down-sampling operation precedes the multiplications and additions, and therefore the arithmetic operations are evaluated at the lower sampling rate. In the structure of Figure 4.2(b), the

Figure 4.1. Factor-of-M decimator with FIR filter

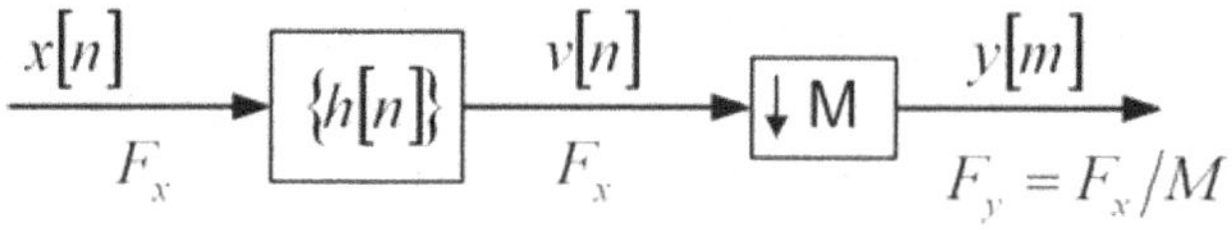

Figure 4.2. Direct implementations of a factor-of-M decimator: (a) Filter and down-sampler in cascade. (b) Efficient direct implementation structure.

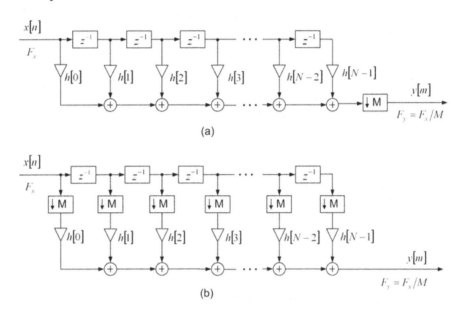

(a)

(b)

data flow through delays as in the case of the conventional FIR filter of Figure 4.2(a). The data are picked up from delays simultaneously every M^{th} instant of time, and then multiplied by the filter coefficients $\{h[n]\}$ and added together to give the output sample $y[m]$.

Formally, the structure of Figure 4.2(b) can be derived also from the structure of Figure 4.2(a) by applying the First Identity, see Chapter II.

In the conventional implementation of Figure 4.2(a), the number of multiplications per input sample in the decimator is equal to the FIR filter length N. The efficient implementation structure of Figure 4.2(b) reduces the number of multiplications per input sample to N/M. This property significantly improves the efficiency of multirate FIR filters.

The number of multiplications can be further reduced in the case of linear-phase FIR filters. By exploiting the coefficient symmetry $h[n] = h[N-n-1]$, $n = 0, 1, \ldots, (N-1)/2$ for N odd, and $n = 0, 1, \ldots, (N-2)/2$ for N even, the number of multiplications can be reduced by 2. Figure 4.3 shows an efficient realization of a linear-phase factor-of-M decimator where the filter length N is an even number. Here, the number of multiplications per input sample is reduced to $N/(2M)$. A similar structure exists for odd-length FIR filters, where the multiplication rate is reduced to $(N+1)/(2M)$.

Direct Implementation Structures for FIR Interpolators

Consider a factor-of-L interpolator consisting of an up-sampler and an FIR filter of the impulse response $\{h[n]\}$ as indicated in Figure 4.4.

The low-rate signal $\{x[n]\}$ is up-sampled by L and then filtered by the antiimaging FIR filter. Our aim is to introduce the up-sampling operation into the filter structure in a manner that the multiplication

Figure 4.3. Direct implementations of a linear-phase factor-of-M decimator with the reduced number of multipliers

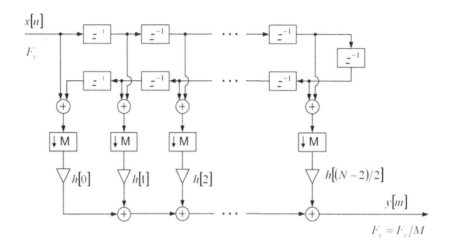

Figure 4.4. Factor-of-L interpolator with FIR filter

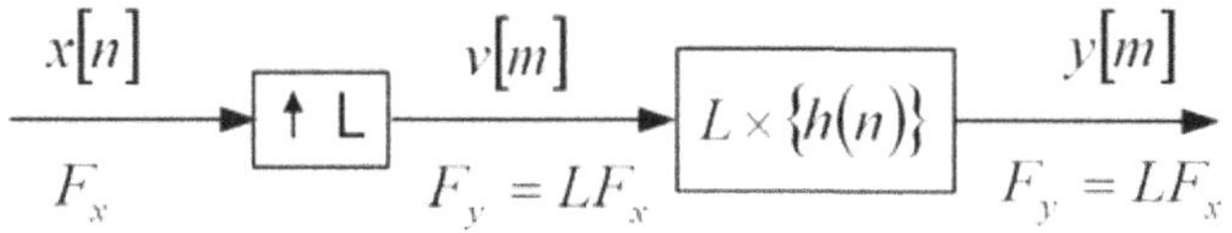

operations precede the sampling rate increase. Let us observe the input to the filter, signal $\{v[m]\}$. This signal is the result of the up-sampling operation performed on the input signal $\{x[n]\}$,

$$v[m]=\begin{cases} x[m/L], & m=0,\pm L,\pm 2L,\ldots \\ 0, & \text{otherwise} \end{cases}. \tag{4.5}$$

Here, we see that only every L^{th} input sample to the filter is non-zero valued. Therefore, in the conventional FIR filter structure $L-1$ out of L input samples are multiplied with the filter coefficients without contributing to the values of the output samples. This is illustrated in Figure 4.5(a) for the transposed direct implementation form.

Since there is no need to multiply the filter coefficients by the zero-valued samples, we can perform multiplications at the sampling rate of the input signal, and then up-sample-by-L the multiplied signals as shown in Figure 4.5(b). The up-sampled samples come to the adders and thus arrive to the chain of adders and delays. This implementation structure reduces the number of multiplications per output sample from N (the filter length) to N/L.

The number of multiplications per output sample can be further reduced. Namely, for interpolators with linear-phase FIR filters, we can exploit the inherent coefficient symmetry and decrease the number of multiplications by 2 as already shown in the case of decimators. An efficient implementation structure

Figure 4.5. Direct implementations of a factor-of-L interpolator: (a) Up-sampler and filter in cascade. (b) Efficient direct implementation structure.

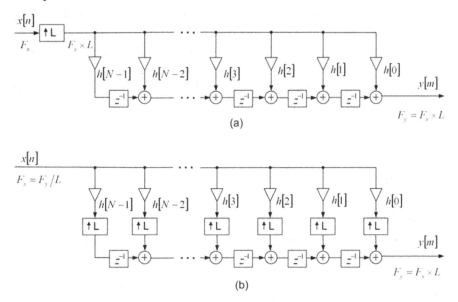

Figure 4.6. Direct implementations of a linear-phase factor-of-L interpolator with the reduced number of multipliers

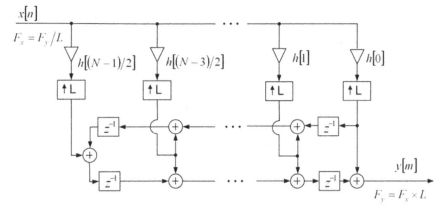

of the linear-phase interpolator is shown in Figure 4.6. The example interpolator structure of Figure 4.6 is given for N being an odd number. A similar structure can be derived for FIR filters of even lengths.

POLYPHASE IMPLEMENTATION OF DECIMATORS AND INTERPOLATORS

An efficient implementation of FIR decimators and interpolators is based on the polyphase decomposition of the filter transfer function. In this section, we consider the polyphase realization structures for FIR filters, and then show efficient implementations of polyphase decimators and interpolators.

Polyphase Realization Structures for FIR Filters

A higher-order FIR filter can be realized in a parallel structure based on the polyphase decomposition of the transfer function. The FIR transfer function is decomposed into M lower-order transfer functions, called the *polyphase components*, which are afterwards added together to compose the original overall transfer function. For the sake of simplicity, we demonstrate next the decomposition of an FIR system $H(z)$ into two polyphase components $E_0(z)$ and $E_1(z)$. Let us express the FIR transfer function $H(z)$ in developed form,

$$H(z) = h[0] + h[1]z^{-1} + h[2]z^{-2} + h[3]z^{-3} + \ldots$$
$$+ h[N-4]z^{-(N-4)} + h[N-3]z^{-(N-3)} + h[N-2]z^{-(N-2)} + h[N-1]z^{-(N-1)}. \tag{4.6}$$

The above transfer function can be expressed as a sum of two terms. We form the first term from the even-indexed coefficients and the second term from the odd-indexed coefficients. Without loss of generality, we can assume that N is an odd number, and express equation (4.6) as follows,

$$H(z) = h[0] + h[2]z^{-2} + h[4]z^{-4} + \ldots + h[N-3]z^{-(N-3)} + h[N-1]z^{-(N-1)} +$$
$$h[1]z^{-1} + h[3]z^{-3} + \ldots + h[N-4]z^{-(N-4)} + h[N-2]z^{-(N-2)}, \tag{4.7}$$

or equivalently,

$$H(z) = h[0] + h[2]z^{-2} + h[4]z^{-4} + \ldots + h[N-3]z^{-(N-3)} + h[N-1]z^{-(N-1)} +$$
$$z^{-1}\left(h[1] + h[3]z^{-2} + \ldots + h[N-4]z^{-(N-5)} + h[N-2]z^{-(N-3)} \right). \tag{4.8}$$

By using the notation

$$E_0(z) = h[0] + h[2]z^{-1} + h[4]z^{-2} + \ldots + h[N-3]z^{-(N-3)/2} + h[N-1]z^{-(N-1)/2}$$

$$E_1(z) = h[1] + h[3]z^{-1} + \ldots + h[N-4]z^{-(N-5)/2} + h[N-2]z^{-(N-3)/2}, \tag{4.9}$$

we express transfer function (4.6) as the sum of polyphase components $E_0(z)$ and $E_1(z)$

$$H(z) = E_0(z^2) + z^{-1}E_1(z^2). \tag{4.10}$$

In a general case, an N-length transfer function $H(z)$ can be decomposed into M polyphase branches $E_0(z), E_1(z), \ldots, E_{M-1}(z)$ in a manner that $H(z)$ is expressible in the form

$$H(z) = \sum_{k=0}^{M-1} z^{-k}E_k(z^M), \tag{4.11}$$

where

$$E_k(z) = \sum_{n=0}^{\lfloor N/M \rfloor} h[Mn+k]z^{-n}, \qquad 0 \le k \le M-1. \tag{4.12}$$

Here one should assume that $h[n]$ is zero outside the interval $0 \le n \le N-1$. Note that the symbol $\lfloor x \rfloor$ means the integer part of x.

It is very easy to determine polyphase components for an FIR filter using the MATLAB function reshape. We take the vector of filter coefficients h as an input and find the polyphase matrix E,

E = reshape(h,M,length(h)/M);

This command returns an M-by-length(h)/M matrix E, which in the M rows stores the M polyphase components for the given coefficient vector h. Here, the ratio length(h)/M should be an integer, if not, the sequence h can be zero padded to reach the necessary length. The kth polyphase component ek is stored in the kth row of matrix E, and can be extracted using the command,

ek = E(k,:);   for k = 1, 2, ..., M

Efficient Polyphase Realizations of FIR Decimators and Interpolators

An FIR filter can be implemented as a parallel connection of M (L) polyphase components, which are added together at the output. The polyphase components are sometimes called polyphase subfilters or polyphase branches. A polyphase component is usually implemented in the direct transversal form. Figure 4.7(a) shows a decimator composed of the cascade of an FIR filter implemented as a parallel connection of M polyphase branches, and factor-of-M down-sampler. Here the arithmetic operations in the polyphase branches are to be performed at the input sampling rate, i.e. at the higher sampling rate of the system. Instead of down-sampling at the filter output, one can shift the down-sampling operation into the polyphase branches before the output adders. This modification opens the opportunity of

Figure 4.7. Polyphase implementation of FIR decimator: (a) Filter and down-sampler. (b) Efficient polyphase decimator.

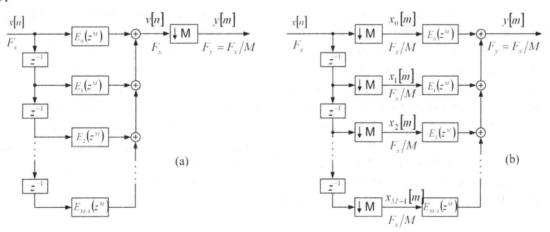

applying the Third Identity (see Chapter II, Figure 2.19) and to arrive to the efficient implementation of Figure 4.7(b).

In the structure of Figure 4.7(b), the down-sampling-by-M occurs at the inputs of the polyphase components $E_0(z)$, $E_1(z)$, ..., $E_{M-1}(z)$ and filtering is performed at the sampling rate F_x/M. The overall computational complexity of the decimator is reduced by M.

The input sequences in the polyphase branches of Figure 4.7(b), $\{x_0[m]\}$, $\{x_1[m]\}$, $\{x_2[m]\}$, ..., $\{x_{M-1}[m]\}$, are delayed and down-sampled versions of the input signal $\{x[n]\}$. We can say that the particular sequence $\{x_k[m]\}$ is obtained when down-sampling-by-M the sequence $\{x[n]\}$ with the phase offset k, $k = 0, 1, ...,$ $M-1$. Hence, from the causal sequence $\{x[n]\} = \{..., 0, 0, x[0], x[1], x[2], ..., x[M-1], x[M], x[M+1], ...\}$ it is straightforward to extract the M sequences,

$$\{x_0[m]\} = \{x[0], x[M], x[2M], ...\}$$
$$\{x_1[m]\} = \{x[-1], x[M-1], x[2M-1], ...\}$$
$$\{x_2[m]\} = \{x[-2], x[M-2], x[2M-2], ...\}$$

$$\begin{matrix} \cdot & & \cdot \\ \cdot & & \cdot \\ \cdot & & \cdot \end{matrix}$$

(4.13)

$$\{x_{M-1}[n]\} = \{x[-M+1], x[1], x[M+1], x[2M+1], ...\}.$$

Evidently, those sequences can be selected from the input signal $\{x[n]\}$ directly by using the commutative structure with the rotator shown in Figure 4.8. The rotator starts at the starting time $n = 0$ and gives the current sample $x[0]$ to $E_0(z)$. The next sample $x[1]$, at the time instant $n = 1$, goes to $E_{M-1}(z)$. The rotator continues in the same manner by moving to the left. Obviously, the rotator operates at the rate of the high-rate input signal $\{x[n]\}$, whereas filtering in the polyphase branches is performed at the rate of the low-rate signal $\{y[n]\}$. MATLAB programs demo_4_1 and demo_4_2 demonstrate the operation of the efficient decimator of Figure 4.8.

The parallel polyphase structure composed of L polyphase components can be used to provide an efficient implementation of a factor-of-L interpolator. Figure 4.9(a) shows the interpolator consisting of a factor-of-L down-sampler and an FIR filter realized in the polyphase form. Notice that the realization form in Figure 4.9(a) is a transpose of the realization form used for decimator in Figure 4.7(a).

In the interpolator structure of Figure 4.9(a), the up-sampling precedes filtering, and according to this, filtering in the polyphase components is performed at the higher sampling rate. Using the Sixth Identity, the structure of Figure 4.9(a) can be modified to the more efficient structure shown in Figure 4.9(b). The positions of up-samplers and polyphase components are interchanged, and filtering in the polyphase branches is to be performed at the lower sampling rate. The filtered signals $\{u_0[n]\}$, $\{u_1[n]\}$, $\{u_2[n]\}$, ..., $\{u_{M-1}[n]\}$ are up-sampled-by-L and fed to the chain of adders and delays. The samples flow through delays at the output sampling rate and finally give the output signal $\{y[m]\}$.

Since the up-sampled signals in polyphase branches have $L-1$ zero valued samples, the set of up-samplers and delays in the interpolator structure of Figure 4.9(b) can be replaced with the commutative structure with the rotator shown in Figure 4.10. The samples at the output $y[m]$ are obtained by picking up sequentially samples from the filtered signals $\{u_0[n]\}$, $\{u_1[n]\}$, $\{u_2[n]\}$, ..., $\{u_{M-1}[n]\}$ at the output sampling rate. The operation of the efficient interpolator of Figure 4.10 is simulated in MATLAB and demonstrated in programs demo_4_1 and demo_4_2.

Figure 4.8. Commutative polyphase structure of a decimator

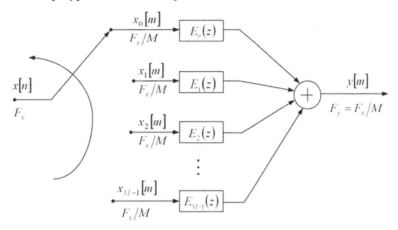

Figure 4.9. Polyphase implementation of FIR interpolator: (a) Up-sampler and. (b) Efficient polyphase interpolator

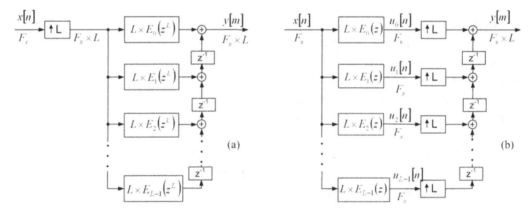

STRUCTURE VERIFICATION AND SIMULATION USING MATLAB

Polyphase decimation and interpolation presented in the preceding subsection can be evaluated in MATLAB. To illustrate computationally efficient decimators and interpolators, we use MATLAB to decimate and interpolate an input signal according to the polyphase structures of Figure 4.8, and Figure 4.10, respectively. We show on examples the application of MATLAB in verification of the developed structure, and in simulation of the sampling-rate conversion process. In this subsection we present two MATLAB programs: demo_4_1 for structure verification, and demo_4_2 which simulates the decimator and interpolator in real-time operation.

The sampling rate conversion process is demonstrated on the example sequence consisting of three sinusoids and the additive noise of normal distribution. This sequence is decimated by using the structure of Figure 4.8, and then the decimated sequence is interpolated using the structure of Figure 4.10.

Figure 4.10. Commutative polyphase structure of an interpolator

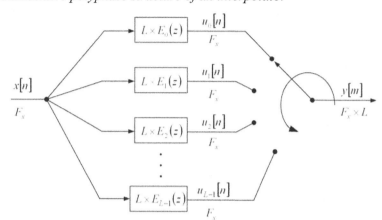

Programs demo_4_1 and demo_4_2 present:

- Decimation and interpolation by integer factor $M = L = 5$
- Design of optimal linear-phase FIR filter
- Creation of the input signal
- Polyphase decomposition
- Polyphase decimation according to Figure 4.8
- Polyphase interpolation according to Figure 4.10

Lowpass FIR filter for decimator and interpolator: Filter specifications according to the tolerance scheme *Case a*, see Figure 3.5. The passband edge frequency: $f_p = 0.08$, the stopband edge frequency: $f_s = 1/M - f_p$. Passband ripple $a_p = 0.01$ dB, minimum stopband attenuation $a_s = 60$ dB. The specifications are met with an optimal FIR filter of the length $N = 60$.

The input signal: The input signal $\{x[n]\}$ is a causal sequence formed as a sum of three sinusoidal sequences corrupted in noise:

$$x[n] = \cos(2\pi F_1 n/F_x) + \cos(2\pi F_2 n/F_x) + 0.7\cos(2\pi F_3 n/F_x) + s[n], \tag{4.14}$$

where $F_1 = 200$ Hz, $F_2 = 400$ Hz, $F_3 = 3800$ Hz, and $s[n]$ is a random signal with the normal distribution. The sampling frequency at the input is $F_x = 10000$ Hz. The sequence length is $N = 511$ samples.

Program demo_4_1 is intended for structure verification. This program is organized in 5 steps:

- *Step 1:* Design of the FIR filter to meet the specifications given above by using MATLAB function firpm from *Signal Processing Toolbox*. This filter is the anti-aliasing filter in decimator, and the anti-imaging filter in interpolator.
- *Step 2:* Decomposition of the filter transfer function into 5 polyphase components using the MATLAB function reshape.
- *Step 3:* Generation of the specified input signal $\{x[n]\}$.

- *Step 4:* Decimation-by-5 according to Figure 4.8. In this step, the input $\{x[n]\}$ is decomposed into the set of 5 subsequences: $\{x_0[m]\}$, $\{x_1[m]\}$, $\{x_2[m]\}$, $\{x_3[m]\}$, $\{x_4[m]\}$ according to (4.13). Here, for the causal $\{x[n]\}$ we have $x[-1] = x[-2] = x[-3] = x[-4] = 0$. The 5 subsequences are filtered in the polyphase branches using the MATLAB function filter, and added together to give the decimated signal $\{y_{dec}[m]\}$.
- *Step 5:* Interpolation-by-5 according to Figure 4.10. In this step, signal $\{y_{dec}[m]\}$ computed in *Step 4* is used as an input to the interpolator of Figure 4.10. The signal $\{x[n]\} = \{y_{dec}[m]\}$ is filtered in the parallel polyphase branches and the set of 5 signals $\{u_0[n]\}$, $\{u_1[n]\}$, $\{u_2[n]\}$, $\{u_3[n]\}$, $\{u_4[n]\}$ is obtained. The matrix **U** is composed of 5 row vectors: $\{u_0[n]\}$, $\{u_1[n]\}$, $\{u_2[n]\}$, $\{u_3[n]\}$, $\{u_4[n]\}$. The samples of the interpolated signal $\{y_{int}[m]\}$ are stored coloumnwise in matrix **U**. The interpolated signal is obtained simply by picking up the samples from matrix **U**.

Program plots the following characteristics: filter magnitude response in dB, spectrum of the input signal $X(F)$, spectrum of the decimated signal $Y_{dec}(F)$, and spectrum of the interpolated signal $Y_{int}(F)$. Here, we denote the resulting decimated signal with $\{y_{dec}[m]\}$, and the interpolated signal with $\{y_{int}[m]\}$.

```
% Program demo_4_1.m
% Structure verification of decimator, Figure 4.8.
% Structure verification of interpolator, Figure 4.10.
clear all, close all

% STEP 1
% Lowpass FIR filter design
h = firpm(59,[0,2*400/10000,1000/5000,1],[1,1,0,0]); % Computation of filter coefficients
[H,F1] = freqz(h,1,1024,10000);  % Computation of filter frequency response
figure (1)
subplot(4,1,1), plot(F1,20*log10(abs(H))), legend('Filter magnitude response'), ylabel('Gain, dB')
axis([0,5000,-80,5])

% STEP 2
% Generating the input signal
n = 0:510; % Time index
x = cos(2*pi*200*n/10000)+cos(2*pi*400*n/10000)+0.7*cos(2*pi*3800*n/10000)+randn(size(n));
[X,F1] = freqz(x,1,1024,10000); % Computing the signal spectrum
subplot(4,1,2), plot(F1,abs(X)), legend('Spectrum: input signal'), ylabel('|X(F)|'), axis([0,5000,0,300])

% STEP 3
 % Polyphase decomposition
E = reshape(h,5,length(h)/5);

% STEP 4
 % Polyphase downsampling and filtering
x0 = x(1:5:length(x)); x1 = [0,x(5:5:length(x))]; x2 = [0,x(4:5:length(x))];
x3 = [0,x(3:5:length(x))]; x4 = [0,x(2:5:length(x))];
```

```
y0 = filter(E(1,:),1,x0); y1 = filter(E(2,:),1,x1); y2 = filter(E(3,:),1,x2); y3=filter(E(4,:),1,x3);
y4=filter(E(5,:),1,x4);
ydec = y0 + y1 + y2 + y3 + y4; % Decimated signal
[Ydec,F2] = freqz(ydec,1,512,2000);
subplot(4,1,3), plot(F2,abs(Ydec)), legend('Spectrum: decimated signal'), ylabel('|Y_d_e_c(F)|'),
axis([0,1000,0,60])


% STEP 5
x = ydec; % Input to the interpolator
% Poliphase filtering and up-sampling
u0 = 5*filter(E(1,:),1,x); u1 = 5*filter(E(2,:),1,x); u2 = 5*filter(E(3,:),1,x);
u3 = 5*filter(E(4,:),1,x); u4 = 5*filter(E(5,:),1,x);
U = [u0;u1;u2;u3;u4];
yint = U(:); % interpolated signal
[Yint,F3] = freqz(yint,1,1024,10000);
subplot(4,1,4), plot(F3,abs(Yint)); legend('Spectrum: interpolated signal'), xlabel('Frequency, Hz'),
ylabel('|Y_i_n_t(F)|')
axis([0,5000,0,300])
```

Figure 4.11 plots the results. Comparing the spectra plotted in Figure 4.11, we can verify the correctness of decimator and interpolator structures of Figures 4.8 and 4.10, respectively. For the decimator, we compare the spectrum of the input signal and the spectrum of the decimated signal. Evidently, the decimated signal contains two sinusoids of frequencies $F_1 = 200$ Hz and $F_2 = 400$ Hz, whereas the aliasing of the third sinusoid ($F_3 = 3800$ Hz) and that of the out-of-band noise is suppressed sufficiently. The correctness of the interpolator structure is evident when we compare the spectrum of the interpolated signal with the spectrum of the input signal. The interpolated signal recovers exactly the two sinusoids of the input signal $F_1 = 200$ Hz and $F_2 = 400$ Hz with a practically unchanged signal-to-noise ratio. In the interpolated signal, high frequencies are significantly attenuated.

It is left to the reader to accomplish the structure verification by implementing decimation as the cascade of FIR filter and up-sampler, and interpolation as the cascade of down-sampler and FIR filter, and then to compare obtained results with those of Figure 4.11, see MATLAB Exercise 4.1.

This is of practical interest to simulate and verify in MATLAB the real-time operation of decimators and interpolators. Using the simulation model, one can detect weak points and faults in an early phase of system design. Here, we describe in MATLAB the real-time operation of the decimator structure of Figure 4.8 and of the interpolator structure of Figure 4.10. Program demo_4_2 is developed to simulate a sample-by-sample operation in the time-domain for a factor-of-5 decimator and for a factor-of-5 interpolator based on the efficient structures of Figures 4.8 and 4.10.

Program demo_4_2 uses the FIR filter and the input signal $\{x[n]\}$ already used in the program demo_4_1, and therefore, the *Steps 1-3* are identical for both programs. *Step 4* (decimation) and *Step 5* (interpolation) are different.

Step 4: Decimation-by-5 according to the structure of Figure 4.8. In this step, the rotator takes the current sample of the input signal $\{x[n]\}$, which is then inputted to the corresponding polyphase component. The rotator turns one step left, and the process is repeated for the next sample of the signal $\{x[n]\}$. When the rotator completes the full cycle, the filtered samples are added together to give the sample of the decimated signal $y_{dec}[m]$.

Figure 4.11. Decimation and interpolation with M =L=5: Filter characteristic. Spectrum of the input signal. Spectrum of the decimated signal. Spectrum of the interpolated signal.

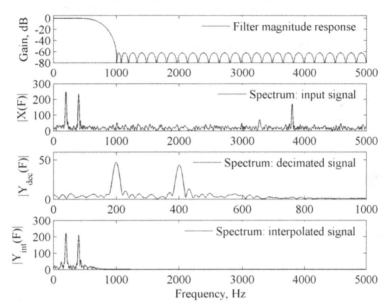

Step 5: Interpolation-by-5 according to the structure of Figure 4.10. In this step, the sample $y_{dec}[m]$ computed in *Step 4* is used as an input to the interpolator of Figure 4.10. The sample $x[n] = y_{dec}[m]$ is filtered in the polyphase branches by means of the program firsim. Filtering is performed at the lower sampling rate. Rotator at the higher sampling rate picks up the samples from the polyphase branches and outputs the samples of the interpolated signal $y_{int}[n]$.

Program demo_4_2 plots the same characteristics as the program demo_4_1: filter magnitude response in dB, spectrum of the input signal $X(F)$, spectrum of the decimated signal $Y_{dec}(F)$, and spectrum of the interpolated signal $Y_{int}(F)$.

```
% Program demo_4_2.m
% Structure simulation of decimator, Figure 4.8.
% Structure simulation of interpolator, Figure 4.10.
clear all, close all

% STEP 1
% Lowpass FIR filter design
h = firpm(59,[0,400/5000,1000/5000,1],[1,1,0,0]); % Computing the filter coefficients
[H,F1] = freqz(h,1,1024,10000); % Computing the signal spectrum
figure (1)
subplot(4,1,1), plot(F1,20*log10(abs(H))), legend('Filter magnitude response'), ylabel('Gain, dB')
axis([0,5000,-80,5])

% STEP 2
% generating the input signal
```

```
n = 0:510;  % Time index
x = cos(2*pi*200*n/10000)+cos(2*pi*400*n/10000)+0.7*cos(2*pi*3800*n/10000)+randn(size(n));
[X,F1] = freqz(x,1,1024,10000); % Computing the spectrum of the input signal
subplot(4,1,2)
plot(F1,abs(X)), legend('Spectrum: input signal'), ylabel('|X(F)|')
axis([0,5000,0,300])
% STEP 3
% Polyphase decomposition
E = reshape(h,5,length(h)/5);

% STEP 4
% Initial states in polyphase components
vk_i = zeros(size(E));
% Decimation
rot = 0; % Initial rotator position
ydec = [];
for n=1:length(x)
xn = x(n);
vk = vk_i(rot+1,:); ek = E(rot+1,:);
vk = [xn,vk(1:length(vk)-1)]; yn = sum(vk.*ek); % filtering
vk_i(rot+1,:)=vk;
if n==1, yy=0; else, end
if rot == 0 ss = 4; yy = yy+yn; ydec = [ydec,yy]; else end
if rot == 1 ss = 0; yy = yy+yn; else end
if rot == 2 ss = 1; yy = yy+yn; else end
if rot == 3 ss = 2; yy = yy+yn; else end
if rot == 4 ss = 3; yy = yn; else end
rot = ss;
end
[Ydec,F2] = freqz(ydec,1,512,2000);
subplot(4,1,3), plot(F2,abs(Ydec)), legend('Spectrum: decimated signal'), ylabel('|Y_d_e_c(F)|')
axis([0,1000,0,60])

% STEP 5
% Interpolation
yi = [0]; xk_i = zeros(size(E));
rot = 0;
k = 1;
for n=1:length(ydec)*5
   xn=ydec(k);
xk = xk_i(rot+1,:); ek = E(rot+1,:);
xk=[xn,xk(1:length(xk)-1)]; yn=sum(xk.*ek); % filtering
xk_i(rot+1,:)=xk; yn=5*yn;
```

```
if rot == 0 ss = 1; yi = [yi,yn]; else end
if rot == 1 ss = 2; yi = [yi,yn]; else end
if rot == 2 ss = 3; yi = [yi,yn]; else end
if rot == 3 ss = 4; yi = [yi,yn]; else end
if rot == 4 ss = 0; k = k+1; yi = [yi,yn]; else end
rot = ss;
end
[Yint,F3] = freqz(yi,1,1024,10000);
subplot(4,1,4)
plot(F3,abs(Yint)); legend('Spectrum: interpolated signal')
xlabel('Frequency, Hz'),ylabel('|Y_i_n_t(F)|')
axis([0,5000,0,300])
```

Figure 4.11 displays also the plots of the program demo_4_2. Reader can easily modify *Steps 1-3* in the program demo_4_2 and examine the sampling rate conversion for various signals with different FIR filters.

MEMORY SAVING STRUCTURES FOR FIR POLYPHASE DECIMATORS AND INTERPOLATORS

The polyphase subfilters in the commutative configuration of decimator or interpolator suffer from the drawback that their impulse responses are not symmetric. Therefore, the coefficient symmetry property used to half the number of multipliers in the direct implementations cannot be exploited for the efficient implementation of the individual polyphase subfilters.

However, the advantage of polyphase configurations is in the possibility to save the number of delays (memory elements). When the direct form or the transposed direct form is used to implement the polyphase subfilters in an interpolator or in a decimator, the number of delays can be reduced by the sampling rate conversion factor M (L). In order to demonstrate the development of memory saving structures let us examine the polyphase implementation of a factor-of-3 decimator and a factor-of-3 interpolator using an 11th order FIR filter.

The transfer function of an 11th order FIR filter is given by

$$H(z) = h[0] + h[1]z^{-1} + h[2]z^{-2} + h[3]z^{-3} + h[4]z^{-4} + h[5]z^{-5} + h[6]z^{-6}$$
$$+ h[7] + z^{-7} + h[8]z^{-8} + h[9]z^{-9} + h[10]z^{-10} + h[11]z^{-11}$$

(4.15)

For the implementation of a factor-of-3 decimator or a factor-of-3 interpolator, we find three polyphase components of $H(z)$,

$$E_0(z) = h[0] + h[3]z^{-1} + h[6]z^{-2} + h[9]z^{-3}$$
$$E_1(z) = h[1] + h[4]z^{-1} + h[7]z^{-2} + h[10]z^{-3},$$
$$E_2(z) = h[2] + h[5]z^{-1} + h[8]z^{-2} + h[11]z^{-3}.$$

(4.16)

The polyphase components $E_0(z)$, $E_1(z)$ and $E_2(z)$ can be used in the commutative structures of Figures 4.8 and 4.10 to implement factor-of-3 decimator and factor-of-3 interpolator. Let us consider first the case of decimator. Figure 4.12 shows the configuration with three polyphase components implemented in the direct transpose form. From Figure 4.12, we observe that the samples from the three delay chains are added together to give the output sample $y[m]$. Hence, the individual delay chains can be combined together into a single one. In this way, the memory saving structure for decimator shown in Figure 4.13 is obtained.

Figure 4.14 shows the commutative structure of the factor-of-3 interpolator with the polyphase components as given in (4.16). Here, the polyphase components $E_0(z)$, $E_1(z)$ and $E_2(z)$ are implemented in the direct form. Observing the structure of Figure 4.14, we see that the same samples of the input signal appear simultaneously at the same time instants at the corresponding nodes of the three delay chains. Obviously, the individual delay chains can be combined together into a single chain. From the adders, the rotator simply picks-up the output samples $y[m]$. The resulting memory saving structure for the interpolator is shown in Figure 4.15.

Program demo_4_3 is developed to simulate the memory saving commutative decimator and interpolator from Figures 4.13 and 4.15. Here, the sampling rate conversion factors are $M = L = 3$, and the FIR filter is of the length $N = 12$.

Similarly to the organization of demo_4_1 and demo_4_2, program demo_4_3 is organized in the following 5 steps:

- *Step 1:* FIR filter design. Optimal FIR filter of the length $N = 12$. Specifications for the tolerance scheme *Case b* of Figure 3.5, and the sampling rate conversion factors $M = L = 3$.
- *Step 2:* Decomposition of the filter transfer function to 3 polyphase components using the MATLAB function reshape.
- *Step 3:* The input signal generation. The input signal $\{x[n]\}$ is a causal sequence consisting of three sinusoidal sequences and a noise, with $F_1 = 350$ Hz, $F_2 = 750$ Hz, $F_3 = 3800$ Hz, see (4.14).

Figure 4.12. Polyphase factor-of-3 decimator. Subfilters in the transposed direct form.

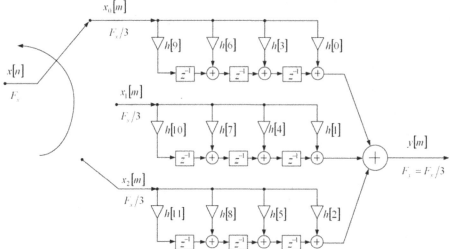

Figure 4.13. Memory saving structure for the polyphase factor-of-3 decimator

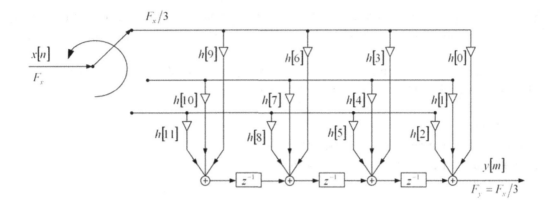

Figure 4.14. Polyphase factor-of-3 interpolator. Subfilters in the direct form.

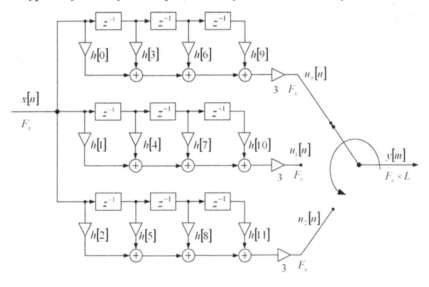

Figure 4.15. Memory saving structure for the polyphase factor-of-3 interpolator.

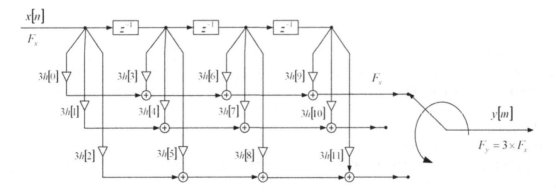

The noise and $s[n]$ is a random signal with the normal distribution. The sampling frequency is F_x = 10000 Hz, and the sequence length is $N = 511$ samples.

- *Step 4:* Simulation of the decimation process for the memory saving structures of Figure 4.13.
- *Step 5:* Simulation of the interpolation process for the memory saving structures of Figure 4.15.

Program demo_4_3 plots: filter magnitude response in dB, spectrum of the input signal $X(F)$, spectrum of the decimated signal $Y_{dec}(F)$, and spectrum of the interpolated signal $Y_{int}(F)$.

```
% Program demo_4_3.m
% Structure simulation of decimator, Figure 4.13.
% Structure simulation of interpolator, Figure 4.15.
clear all, close all

% STEP 1:
% Lowpass FIR filter design
h = firpm(11,[0,800/5000,2*1666/5000-800/5000,1],[1,1,0,0]);
[H,F1] = freqz(h,1,1024,10000);
figure (1)
subplot(4,1,1), plot(F1,20*log10(abs(H))), legend('Filter magnitude response'), ylabel('Gain, dB')
axis([0,5000,-60,5])

% STEP 2:
% Polyphase decomposition
E = reshape(h,3,length(h)/3); E = fliplr(E);

% STEP 3:
% Input signal
n = 0:510; % Time index
x = cos(2*pi*350*n/10000) + cos(2*pi*750*n/10000) + 0.7*cos(2*pi*3800*n/10000) + randn(size(n));
[X,F1] = freqz(x,1,1024,10000); % Computing the spectrum of the input signal
subplot(4,1,2)
plot(F1,abs(X)), legend('Spectrum: input signal'), ylabel('|X(F)|')
axis([0,5000,0,300])

% STEP 4:
% Setting the initial states
xk = zeros(size(1:length(h)/3));
% Decimation
rot = 0; % Initial rotator position
ydec = [];
for n=1:length(x)
xn = x(n); xk = xn*E((rot+1),:)+xk;
if rot == 0 ss = 2; yn = xk(length(xk)); ydec = [yn,ydec]; xk = [0,xk(1:length(xk)-1)]; else end
if rot == 1 ss = 0; else end
```

```
if rot == 2 ss = 1; else end
rot = ss; end
[Ydec,F2] = freqz(ydec,1,512,3332);
subplot(4,1,3), plot(F2,abs(Ydec)), legend('Spectrum: decimated signal'), ylabel('|Y_{dec}(F)|')
axis([0,1662,0,100])

% STEP 5:
% Interpolation
E = 3*fliplr(E);
yi = [];
xk = zeros(size(1:length(h)/3));;
rot = 0; % Initial rotator position
k = 1;
for n=1:length(ydec)*3
    xn = ydec(k); ek = E(rot+1,:);
    if k == 1,  xk = [xn,xk(1:length(xk)-1)]; else end
    yn = sum(xk.*ek); yi = [yi,yn];
if rot == 0 ss = 1; else end
if rot == 1 ss = 2; else end
if rot == 2 ss = 0; k = k+1; xk = [xn,xk(1:length(xk)-1)]; else end
rot = ss;
end
[Yint,F3] = freqz(yi,1,1024,10000);
subplot(4,1,4)
plot(F3,abs(Yint)); legend('Spectrum: interpolated signal')
xlabel('Frequency, Hz'),ylabel('|Y_{int}(F)|')
axis([0,5000,0,300])
```

Figure 4.16 plots the results. In this example, decimation (interpolation) is performed with 5 multiplications per input (output) sample, and the structure of the decimator (interpolator) contains only 3 delays (memory elements).

The implementation structures for decimators and interpolators discussed in this chapter show the efficiency of FIR filters when used in multirate systems. The savings to the factor-of-M (L) in the number of multiplications and the savings in memory elements become possible in the structures that incorporate sampling rate alteration into the filtering process. The next section considers the computation savings in more detail.

COMPUTATIONAL EFFICIENCY OF FIR DECIMATORS AND INTERPOLATORS

Computing the Multiplication Rate

The computational efficiency is determined by the number of multiplications per input sample for the decimator, and by the number of multiplications per output sample for the interpolator. Alternatively, the

Figure 4.16. Decimation and interpolation with memory saving structures of Figures 4.13 and 4.15: Filter characteristic, spectrum of the input signal, spectrum of the decimated signal, spectrum of the interpolated signal.

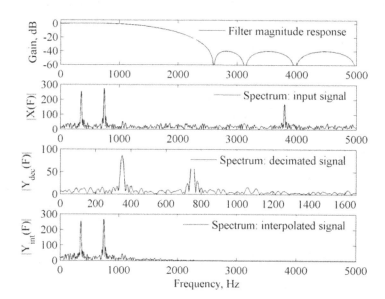

computational efficiency can be expressed as the number of multiplications per second. In the single-rate FIR filter with N non-zero coefficients, N multiplications are needed for computing one output sample. It was shown in the preceding section that employing multirate techniques in implementation of a factor-of-M decimator (factor-of-L interpolator) leads to a factor-of-M (L) savings in computational complexity. In this section, we consider the computational efficiency of the FIR decimator and interpolator structures in more detail.

Let us denote with *MPIS* the number of multiplications per input sample in the case of decimator, and with *MPOS* the number of multiplications per output sample in the case of interpolator. An FIR decimator can be implemented efficiently by using the direct structures of Figure 4.2 (b) and Figure 4.3, or polyphase configuration of Figure 4.8. The computational requirements for a factor-of-M decimator of Figure 4.2(b) expressed in *MPIS* are given by

$$MPIS = \frac{N}{M}.$$
(4.17)

where N is the filter length. The computational requirements can be halved in the case of a linear-phase FIR filter. By exploiting the coefficient symmetry according to Figure 4.3, *MPIS* reduces to

$$MPIS = \frac{N}{2M}.$$
(4.18)

The number of delays (memory elements) is equal to the filter order, $N_{ord} = N - 1$.

In the polyphase decomposition, the coefficient symmetry in the individual polyphase components generally doesn't exist. At least, one component can exhibit the coefficient symmetry. Thereby, when the decimator is implemented according to the polyphase configuration of Figure 4.8, the *MPIS* is roughly the same as given for the direct implementation structure in (4.17). The polyphase configuration is suitable for the savings in memory as shown in Figure 4.13 and demonstrated in the program demo_4_3. The minimal required number of delay elements in the overall decimator is equal to the number of delays in the highest-order polyphase subfilter.

The computational complexity of an interpolator can be considered in a similar manner. Hence for the direct implementation of a factor-of-L interpolator according to Figure 4.5(b), the computational complexity expressed in the number of multiplications per output sample $MPOS$ is given by

$$MPOS = \frac{N}{L}. \tag{4.19}$$

As in the case of decimator, the computational requirements can be halved in the case of a linear-phase FIR filter. For the linear-phase interpolator, the structure of Figure 4.6 can be used reducing the $MPOS$ to

$$MPOS = \frac{N}{2L}. \tag{4.20}$$

The number of delays (memory elements) is equal to the filter order, $N_{ord} = N - 1$.

When the interpolator is implemented according to the polyphase configuration of Figure 4.10, the $MPOS$ is roughly the same as given for the direct implementation structure in (4.17). The minimal required number of delay elements in the overall interpolator is equal to the number of delays in the highest-order polyphase subfilter.

Sometimes, the computational requirements are expressed as the number of multiplications per second and denoted by R_M, called also the *multiplication rate*. Hence, a single-rate FIR filter with N non-zero coefficients operating at the sampling frequency F_x, requires $R_M = N \times F_x$ multiplications per second, or $R_M = N \times F_x / 2$ when exploiting the coefficient symmetry. For multirate filters, R_M can be reduced by a sampling-rate conversion factor. In the case of a factor-of-M decimator, when neglecting the coefficient symmetry, we have

$$R_{M\_DEC} = \frac{N \times F_x}{M} = N \times F_y. \tag{4.21}$$

Here, F_x denotes the input, and F_y the output sampling rates.

For an efficient factor-of-L interpolator, the number of multiplications per second can be computed in a similar manner, i.e.,

$$R_{M\_INT} = \frac{N \times F_y}{L} = N \times F_x. \tag{4.22}$$

Computational Efficiency of Single-Stage and Multistage Systems

Computational cost of a sampling-rate conversion system needs to be minimized in practice. Until now, we have been focused on the efficient implementations of single-stage decimators and interpolators. In this subsection, we show that the significant savings in computational requirements can be achieved when the sampling rate alteration is accomplished in several stages. The multi-stage decimators and interpolators are possible if $M(L)$ is factorable into a product of integers, as indicated in Chapter II. In the following, we demonstrate the efficiency of multi-stage design on the example of two-stage decimator with the overall conversion factor $M = 15$.

Let us examine the example factor-of-15 decimator which converts the input sampling rate of $F_x = 30$ kHz to the output sampling rate of $F_y = 2$ kHz. We consider the *Case a* tolerance scheme of Figure 3.5 (a) with the passband edge frequency $F_p = 0.5$ kHz. For the decimation factor $M = 15$ and the *Case a* tolerance scheme, the stopband edge frequency is $F_s = 1$ kHz. The requested peak passband ripple is specified to $\delta_p = 0.01$, and the maximal stopband ripple is $\delta_s = 0.001$ ($a_s = 60$ dB).

Single-Stage Solution for M = 15:

For the single-stage decimator, an optimal FIR filter of the length $N = 180$ meets the design requirements specified above. The polyphase decomposition for $M = 15$ gives 15 polyphase subfilters, and the length of each subfilter is 12. When the polyphase decimator for the factor-of-15 is implemented according to Figure 4.8, the arithmetic operations are evaluated at the lower sampling rate. For this design, the multiplication rate R_{M_DEC} is given by $R_{M\_DEC} = 180 \times 2000 = 360000$ multiplications per second.

The polyphase decimator can exploit the memory saving structure in a similar manner as shown in Figure 4.13 for the case $M = 3$. By extending the configuration from Figure 4.13 to $M = 15$ polyphase branches, the single-stage decimator of this example can be implemented with 11 delays since the length of the polyphase subfilters is 12.

Two-Stage Solution for M = 15:

The decimation factor $M = 15$ is expressible as a product of two integers, i.e., $M = 5 \times 3$, and instead of the single-stage decimation, the decimation can be performed in two steps. We can decimate the input signal by the factor-of-5, and then by the factor-of-3, or vice versa.

Let us consider first the two stage decimator as indicated in Figure 4.17(a). In the first stage, the input signal is decimated by 5, and in the second stage by 3. Figure 4.17(b) presents the single-stage equivalent of the two-stage decimator.

The role of filters $H_1(z)$ and $H_2(z)$ is to provide that the overall design requirements specified for the decimator are met. Hence, the overall decimator characteristics are the design goal when specifying the particular design requirements for $H_1(z)$ and $H_2(z)$. The single-stage equivalent representation shown in Figure 4.17(b) serves appropriately in determining the specifications for $H_1(z)$ and $H_2(z)$. The single-stage equivalent of Figure 4.17(b) is obtained when the Third Identity is applied to the cascade of the factor-of-5 down-sampler and filter $H_2(z)$. Therefore, for the two-stage decimator with $M = 5 \times 3$, we design the individual filters $H_1(z)$ and $H_2(z)$ under the condition that $H(z) = H_1(z) \times H_2(z^5)$ meets the specifications.

Schematic representations of the requested magnitude responses for $H_1(z)$ and $H_2(z)$ are indicated in Figure 4.18. Filter $H_2(z)$ operating at the sampling rate 6 kHz is designed for the passband boundary frequency of 0.5 kHz, and the stopband edge frequency of 1 kHz, Figure 4.18(a). Periodic characteristic, which follows for $H_{2u}(z) = H_2(z^5)$ is shown in Figure 4.18(b). Obviously, the role of $H_1(z)$ is to provide the passband in the range 0-0.5 kHz and to provide the stopband attenuation of 60 dB in the two passbands of the periodic filter $H_{2u}(z) = H_2(z^5)$, see Figure 4.18(c). Here, $H_1(z)$ ensures the stopband attenuation and is requested to satisfy specifications of the *Case c* tolerance scheme. The resulting overall response $H(z) = H_1(z) H_2(z^5)$ is shown in Figure 4.18(d). With this approach, the overall filtering task is shared between two lower-order filters as will be illustrated in detail later on.

Figure 4.17. Two-stage decimator. (a) Two-stage implementation for M =5×3. (b) Single-stage equivalent for M =5×3.

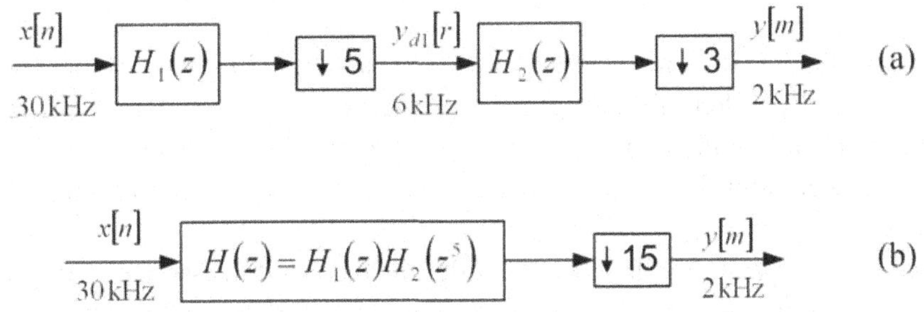

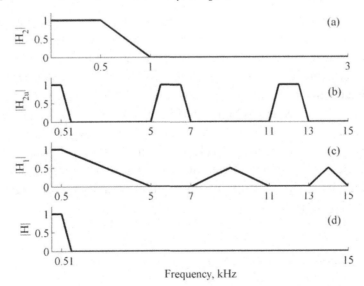

When specifying the passband ripples δ_{p1} and δ_{p2} for $H_1(z)$ and $H_2(z)$, respectively, we assume that each filter in cascade contributes to the resulting ripple of the multistage system. If the number of stages is small (<4), and $\delta_p \ll 1$, we can determine the individual ripples by dividing the specified ripple of the system (δ_p) with the number of stages. For the two-stage system and $\delta_p = 0.01$, we compute,

$\delta_{p1} = \delta_{p2} = \delta_p/2 = 0.005$.

The design specifications for $H_1(z)$ and $H_2(z)$ in the two-stage decimator of Figure 4.17(a) follow directly from the characteristics shown in Figure 4.18:

- The lowpass filter $H_2(z)$ as specified in Figure 4.18(a) has the boundary frequencies $F_p = 0.5$ kHz and $F_s = 1$ kHz and the sampling frequency is 6 kHz. The passband ripple is $\delta_p = 0.005$, and the stopband ripple $\delta_s = 0.001$.

Figure 4.18. Magnitude responses for the single-stage equivalent of the two stage decimator, M =5×3. (b) Magnitude response for H₂(z). (a) Magnitude response for H₂ᵤ(z) = H₂(z⁵). (c) Magnitude response for H₁(z). (d) Magnitude response for H (z) = H₁(z) H₂(z⁵).

- The lowpass filter $H_1(z)$ according to Figure 4.18(c) has the passband boundary frequencies $F_p = 0.5$ kHz and two stopband regions [5–7 kHz] and [11–13 kHz] The sampling frequency is 30 kHz. The passband ripple is $\delta_p = 0.005$, and the stopband ripple $\delta_s = 0.001$.

The specifications for $H_1(z)$ are met with an optimal linear-phase FIR filter of the length $N = 19$, and an optimal linear-phase FIR filter of the length $N = 36$ meets the specifications for $H_2(z)$. Thereby, the total number of coefficients in the two-stage decimator for $M = 5\times3$ amounts to $19 + 36 = 55$. Recall that the single-stage decimator for $M = 15$ considered earlier in this section requires the FIR filter of 180 coefficients.

The first stage consisting of the filter $H_1(z)$ and the factor-of-5 down-sampler can be implemented with 5 polyphase subfilters operating at the intermediate sampling rate of 6 kHz. The multiplication rate R_{M_DEC1} for the first stage with the total of 19 coefficients amounts to,

$R_{M\_DEC1} = 19 \times 6000 - 114000$ multiplications per second.

The second stage consisting of the filter $H_2(z)$ and the factor-of-3 down-sampler can be implemented with 3 polyphase subfilters operating at the output sampling rate of 2 kHz. Hence, for the second stage, which has 36 coefficients, the multiplication rate is

$R_{M\_DEC2} = 36 \times 2000 - 72000$ multiplications per second.

The total multiplication rate of the two-stage decimator is therefore

$R_{M\_DEC} = R_{M\_DEC1} + R_{M\_DEC2} = 186000$ multiplications per second.

Comparing this result with 360000 multiplications per second required for the single-stage decimator, the computational efficiency in the two-stage design is nearly doubled.

The two-stage decimation for the specifications given above is evaluated in the program demo_4_4. This program designs two optimal FIR filters: $H_1(z)$ of the length $N = 19$ and $H_2(z)$ of the length $N = 19$. The MATLAB function firpm from *Signal Processing Toolbox* is used to compute the filter coefficients. In order to distinguish the low-frequency signal components from the high-frequency spectrum, the input signal $\{x[n]\}$ is composed of two sinusoidal sequences of the frequencies 300 and 400 Hz, and of the additional sequence whose spectrum has the prescribed shape at high frequencies.

Program demo_4_4 performs in the first stage the decomposition of the filter $H_1(z)$ into 5 polyphase components and decimates the input signal $\{x[n]\}$ by 5 giving the intermediate signal $\{y_{d1}[r]\}$. In the second stage, program performs decomposition of the filter $H_2(z)$ into 3 polyphase components and decimates by 3 the intermediate signal $\{y_{d1}[r]\}$. The result of the second stage is the output signal $\{y[m]\}$.

The program returns two figures. One figure plots of the magnitude responses of $H_2(z)$, $H_{2u}(z)$, $H_1(z)$, and the overall magnitude response of the single-rate equivalent $H(z) = H_1(z)\times H_2(z^5)$. The second figure plots the spectrum of the input signal $\{x[n]\}$, the spectrum of the intermediate signal $\{y_{d1}[r]\}$, and the spectrum of the output signal $\{y[m]\}$.

```
% Program demo_4_4.m
% Efficient 2-stage decimator: M = M1*M2, M1 = 5, M2= 3
```

```
close all, clear all
h1 = firpm(18,[0,0.5/15,5/15,7/15,11/15,13/15,1,1],[1,1,0,0,0,0,0,0],[1,5,5,5]); % Filter H1(z)
[H1,F1] = freqz(h1,1,1024,30); % Frequency response  of H1(z)
h2 = firpm(35,[0,5*0.5/15,5*1/15,1],[1,1,0,0],[1,5]); % Filter H2(z)
[H2,F2] = freqz(h2,1,1024,6); % Frequency response of H2(z)
h2u = upsample(h2,5); % Periodic filter H_2(z^5)
[H2u,F1] = freqz(h2u,1,1024,30); % Spectrum of the periodic filter H_2(z^5)
H = H1.*H2u; % The overall frequency response
figure (1)
subplot(4,1,1), plot(F2,20*log10(abs(H2))), ylabel('|H_2(F)|'), axis([0,3,-80,2])
subplot(4,1,2), plot(F1,20*log10(abs(H2u))), ylabel('|H_2_u(F)|'),axis([0,15,-80,2])
subplot(4,1,3), plot(F1,20*log10(abs(H1))), ylabel('|H_1(F)|'), axis([0,15,-80,2])
subplot(4,1,4), plot(F1,20*log10(abs(H))), ylabel('|H(F)|'), axis([0,15,-80,2])
xlabel('Frequency, kHz')

% Input signal
ff = [0,1/15,1.5/15,0.6,0.8,1]; a = [0,0,0.8,0.7,0.9,0];
x = fir2(1000,ff,a);
x = 5*x+0.01*sin(2*pi*(1:1001)*300/30000) + 0.01*sin(2*pi*(1:1001)*400/30000);
[X,F1] = freqz(x,1,2048,30); % Soectrum of the input signal
figure (2)
subplot(3,1,1), plot(F1,abs(X)), ylabel('|X(F)|'), legend('(a)')

% Polyphase decomposition, M1 = 5
h1 = [h1,0];
E1 = reshape(h1,5,length(h1)/5);
% Polyphase downsampling and filtering, M1 = 5
v0 = x(1:5:length(x)); v1 = [0,x(5:5:length(x))]; v2 = [0,x(4:5:length(x))];
v3 = [0,x(3:5:length(x))]; v4 = [0,x(2:5:length(x))];
y0 = filter(E1(1,:),1,v0); y1 = filter(E1(2,:),1,v1); y2 = filter(E1(3,:),1,v2);
y3 = filter(E1(4,:),1,v3); y4 = filter(E1(5,:),1,v4);
yd1 = y0 + y1 + y2 + y3 + y4; % Decimated signal
[Yd1,F2]=freqz(yd1,1,2048,6);
subplot(3,1,2), plot(F2,abs(Yd1)), legend('(b)'), ylabel('|Y_d_1(F)|'), %axis([0,1000,0,60])

% Polyphase decomposition, M2 = 3
E2 = reshape(h2,3,length(h2)/3);
% Polyphase downsampling and filtering, M2 = 3
x = [yd1,0];
v0 = x(1:3:length(x)); v1 = [0,x(3:3:length(x))]; v2 = [0,x(2:3:length(x))];
y0 = filter(E2(1,:),1,v0); y1 = filter(E2(2,:),1,v1); y2 = filter(E2(3,:),1,v2);
y = y0 + y1 + y2; % decimated signal
[Y,F3] = freqz(y,1,2048,2);
subplot(3,1,3), plot(F3,abs(Y)), legend('(c)'), ylabel('|Y(F)|')
xlabel('Frequency, kHz')
```

Figure 4.19. Magnitude responses of the two-stage decimator, in [dB]: (a) Second-stage filter $H_2(z)$. (b) Periodic filter $H_2(z^5)$. (c) First-stage filter $H_1(z)$. (d) The single-stage equivalent of the overall decimation filter $H(z)$.

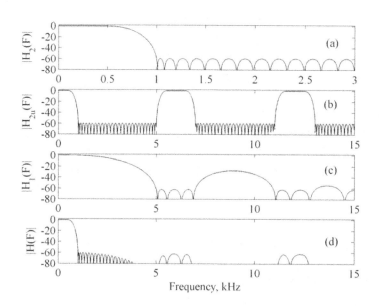

Figures 4.19 and 4.20 expose the results computed by the program demo_4_4.

Figure 4.19 presents the magnitude responses of the system. The magnitude response of the second-stage filter $H_2(z)$ is shown in Figure 4.19(a), and Figure 4.19(b) presents the magnitude response of the periodic filter $H_{2u}(z) = H_2(z^5)$. Figure 4.19(c) plots the magnitude response of $H_1(z)$. Finally, Figure 4.19(d) shows the magnitude response of the single-stage equivalent $H(z) = H_1(z) \times H_2(z^5)$, which demonstrates that the overall design specifications are met.

Figure 4.20 illustrates the two-stage decimation of the input signal. Figure 4.20(a) plots the spectrum of the input signal, which shows the two sinusoids in the low frequency range and the spectrum of a wide-band signal covering the frequency band 1–15 kHz. Figure 4.20(b) shows the spectrum of the signal decimated-by-5 in the first stage, and Figure 4.20(c) plots the resulting spectrum obtained in the second stage. Evidently, the two-stage decimation is performed without destroying the signal in the frequency band of interest, which in this case is limited to the range 0–0.5 kHz.

It is interesting to examine the savings in the number of memory elements in the two-stage decimator presented above. Here, the number of delays has to be computed for each stage and added afterwards. The number of delays in the memory-saving structure is determined according to the number of delays in the polyphase component of the highest order. In the first stage, filter $H_1(z)$ is of the length $N = 19$, and therefore, for being decomposed into 5 polyphase components, has to be extended in length with a single zero sample to provide length$\{h_2[n]\} = 20$. The polyphase structure consists of 5 subfilters: 4 subfilters of the order 3, and one has the order 2. Using the memory-saving structure, the decimation-by-5 in the first stage can be performed with only 3 delay elements. In the second stage, the filter $H_2(z)$ is of the length

Figure 4.20. Two-stage decimation: (a) Spectrum of the input signal. (b) Decimation-by-5, resulting spectrum after the first stage. (c) Decimation-by-15, resulting spectrum after the second stage.

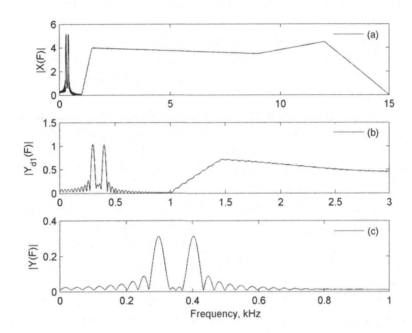

Table 4.1. Implementations of the factor-of-15 decimator

Design	Number of constants	Multiplication rate, $R_{M\text{-DEC}}$	Min. No of delays
Single-stage	180	360000	11
Two-stage, 5×3	55	186000	14
Two-stage, 3×5	70	220000	12

$N = 36$. The decomposition into 3 polyphase components results in three 12-length subfilters. For the memory-saving structure, the number of delays in the second stage reduces to 11. The total number of delays in the example two-stage factor-of-15 decimator is 14. Notice that the minimal number of delays in the single stage decimator was only 11.

This example shows that the two-stage decimator provides significant savings in the multiplication rate when compared with the single-stage design. But somewhat better savings in memory have been achieved with the single-stage design. The advantages of the two-stage decimator are evident when comparing the filter characteristics with those requested for the single-stage decimator. The relaxed design specifications bring the lower-order filters and therefore decrease the effects of the finite precision arithmetic.

In the two-stage decimator discussed above, we have examined the implementation when the decimation-by-5 precedes the decimation-by-3. Obviously, the two-stage decimation-by-15 can be performed in the reversed order. It is simple to verify by solving MATLAB Exercise 4.4 that the decimation-by-3 followed with decimation-by-5 is inferior.

Table 4.1 summarizes the computational efficiency of three designs: single-stage decimator, two-stage 5×3, and two-stage 3×5.

In this subsection, we have considered the implementation and the computational complexity of the two-stage decimator. Because of the duality of decimators and interpolators, the similar results are obtainable in the interpolation case. This can be investigated and verified by solving MATLAB Exercises 4.4, 4.5 and 4.6.

When the sampling rate alteration factor is a product of more than two integers, the decimator (interpolator) is realizable in several stages as shown in Chapter II, Figure 2.35 and Figure 2.36. If the decimation factor M is expressible as a product of K integers, $M = M_1 \times M_2 \times \ldots M_K$, the decimator can be implemented in K decimation stages as indicated in Figure 2.35, whereas K interpolation stages can be used for the interpolation-by-L with $L = L_1 \times L_2 \times \ldots L_K$, see Figure 2.36.

We have seen in the design example considered above, that even in the two-stage system with $M = M_1 \times M_2$, the multiplication rate depends significantly on the ordering of the decimation stages M_1 and M_2. In the multistage sampling rate conversion, the selection of the optimum ordering is not simple. The best solution that minimizes the required number of multiplications per second depends on the selection of the number of stages K and on the combination and ordering of $M_1, M_2, \ldots, M_K$ ($L_1, L_2, \ldots, L_K$).

The design procedure of the two-stage system considered in this subsection can be extended to more than two stages. Experience has shown that in most cases the largest savings in the computational complexity are reached in going from a single-stage to a two-stage implementation (Ansari, 1993).

Advanced Methods for Improving the Computational Efficiency of FIR Decimators and Interpolators

The design procedure described in the previous subsection is carried out using the standard MATLAB functions for filter design, and consequently, the resulting solutions are suboptimal. Looking in the plots of Figure 4.19, where the frequency responses of the two-stage decimator are shown, we observe that in some stopband regions the attenuation unnecessarily exceeds the requested minimum of 60 dB. This indicates that the minimal requested attenuation in the whole stopband can be reached with filters $H_1(z)$ and $H_2(z)$ of the lower order if the appropriate optimization algorithm is used.

An effective approach for reducing the computational complexity of FIR decimators and interpolators based on an additional low-order FIR filter was proposed in (Saramäki, 1984). The sampling rate conversion system includes an additional filter $A(z)$ as depicted in Figure 4.21. In decimator, filter $A(z)$ is placed at the output and operates at the output sampling rate as shown in Figure 4.21(a), whereas in the case of interpolator filter $A(z)$ is placed at the input and operates at the input sampling rate, Figure 4.21(b).

Actually, the configurations of Figure 4.21 represent the two-stage implementations. The first stage of the decimator consists of the filter $H_1(z)$ followed with an M-fold-down-sampler, and the second stage is the additional filter $A(z)$. The interpolator in the first stage has the filter $A(z)$, and the second stage consists of an L-fold up-sampler followed with the filter $H_1(z)$. Therefore, the additional filter $A(z)$ in both realizations operates at the lower sampling rate. The single-stage equivalent $H(z)$ is given by

$$H(z) = H_1(z)A(z^M), \text{ and } H(z) = H_1(z)A(z^L) \qquad (4.23)$$

Figure 4.21. (a) Decimator with the extra filter A(z) at the output. (b) Interpolator with the extra filter A(z) at the input.

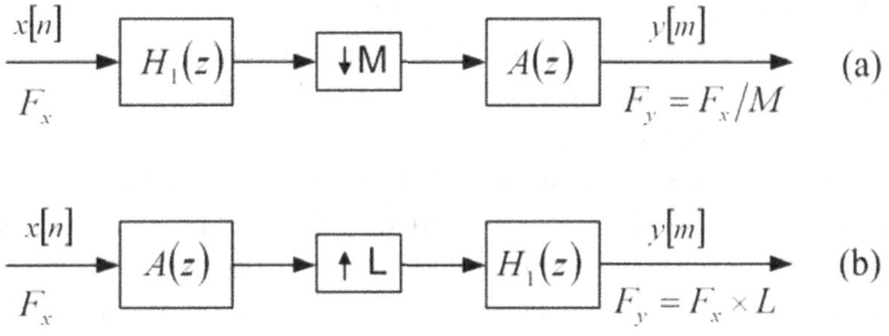

for decimator and for interpolator, respectively.

Filter $H_1(z)$ is a decimation (interpolation) filter that prevents aliasing (suppress images), and $A(z)$ is dedicated to compensating the passband characteristic of $H_1(z)$. With this approach, the filter $H_1(z)$ can be designed with significantly relaxed passband specifications, and $A(z)$ is adjusted to provide the overall equiripple magnitude response in the passband. An efficient iteration procedure for the design of filters of this type is presented in (Saramäki, 1984). It was shown that the significant savings in the overall computational efficiency of decimators and interpolators can be achieved with this approach.

Furthermore, the computational efficiency of multistage decimators and interpolators can be improved when optimizing the cascade of several stages of the configurations of Figures 4.21(a) and (b), respectively. The description of the optimization algorithm illustrated with examples can be found in (Saramäki, 1984, 1986, 2001).

The computational efficiency of an optimal sampling rate conversion system with a minimal number of multiplication constants can be further improved by simplifying the arithmetic operation in subfilters. This demand is of particular importance for VLSI implementations.

It was shown recently (Gustafsson 2004, and 2006; Eghbali, 2007) that multiple constant multiplication (MCM) technique is an efficient way to reduce the computational workload in the polyphase decimators and interpolators. The basic idea in the MCM technique is to multiply one data with several coefficients. Each constant is expressed using shifts, additions and subtractions (Dempster 1995, and 2000). The number of adders and subtractors can be significantly reduced by using common partial results. Generally, MCM technique replaces a general multiplier with a combination of a small number of shifters, adders and subtractors. In this way, the computational efficiency is increased.

An efficient approach for reducing the computational complexity of polyphase decimation filter suitable for VLSI implementations was proposed in (Yeary, 2006).

MATLAB EXERCISES

4.1 Study the polyphase decimation and interpolation by modifying program demo_4_1.

(a)   Design an optimal linear-phase FIR filter to satisfy the *Case b* tolerance scheme of Figure 3.5 (b) for the sampling-rate conversion factor $M = L = 4$. The peak passband ripple is $a_p = 0.1$ dB, and the minimal stopband attenuation amounts to $a_s = 60$ dB. Choose the passband edge frequency $\omega_p < \pi/M$. The sampling frequency at the input of decimator and at the output of interpolator is 12 kHz.

(b) Generate the input signal $\{x[n]\}$ according to your own choice.

(c) Compute four polyphase components for the filter designed in point (a).

(d) Use the commutative polyphase structure of Figure 4.8 to decimate-by-4 signal $\{x[n]\}$. To perform decimation, modify *Step 4* in program demo_4_1.

(e) Use the commutative polyphase configuration of Figure 4.10 to interpolate-by-4 decimated signal obtained in point (d). To perform interpolation, modify *Step 5* in program demo_4_1.

(f) Compute the multiplication rate in decimator and in interpolator, respectively.

(g) Compute and plot: filter magnitude response, spectrum of the input signal, spectrum of the decimated signal, and spectrum of the interpolated signal.

(h) Verify the commutative configurations used above: (i) Evaluate decimation by using the cascade connection of FIR filter from point (a) and the factor-of-4 downsampler. (ii) Evaluate interpolation with the cascade of the factor-of-4 up-sampler and FIR filter from point (a). (iii) Compute and plot spectra of the decimated signal and that of the interpolated signal and compare with the results obtained in point (f).

4.2 Repeat Exercise 4.1 for the sampling rate conversion factor $M = L = 6$. This time simulate decimation and interpolation by modifying program demo_4_2.

4.3 Study the application of memory-saving structures of polyphase decimators and interpolators by modifying program demo_4_3.

(a) Design an optimal linear-phase FIR filter to satisfy the *Case a* tolerance scheme of Figure 3.5 (a), for the sampling-rate conversion factor $M = L = 4$. The peak passband ripple is $a_p = 0.1$ dB, and the minimal stopband attenuation is $a_s = 50$ dB. Choose the passband edge frequency $\omega_p < \pi/M$. The sampling frequency at the input of decimator and at the output of interpolator is 12 kHz.

(b) Generate the input signal $\{x[n]\}$ according to your own choice.

(c) Compute four polyphase components for the filter designed in point (a).

(d) On the basis of the memory-saving structure of Figure 4.13 develop the corresponding structure for $M = 4$. Decimate-by-4 signal $\{x[n]\}$ by modifying *Step 4* in the program demo_4_3.

(e) On the basis of the memory-saving structure of Figure 4.15 develop the corresponding structure for $M = 4$. Interpolate-by-4 decimated signal by modifying *Step 5* in the program demo_4_3.

(f) Compute the multiplication rate in decimator and in interpolator, respectively.

(g) Compute the number of delays in decimator and in interpolator, respectively.

(h) Compute and plot: filter magnitude response, spectrum of the input signal, spectrum of the decimated signal, and spectrum of the interpolated signal.

4.4 Investigate the efficiency of two-stage decimator by modifying program demo_4_4 in such a manner that decimation-by-3 is followed by decimation-by-5. Generate the corresponding plots to those given in Figures 4.19 and 4.20.

4.5 Modify program demo_4_4 to study the two-stage implementation of factor-of-15 interpolator.

(a) Implement the interpolator with $L_1 = 5$ in the first stage, and $L_2 = 3$ in the second stage.

(b)   Implement the interpolator with $L_1 = 3$ in the first stage, and $L_2 = 5$ in the second stage.

(c)   Compare the computational efficiency of two solutions from (a) and (b).

4.6 Design and implement the factor-of-6 decimator, which converts the input frequency of 24 kHz to the frequency of 4 kHz.

(a)   Design and implement the single-stage decimator.

(b)   Design and implement two-stage decimator with $M_1 = 3$, and $M_2 = 2$.

(c)   Design and implement two-stage decimator with $M_1 = 2$, and $M_2 = 3$.

(d)   Plot and comment on the results.

(e)   Compute and compare the multiplication rates of the solutions from (a), (b) and (c).

4.7 Design and implement the factor-of-6 interpolator, which converts the input frequency of 4 kHz to the frequency of 24 kHz.

(a)   Design and implement the single-stage interpolator.

(b)   Design and implement two-stage interpolator with $L_1 = 3$, and $L_2 = 2$.

(c)   Design and implement two-stage interpolator with $L_1 = 2$, and $L_2 = 3$.

(d)   Plot and comment on the results.

(e)   Compute and compare the multiplication rates of the solutions from (a), (b) and (c).

REFERENCES

Ansari,R., & Liu,B., (1993). Multirate signal processing. In Sanjit. K. Mitra and James F. Kaiser (ed.), *Handbook for Digital Signal Processing*. New York, NY: John Wiley-Interscience, 981-1084.

Bellanger, M.G., Bonnerot, G., & Coudreuse, M. (1976). Digital filtering by polyphase network: application to sample-rate alteration and filter banks. *IEEE Transactions on Acoustics, Speech, and Signal Processing, 24*(2), 109-114.

Bellanger, M. (2000). *Digital processing of signals: Theory and practice*. 3rd edition. New York, NY: John Wiley.

Burrus, C.S., McClellan, J.H., Oppenheim, A.V, Parks, T.W., Schaffer, R.W. & Schussler, H.W. (1994). *Computer-based exercises for signal processing using MATLAB.* Englewood Clifs: Prentice-Hall.

Crochiere, R.E., & Rabiner, L.R., (1981, March). Interpolation and decimation of digital signals - A Tutorial Review. *Proceedings of the IEEE, 69*(3), 300-331.

Crochiere, R.E., & Rabiner, L.R. (1983). *Multirate digital signal processing*. Englewood Cliffs, NJ: Prentice-Hall.

Dempster, A. G., & Macleod, M.D. (1995). Use of minimum-adder multiplier block in FIR digital filters. *IEEE Trans. Circuits and Systems II: Analog and Digital Signal Processing, 42*(9), 569-577.

Dempster, A. G., & Marphy, N. R. (2000). Efficient interpolators and filter banks using multiplier blocks. *IEEE Trans. Signal Processing, 48*(1), 257-261.

Diniz, P., Netto, S., & Da Silva, E. (2002). *Digital Signal Processing: System Analysis and Design*. New York, NY: Cambridge University Press.

Eghbali, A., Gustafsson, O., Johansson, H., & Lövenborg, P. (2007). On the complexity of multiplierless direct and polyphase FIR filter structures. *Proc. of the 5ᵗʰ International Symposium on Image and Signal Processing and Analysis, ISPA 2007*, 200-205.

Filter design toolbox for use with MATLAB. User's guide. Version 6. (2006). Natick: MathWorks.

Fliege, N. J. (1994). *Multirate digital signal processing*. New York, NY: John Wiley.

Gustafsson, O., & Dempster, A.G., (2004). On the use of multiple constant multiplication in polyphase FIR filters and filter banks. *Proc. Nordic Signal Proc. Symp.* 53-56.

Gustafsson, O. Johansson, K., Johansson, H. and Wanhammar, L. (2006). Implementation of polyphase decomposed FIR filters for interpolation and decimation using multiple constant multiplication techniques. *Proc. 2006 Asia Pacific Conference on Circuits and Systems*. 926-923.

Harris, F. J., (2004). *Multirate signal processing for communication systems*. Upper Saddle River, NJ: Prentice Hall PTR.

Hentchel, T. (2002). *Sample rate conversion in software configurable radios*. Morwood, MA: Artech House.

McClellan, J. H., & Parks, T. W., & Rabiner, L. R. (1973). A computer program for designing optimum FIR linear-phase digital filters. *IEEE Transactions on Audio Electroacoustics*, *21*(6), 506-526.

Milić, Lj., & Lutovac, M.D. (2002). Efficient multirate filtering. In Gordana Jovanović-Doleček, (ed.), *Multirate Systems: Design & Applications*. Hershey, PA: Idea Group Publishing, 105-142.

Mitra, S. K. (2006). *Digital signal processing: A computer based approach*. 3rd edition. New York, NY: The McGraw-Hill Companies, Inc.

Proakis J. G., & Manolakis D.G. (1996). *Digital signal processing: Principles, algorithms, and applications*. London: Prentice Hall.

Saramäki, T. (1984). A class of linear-phase FIR filters for decimation, interpolation and narrow-band filtering. *IEEE Transactions Acoustics, Speech, Signal Processing*, *32*(5), pp. 1024-1036.

Saramäki, T. (1986). Design of optimal multistage IIR and FIR filters for sampling rate alteration. *Proc. 1986 IEEE International Symposium on Circuits and Systems*. 227-230.

Saramäki, T. (1993). Finite impulse response filter design., Chapter 4 in *Handbook for Digital Signal Processing*. Edited by S. K. Mitra and J. F. Kaiser, New York, NY: John Wiley Interscience, pp. 155-277.

Saramäki, T. *Multirate Signal Processing*. (2001). Lecture notes for a graduate course, the Institute of Signal Processing, Tampere University of Technology, Finland.

Signal processing toolbox for use with MATLAB. User's guide. Version 6. (2006). Natick: MathWorks.

Shpak, D.J., & Antoniou, A. (1990, February). A generalized Remez method for the design of FIR digital filters. *IEEE Transactions on Circuits and Systems*, 37(2), 161-174.

Vaidyanathan, P.P., (1990). Multirate digital filters, filter banks, polyphase networks, and applications: A Tutorial. *Proceedings of the IEEE*, 78(1), 56-93.

Vaidyanathan, P.P., (1993). *Multirate systems and filter banks*. Englewood Cliffs, NJ: Prentice Hall.

Yeary, M.B., Zhang, W., Trelewicz, J. O., Zhai, Y., & McGuire, B. (2006). Theory and implementation of computationally efficient decimation filter for power-aware embedded systems. *IEEE Trans. on Instrumentation and Measurement.* 55(5), 1839-1849.

Chapter V
IIR Filters for Sampling Rate Conversion

INTRODUCTION

Infinite impulse response (IIR) filters are used in applications where the computational efficiency is the highest priority. It is well known that an IIR filter transfer function is of a considerably lower order than the transfer function of an FIR equivalent. The drawbacks of an IIR filter are the nonlinear phase characteristic and sensitivity to quantization errors.

In multirate applications, the computational requirements for FIR filters can be reduced by the sampling rate conversion factor as demonstrated in Chapter IV. However, such a degree of computation savings cannot be achieved in multirate implementations of IIR filters. This is due to the fact that every sample value computed in the recursive loop is needed for evaluating an output sample. Based on the polyphase decomposition, several techniques have been developed which improve the efficiency of IIR decimators and interpolators as will be shown later on in this chapter.

In this chapter, we consider first the direct implementation structures for IIR decimators and interpolators. In the sequel, we demonstrate the computational requirements for direct form IIR decimators and interpolators. The polyphase decomposition of an IIR transfer function is explained with its application to decimation and interpolation. Then, we demonstrate an efficient IIR polyphase structure based on all-pass subfilters, which is applicable to a restricted class of decimators and interpolators. In this chapter, we discuss the application of the elliptic minimal Q factor (EMQF) filter transfer function in constructing high-performance decimators and interpolators. The chapter concludes with a selection of MATLAB exercises for the individual study.

DIRECT IMPLEMENTATION STRUCTURES FOR IIR FILTERS FOR DECIMATION AND INTERPOLATION

For studying the efficiency of IIR decimators and interpolators let us consider the factor-of-M decimator depicted in Figure 5.1, where the decimation filter $H(z)$ is an Nth order IIR transfer function.

For an IIR decimation filter, the $H(z)$ is the transfer function of the form

$$H(z) = \frac{V(z)}{X(z)} = \frac{B(z)}{A(z)}, \tag{5.1}$$

where $A(z)$ and $B(z)$ are N^{th} order polynomials in z^{-1},

$$B(z) = \sum_{n=0}^{N} b_n z^{-n}, \quad A(z) = 1 + \sum_{n=1}^{N} a_n z^{-n}. \tag{5.2}$$

The input-output relations (5.1) can be rearranged in a product of the recursive section $1/A(z)$ and nonrecursive section $B(z)$,

$$V(z) = \frac{1}{A(z)} X(z) B(z). \tag{5.3}$$

Denoting the output of the recursive part with $W(z)$, i.e.,

$$W(z) = \frac{1}{A(z)} X(z) \tag{5.4}$$

we present relation (5.3) as the output of the nonrecursive part inputted by $W(z)$,

$$V(z) = W(z) B(z). \tag{5.5}$$

In time domain, input-output relations (5.4) and (5.5) are expressed by the difference equations,

$$w[n] = x[n] - a_1 w[n-1] - a_2 w[n-2] \quad \dots \quad - a_N w[n-N], \tag{5.6}$$

$$v[n] = b_0 w[n] + b_1 w[n-1] + b_2 w[n-2] \quad \dots \quad + b_N w[n-N]. \tag{5.7}$$

The recursive equation (5.6) computes the intermediate sequence $\{w[n]\}$, which is used in the nonrecursive equation (5.7) for computing the output samples $v[n]$.

Figure 5.1. General structure of factor-of-M decimator

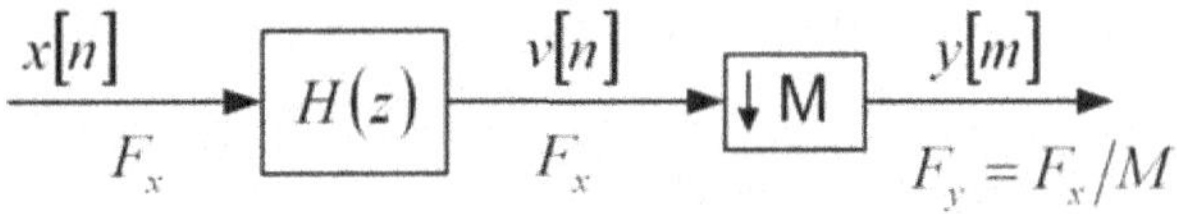

For computing the decimated sequence $\{y[m]\}$, we compute only every M^{th} sample of $\{v[n]\}$, i.e.,

$$y[m] = v[mM] = b_0 w[mM] + b_1 w[mM - 1] + b_2 w[mM - 2] \quad \ldots \quad + b_N w[mM - N]. \tag{5.8}$$

Equation (5.8) shows that for computing a single sample in $\{y[m]\}$ we use $N+1$ consecutive samples of $\{w[n]\}$. Therefore, the computations can be evaluated at the lower sampling rate only in the nonrecursive part of the IIR decimator. Figure 5.2 shows the efficient direct form implementation of a factor-of-M IIR decimator. In the recursive part, the computations are evaluated at the input sampling rate, and in the nonrecursive part only every M^{th} sample is computed, i.e., the computations in the nonrecursive part are evaluated at the output sampling rate.

The computationally efficient direct-form structure for an IIR interpolator can also be developed starting from the general interpolator structure of Figure 5.3.

With the use of the procedure explained above in the case of decimator, a structure similar to that shown in Figure 5.2 can be obtained for the interpolator. As a result, Figure 5.4 shows the computationally efficient factor-of-L IIR interpolator based on the transpose direct form. In the nonrecursive part, the computations are evaluated at the input sampling rate F_x, whereas in the recursive part, the computations are evaluated at the output sampling rate, $F_y = F_x \times L$. Notice that scaling-by-factor-of-L should be introduced in the interpolator structure of Figure 5.4.

COMPUTATIONAL REQUIREMENTS FOR IIR DECIMATORS AND INTERPOLATORS

We have seen in Chapter IV that in the case of an FIR decimator and interpolator, the computational requirements are reduced by the factor of M, and by the factor of L, respectively. The same order of computational savings cannot be achieved with IIR decimation and interpolation filters. This is due to

Figure 5.2. Direct-form implementation of factor-of- M IIR decimator

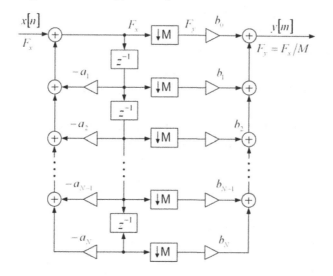

the fact that computations at the lower sampling rate can be evaluated only in the nonrecursive part of the system as demonstrated in the preceding section and illustrated in Figures 5.2 and 5.4.

Considering the structure of decimator in Figure 5.2, we derive easily the corresponding multiplication rate, which we denote by $R_{M,\text{IIR-DEC}}$. The multiplication rate $R_{M,\text{IIR-DEC}}$ is expressed in multiplications per second. We conclude that the computation rate for a factor-of-M decimator consisting of an Nth order IIR filter followed by a factor-of-M down-sampler is determined by the expression,

$$R_{M,\text{IIR-DEC}} = N \times F_x + (N+1) \times F_x / M = N \times F_x + (N+1) \times F_y \text{ multiplications per second} \qquad (5.9)$$

On the other hand, considering the interpolator structure in Figure 5.4, we derive the multiplication rate for the L-fold interpolator denoted with $R_{M\text{-IIR\_INT}}$,

$$R_{M,\text{IIR-INT}} = N \times F_y + (N+1) \times F_y / L = N \times F_y + (N+1) \times F_x \text{ multiplications per second} \qquad (5.10)$$

The following example illustrates the computational complexity of an IIR factor-of-5 decimator implemented in the direct form according to Figure 5.2. We compute the multiplication rate of the IIR decimator and compare the result with the multiplication rate of the corresponding factor-of-5 FIR decimator.

Figure 5.3. General structure of factor-of-L interpolator

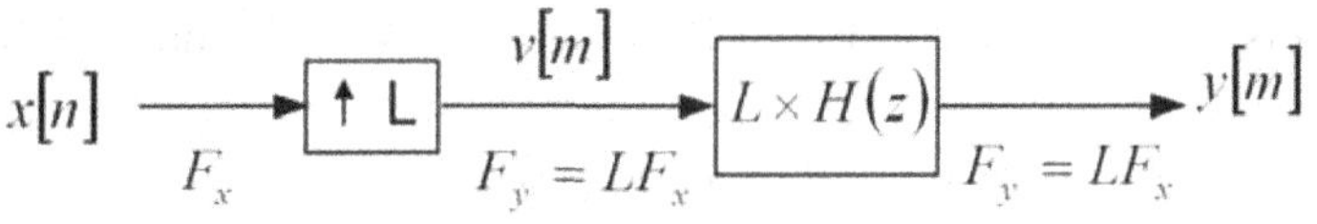

Figure 5.4. Direct-form implementation of factor-of-L IIR interpolator

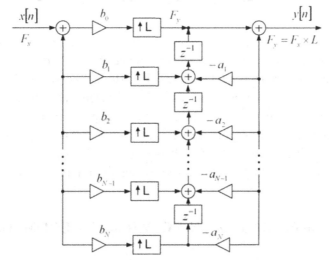

Example 5.1

Compare the multiplication rates of the following decimators:

(a) Factor-of-5 decimator with the elliptic IIR filter of *Example 3.6* designed to satisfy *Case a* tolerance scheme for the decimation factor $M = 5$, see Figure 3.5(a).

(b) Factor-of-5 decimator with the optimal FIR of *Example 3.1* designed to satisfy *Case a* tolerance scheme for the decimation factor $M = 5$, see Figure 3.5(a).

The input sampling frequency for both decimators is $F_x = 10000$ Hz.

Solution:

We compute the multiplication rate $R_{M,\text{IIR-DEC}}$ for the elliptic filter of order $N = 5$ presented in detail in *Example 3.6*, and the multiplication rate $R_{M,\text{FIR-DEC}}$ for the optimal FIR filter of the length $N = 54$ presented in *Example 3.1*. Both filters have been designed to satisfy *Case a* specifications with $M = 5$. Figure 5.5 displays the magnitude and phase responses for two designs.

(a) Multiplication rate of the IIR decimator implemented in the direct form (Figure 5.2) is computed according to (5.9) for $M = 5$, $N = 5$, and $F_x = 10000$ Hz,

$$R_{M,\text{IIR-DEC}} = 5 \times 10000 + (5+1) \times 10000/5 = 62000 \text{ multiplications per second.}$$

(b) Multiplication rate of the FIR decimator implemented in the direct form when neglecting the coefficient symmetry is computed according (4.19) for $M = 5$, $N = 54$, and $F_x = 10000$ Hz,

$$R_{M,\text{FIR-DEC}} = 54 \times 10000/5 = 108000 \text{ multiplications per second.}$$

When the coefficient symmetry is taken into account, see Figure 4.3, the multiplication rate of the FIR decimator reduces to

$$R_{M,\text{FIR-DEC}} = 54 \times 10000/(2 \times 5) = 54000 \text{ multiplications per second.}$$

This example illustrates that the multirate techniques enable a multirate FIR filter to achieve the computational efficiency of a multirate IIR filter. The advantage of the FIR decimator is the linear phase response as illustrated in Figure 5.5. The disadvantage of the IIR decimator, besides the nonlinear phase response, is in the finite-word-length effects such as high coefficient sensitivity and round-off-errors. For the IIR filter considered in this example, the transfer function poles are placed very near to the unit circle of the z-plane making this filter very impractical for the direct form implementation. Notice that the pole radii have the following values: 0.8455, 0.8859, and 0.9618.

Later on, in this chapter we will show that in certain cases IIR filters become very efficient.

IIR FILTER STRUCTURES BASED ON POLYPHASE DECOMPOSITION

The efficient implementation of IIR decimation and interpolation filters can be achieved with polyphase structures. Developing a polyphase structure requires the decomposition of the filter transfer function

Figure 5.5. Magnitude and phase responses of the filters of Example 5.1: Solid line is for FIR filter, dotted line is for IIR filter

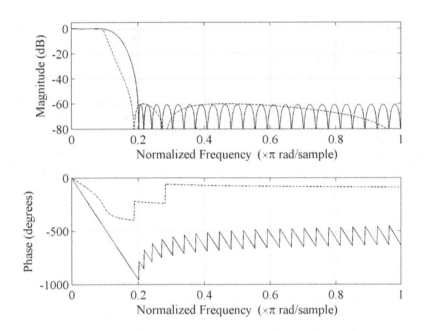

into a set of M (L) polyphase components as already demonstrated in Chapter IV for FIR filters. The overall filter transfer function of a decimation filter $H(z)$ is represented in the form

$$H(z) = \sum_{k=0}^{M-1} z^{-k} H_k\left(z^M\right)$$ (5.11)

where $H_k(z)$ is the kth polyphase IIR subfilter. For an interpolator, the representation of (5.11) is also used including the multiplication by L,

$$H(z) = L\sum_{k=0}^{L-1} z^{-k} H_k\left(z^L\right)$$ (5.12)

In the case of FIR filters the transfer function is a polynomial in terms of z^{-1}, and consequently the polyphase decomposition is very simple as shown in Chapter IV. However, the transfer function of an IIR filter is the ratio of two polynomials, and therefore, the representation of such a function in the form of equations (5.11) and (5.12) requires modifications of the original transfer function in such a way that the denominator contains only powers of z^M (z^L).

There exist in the literature several methods for the polyphase decomposition of the IIR filter transfer function. The well known technique based on the rearrangement of the original IIR transfer function to obtain the form of (5.11) and (5.12) was proposed by Belanger, Bonnerot, and Coudreuse (1976). This technique develops the polyphase subfilters $H_k(z^M)$ starting from the given transfer function $H(z)$. All resulting polyphase subfilters $H_k(z^M)$ have the same denominator, and the total number of the multiplication constants in the overall filter is increased compared with the direct-form implementation of $H(z)$. The benefit to be noticed is that the filter poles to be implemented in the polyphase subfilters are the powers-of-M (L) of the original filter poles. The pole radii become smaller and therefore more distant from the unit circle. This fact decreases the poles' sensitivities and reduces the finite word-length effects in the implementation structure.

The second technique developed for the implementation of polyphase IIR decimators and interpolators uses the distinct all-pass subfilters as polyphase components (Valenzuela & Constantinides, 1983; Drews & Gazsi,1986; Renfors & Saramäki, 1987a&b). With this approach, very efficient IIR decimators and interpolators can be achieved. However, this technique implies certain constraints in the filter frequency response. The properties of this filter class permit that only *Case c* specification of Figure 3.5 can be met for the sampling rate conversion factor of the value M (L) > 2. The exception is the solution for the conversion factor $M = L = 2$, which meets the *Case b* specification.

In this section, we concentrate on the solutions of polyphase IIR decimators and interpolators where the polyphase IIR filter H (z) is constructed as a parallel connection of M (L) distinct all-pass subfilters $A_k(z)$ of the form,

$$A_k(z) = \frac{a_{K,k} + a_{K-1,k}z^{-1} + \ldots + a_{1,k}z^{-(K-1)} + z^{-K}}{1 + a_{1,K}z^{-1} + \ldots + a_{K-1,k}z^{-(K-1)} + a_{K,k}z^{-K}}. \tag{5.13}$$

In practice, a Kth order all-pass transfer function is implemented as a cascade connection of the first-order and the second-order all-pass sections. The first-order all-pass section denoted by $S^{(1)}(z)$ implements the real pole and generally has the form

$$S^{(1)}(z) = \frac{a_1 + z^{-1}}{1 + a_1 z^{-1}}. \tag{5.14}$$

The second-order all-pass section, which we denote by $S^{(2)}(z)$ is used to implement the conjugate complex pole pair, and its general form is given by

$$S^{(2)}(z) = \frac{a_1 + a_2 z^{-1} + z^{-2}}{1 + a_1 z^{-1} + a_2 z^{-2}}. \tag{5.15}$$

The higher order all-pass transfer function $A_k(z)$ is expressible as a product of the corresponding first-order and second-order sections of the forms given by (5.14) and (5.15).

The overall transfer function $H(z)$ of the filter constructed as a parallel connection of M (L) polyphase components is given in (5.11) for the decimation filter, and by (5.12) for the interpolation filter. When the all-pass subfilters $A_k(z)$ are used as polyphase components, equations (5.11) and (5.12) should be modified in the following manner: (i) The functions $H_k(z)$ are replaced with the all-pass functions $A_k(z)$. (ii) Since the $A_k(z)$ is the all-pass transfer function of the unity absolute value, the right-hand sides of equations (5.11) and (5.12) should be scaled with the factor $1/M$ ($1/L$). Therefore, for the decimation filter, we express the transfer function $H(z)$ in the following form

$$H(z) = \frac{1}{M} \sum_{k=0}^{M-1} z^{-k} A_k(z^M), \tag{5.16}$$

and consequently for the interpolation filter we write

$$H(z) = \sum_{k=0}^{L-1} z^{-k} A_k(z^L). \tag{5.17}$$

With the assumption that subfilters $A_k(z^M)$ $[A_k(z^L)]$ are the all-pass transfer functions, the overall frequency responses $H(e^{j\omega})$ can be written as

$$H\left(e^{j\omega}\right) = \frac{1}{M}\sum_{k=0}^{M-1} e^{j\phi_k(\omega)} \tag{5.18}$$

in the case of decimator, and also for the interpolator

$$H\left(e^{j\omega}\right) = \sum_{k=0}^{L-1} e^{j\phi_k(\omega)} \tag{5.19}$$

where

$$\phi_k(\omega) = -k\omega + \arg\left[H_k\left(e^{jM\omega}\right)\right] \tag{5.20}$$

is the phase response of the kth subfilter.

It is obvious from equations (5.18) – (5.20) that the passband and stopband regions are determined by the phase characteristics of the all-pass subfilters. The module of the sum in equations (5.18) and (5.19) should approximate $M(L)$ in the passband, and zero in the stopband, i.e.,

$$\left|\sum_{k=0}^{M-1} e^{j\phi_k(\omega)}\right| \approx \begin{cases} M(L), & \text{passband} \\ 0, & \text{stopband} \end{cases}. \tag{5.21}$$

The passband is obtained in the frequency band in which the phase responses $\phi_k(\omega)$ are almost the same. To provide the stopband in the desired range of frequencies, the phase differences between the consecutive polyphase branches should be approximately $2\pi/M$, $4\pi/M$, … $(2\pi/L, 4\pi/L, …)$ in that range.

Since the polyphase subfilters are the all-pass transfer functions with the magnitude responses equal to unity, the magnitude response for the overall decimation filter is limited by

$$\left|H\left(e^{j\omega}\right)\right| \leq 1, \tag{5.22}$$

and for the interpolator this limit is,

$$\left|H\left(e^{j\omega}\right)\right| \leq L. \tag{5.23}$$

The square magnitude response of this filter class is characterized by the general property (Renfors & Saramäki, 1987a)

$$\sum_{r=0}^{M-1} \left|H\left(e^{j(\omega+2\pi r/M)}\right)\right|^2 = 1. \tag{5.24}$$

It is to be observed that due to the property of the squared magnitude response defined in (5.24), the passband ripple should be very small.

As stated above, the class of IIR filters, which is based on the parallel connection of $M(L)$ all-pass branches, has certain limitations. Namely, their stopband cutoff frequency must be selected higher than the half of the lower sampling rate of the decimator (interpolator). Additionally, for $M(L) > 2$ only *Case c* filter specification can be satisfied. In decimation, those properties cause the aliasing in the transition band, and some imaging effects in interpolation. If those drawbacks can be tolerated, very efficient decimators and interpolators can be obtained. Furthermore, by introducing a very low-order extra filter in the lower sampling rate side of the system, the drawbacks mentioned above can be eliminated (Renfors & Saramäki, 1987a&b).

Usually, the polyphase IIR filters are the nonlinear phase systems, but they can be designed to exhibit approximately linear phase characteristics. The Remez-type algorithm for the optimized design of both nonlinear phase and approximately linear phase polyphase IIR filters with the all-pass polyphase branches was developed by Renfors and Saramäki and the description of the algorithm can be found in (Renfors & Saramäki, 1987a&b).

The polyphase factor-of-M decimator, which represents equation (5.16) can be implemented with the efficient commutative structure shown in Figure 5.6. The configuration of Figure 5.6 is actually the configuration of Figure 4.8 developed for the efficient polyphase implementation of FIR decimators. The main difference is the type of polyphase subfilters: in Figure 4.8, the polyphase branches are the FIR subfilters $E_k(z)$, whereas in Figure 5.6 the polyphase branches are the IIR all-pass subfilters $A_k(z)$. The efficient structure of a factor-of-L interpolator that follows from the polyphase representation of equation (5.17) is given in Figure 5.7. As in the case of decimator, the FIR subfilters $E_k(z)$ are to be replaced with all-pass subfilters $A_k(z)$.

Our goal is to describe by the aid of MATLAB the properties and the operation of such a system and to compare the computational complexity with the corresponding FIR filter. An interested reader is recommended to read the articles of Renfors & Saramäki (1987a&b) where the design algorithm is given in detail.

In this section, we present an example of IIR factor-of-5-decimator constructed as a parallel connection of 5 polyphase all-pass branches. For the all-pass subfilters, we use the design data given in the Lecture Notes written by T. Saramäki (2001). In the following example, we form the all-pass sections

Figure 5.6. Commutative polyphase structure of an IIR factor-of-M decimator with all-pass polyphase branches

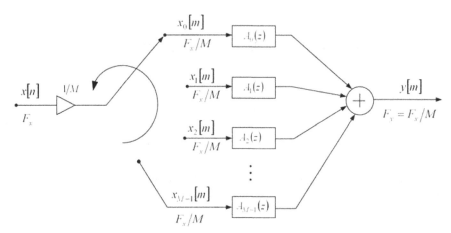

Figure 5.7. Commutative polyphase structure of an IIR factor-of-L interpolator with all-pass polyphase branches

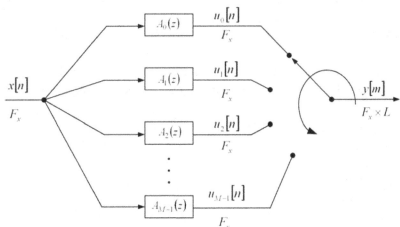

from the given poles of the polyphase subfilters and construct their transfer functions. This provides us the opportunity to study by means of an example the behaviors of the phase responses of the polyphase branches, and the resulting overall magnitude response of the decimation filter. We verify the operation of the decimator by introducing the all-pass polyphase branches into the commutative decimator structure of Figure 5.6 for $M = 5$. We compare the properties and efficiency of the IIR decimator with the corresponding FIR decimator.

Example 5.2

Design specifications:

Decimation filter for $M = 5$ and *Case c* tolerance scheme.
The passband edge frequency $\omega_p = 0.8\pi/5$.
Stopbands: $\{1.2\pi/5 - 2.8\pi/5\}$ and $\{3.2\pi/5 - 4.8\pi/5\}$
Minimal stopband attenuation is 60 dB.

Solution:

The factor-of-5-decimator is constructed as a parallel connection of 5 all-pass polyphase branches $[A_0(z), A_1(z), A_2(z), A_3(z), A_4(z)]$, which are determined as follows (Saramäki, 2001):

$A_0(z)$ has three real poles and therefore consists of three first-order sections
        z01 = -0.03247627480; z02 = -0.4519480048; z03 = -0.9477051753;
$A_1(z)$ has two real poles and therefore consists of two first-order sections
        z11 = -0.08029157130; z12 = -0.5548998293;
$A_2(z)$ has two real poles and therefore consists of two first-order sections
        z21 = -0.1417079348; z22 = -0.6883346404;
$A_3(z)$ has two real poles and therefore consists of two first-order sections
        z31 = -0.2320513100; z32 = -0.7961481351;
$A_4(z)$ has two real poles and therefore consists of two first-order sections
        z41 = -0.3532045984; z42 = -0.8755417392;

In the 6-step procedure presented below, we construct the polyphase IIR decimator using all-pass branches $A_0(z) - A_4(z)$, and perform decimation-by-5 using the efficient configuration of Figure 5.6

Step 1: We first compute in MATLAB the coefficients of the polyphase subfilters:

Branch $A_0(z)$ implements three real poles and therefore is formed as a cascade connection of three first-order sections. Branches $A_1(z) - A_4(z)$ are formed as a cascade connection of two first-order sections.

```
a0 = conv([1,-z01],[1,-z02]); a0 = conv(a0,[1,-z03]);
a1 = conv([1,-z11],[1,-z12]);
a2 = conv([1,-z21],[1,-z22]);
a3 = conv([1,-z31],[1,-z32]);
a4 = conv([1,-z41],[1,-z42]);
```

Step 2: Computation of $A_k(z^5)$, and $z^{-k}A_k(z^5)$.

```
A0 = upsample(a0,5); A1 = upsample(a1,5); A2 = upsample(a2,5); A3 = upsample(a3,5);
A4 = upsample(a4,5);
B0 = fliplr(A0); B1 = [0,fliplr(A1)]; B2 = [0,0,fliplr(A2)]; B3 = [0,0,0,fliplr(A3)]; B4 = [0,0,0,0,fliplr(A4)];
```

Step 3: Compute and plot phase responses of polyphase branches.

```
[PH0,f] = freqz(B0,A0,1024,2); [PH1,f] = freqz(B1,A1,1024,2); [PH2,f] = freqz(B2,A2,1024,2);
[PH3,f] = freqz(B3,A3,1024,2); [PH4,f] = freqz(B4,A4,1024,2);
figure (1)
plot(f,unwrap(angle(PH0)),f,unwrap(angle(PH1)),'b',f,unwrap(angle(PH2)),'b',f,unwrap(angle(PH3)),'b',...
f,unwrap(angle(PH4)),'b')
xlabel('Normalized frequency [\omega/\pi]'), ylabel('Phase [rad]')
```

Step 4: Compute the frequency response of the IIR polyphase filter constructed according to (5.16). Design the equivalent FIR filter. Plot the magnitude and phase responses for both designs.

```
% Overall IIR filter frequency response
H = PH0 + PH1 + PH2 + PH3 + PH4;
figure(2)
subplot(2,1,1), plot(f,20*log10(abs(H)/5),'LineWidth',2), grid
hold on
subplot(2,1,2), plot(f,unwrap(angle(H)),'LineWidth',2)
hold on

% Equivalent FIR filter
ap = 0.000015; as = 60; Fp = 0.8/5;
dev = [(10^(ap/20)-1)/(10^(ap/20)+1),10^(-as/20),10^(-as/20),10^(-as/20)];
F = [Fp,2/5-Fp,2/5+Fp,4/5-Fp,4/5+Fp,1];   % Cutoff frequencies
```

Figure 5.8. Phase responses of the polyphase branches $z^{-k}A_k(z)$, $k = 0, 1, ..., 4$

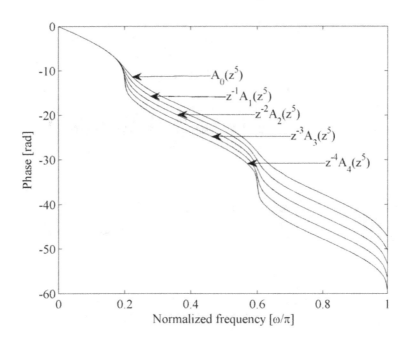

```
A = [1,0,0,0];      % Desired amplitudes
[Nord,Fo,Ao,W] = firpmord(F,A,dev); % Estimating the filter order
b = firpm(126,Fo,Ao,W); % Computing the filter coefficients
[Hfir,f] = freqz(b,1,1024,2); % Computing the filter frequency response
figure (2)
subplot(2,1,1)
plot(f,20*log10(abs(Hfir))), ylabel('Magnitude [dB]'), axis([0,1,-80,2])
subplot(2,1,2), plot(f,unwrap(angle(Hfir)))
xlabel('Normalized frequency [\omega/\pi]'), ylabel('Phase [rad]'),axis([0,1,-50,0])

% Plot the passband details: IIR and FIR
figure (3)
subplot(2,1,1)
plot(f(1:164),20*log10(abs(H(1:164))/5),'LineWidth',2)
hold on
figure (3)
plot(f(1:164),20*log10(abs(Hfir(1:164)))),grid
xlabel('Normalized frequency [\omega/\pi]'), ylabel('Magnitude, dB')
```

Step 5: Generate the input signal $\{x[n]\}$ consisting of three sinusoidal sequences and additive noise of the normal distribution.

```
n = 0:510;
x = cos(2*pi*200*n/10000) + cos(2*pi*400*n/10000) + 0.7*cos(2*pi*3800*n/10000) + randn(size(n));
```

147

Figure 5.9. Magnitude and phase responses: IIR filter – thick line, FIR filter – thin line

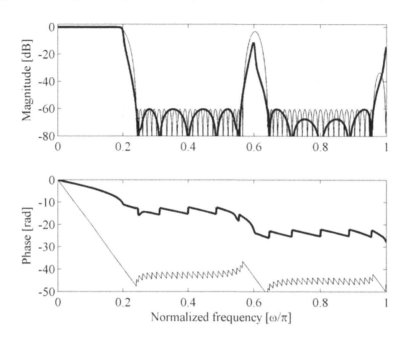

```
[X,F1] = freqz(x,1,1024,10000);
```

Step 6: Decimate-by-5 the input signal {*x*[*n*]} by using the efficient decimator structure of Figure 5.6 and IIR filter of this example.

```
% IIR filter, structure verification: polyphase downsampling and filtering
x0 = x(1:5:length(x)); x1 = [0,x(5:5:length(x))]; x2 = [0,x(4:5:length(x))];
x3 = [0,x(3:5:length(x))]; x4 = [0,x(2:5:length(x))];
y0 = filter(fliplr(a0)/5,a0,x0); y1 = filter(fliplr(a1)/5,a1,x1); y2 = filter(fliplr(a2)/5,a2,x2);
y3 = filter(fliplr(a3)/5,a3,x3); y4 = filter(fliplr(a4)/5,a4,x4);
ydec = y0 + y1 + y2 + y3 + y4; % Decimated signal
[Ydec,F2] = freqz(ydec,1,512,2000);
```

Step 7: Display the spectrum of the input signal and the spectrum of the decimated signal.

```
figure (4)
subplot(2,1,1), plot(F1,abs(X)), legend('Spectrum: input signal'), ylabel('|X(F)|'), axis([0,5000,0,300])
subplot(2,1,2),
plot(F2,abs(Ydec)), legend('Spectrum: decimated signal'), ylabel('|Y_d_e_c(F)|'), xlabel('Frequency [Hz]')
```

Figures 5.8 – 5.11 illustrate properties of the IIR polyphase decimator with all-pass polyphase branches. Figure 5.8 displays the phase characteristics of the branch filters. In the frequency band [0,

Figure 5.10. Passband details: IIR filter – thick line, FIR filter – thin line

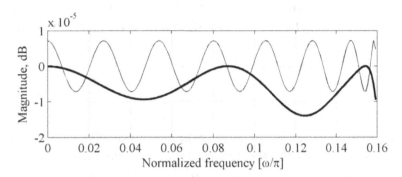

Figure 5.11. Spectrum of the input signal and spectrum of the decimated signal

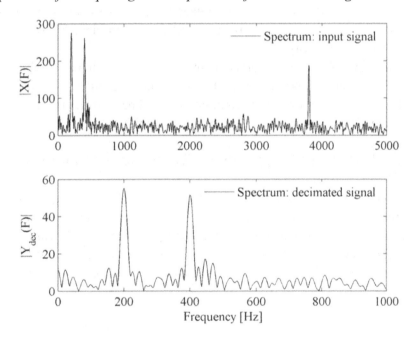

$0.8\pi/5]$ the phases are nearly the same, and in this range the filter magnitude response $|H(e^{j\omega})|$ approximates unity, i.e.,

$$|H(e^{j\omega})| = \frac{1}{M}\left|\sum_{k=0}^{M-1} e^{j\phi_k(\omega)}\right| \approx 1 \quad \text{for} \quad 0 \le \omega \le 0.8\pi/5. \tag{5.25}$$

Thereby, the filter has the passband in that range as shown in Figures 5.9 and 5.10. As shown in Figure 5.10, the passband ripple is extremely small, about 10^{-5}. It has to be pointed out that for a given stopband ripple, the peak passband ripple is restricted by the magnitude response property expressed by equation (5.24). As a consequence, the passband ripple cannot be arbitrarily chosen by designer.

In the frequency band $[1.2\pi/5, 2.8\pi/5]$, the differences between the consecutive phases are approximately $2\pi/5$, and in the frequency band $[3.2\pi/5, 4.8\pi/5]$ the differences are approximately $4\pi/5$. Consequently, in those two ranges, the filter magnitude response $|H(e^{j\omega})|$ approximates zero giving the two stopbands,

$$\left|H\left(e^{j\omega}\right)\right| = \frac{1}{M}\left|\sum_{k=0}^{M-1} e^{j\phi_k(\omega)}\right| \approx 0 \quad \text{for} \quad 1.2\pi/5 \leq \omega \leq 2.8\pi/5, \quad 3.2\pi/5 \leq \omega \leq 4.8\pi/5. \tag{5.26}$$

Figures 5.9 and 5.10 display also the characteristics of the linear phase FIR filter designed to exhibit the similar magnitude response as the IIR filter of this example. It is important to observe that the length of FIR filter is $N = 165$, and that the minimal number of multiplication constants in the implementation exploiting coefficient symmetry amounts to 83. The IIR filter of this example has 11 z-plain poles and therefore the filter order is 11. Since the only one multiplication constant implements the first-order section, the IIR filter is implemented with 11 constants. This comparison illustrates the high efficiency of polyphase IIR filters. The FIR filter has the advantage of linear phase and simultaneously the disadvantage of high phase delay. The phase delay of the IIR filter is considerably smaller, but the main disadvantage is the phase nonlinearity.

Finally, Figure 5.11 shows the spectrum of the input signal $\{x[n]\}$ with the sampling rate 10000 Hz, and the spectrum of the decimated signal $\{y_{\text{dec}}[m]\}$ whose sampling rate is reduced to 2000 Hz. The plots of Figure 5.11 verify the operation of the commutative decimator structure of Figure 5.6 with the all-pass polyphase branches $[A_0(z), A_1(z), ..., A_4(z)]$ of this example.

Computational Efficiency of Polyphase IIR Decimators and Interpolators

As shown above, the commutative configurations of Figures 5.6 and 5.7 have been applied to implement IIR polyphase decimators and interpolators, respectively. The computations in the all-pass polyphase branches are evaluated at the lower sampling rate. Since the polyphase branches are distinct subfilters, the computational efficiency can be computed in the same manner as already used for FIR decimators and interpolators in Chapter IV. Hence, the average multiplication rate for an IIR polyphase factor-of-M decimator is given by

$$R_{M\_\text{DEC}} = \frac{N \times F_x}{M} = N \times F_y, \tag{5.27}$$

and similarly, for an IIR polyphase factor-of-L interpolator

$$R_{M\_\text{INT}} = \frac{N \times F_y}{L} = N \times F_x. \tag{5.28}$$

Recall that F_x denotes the input sampling frequency, and F_y is the sampling frequency at the output.

The IIR polyphase decimator of *Example 5.2* has five all-pass branches. Since the filter transfer function has 11 real poles, only 11 multiplication constants are needed to implement the all-pass subfilters in the polyphase branches:

- $A_0(z)$ implements 3 real poles in 3 all-pass sections in the cascade connection. Each of the first-order sections of the form (5.14) is implementable with only one multiplication constant. Therefore, branch $A_0(z)$ is implemented with 3 multiplication constants.

- Each of the all-pass branches $A_1(z)$, $A_2(z)$, $A_3(z)$ and $A_4(z)$ implement two real poles, and therefore only 2 multiplication constants are needed per branch.

Therefore, the multiplication rate for the IIR decimator of *Example 5.2* when used to convert the input sampling rate of 10000 Hz to the rate of 2000 Hz is given by $R_{M,\text{IIR}-\text{DEC}} = 11 \times 10000/5 = 22000$ multiplications per second.

The efficiency of IIR polyphase filter is visible when compared with the efficiency of a corresponding FIR filter. As shown in Figures 5.9 and 5.10, an FIR filter of the length of $N = 127$ exhibits a similar magnitude response as the examined IIR filter. When neglecting the coefficient symmetry, the multiplication rate for this FIR filter amounts to $R_{M,\text{FIR-DEC}} = 127 \times 10000/5 = 254000$ multiplications per second, and when exploiting the coefficient symmetry the multiplication rate reduces to 127000 mult/sec.

The last figures illustrate the extremely high computational efficiency achieved with the polyphase IIR filter. When the linear phase response is not of the highest priority, this class of filters provides very efficient filtering with excellent magnitude characteristics.

As the consequence of the constraint of equation (5.24), the inherent property of the polyphase decimators and interpolators with the all-pass subfilters is an extremely low passband ripple, as illustrated in Figures 5.9 and 5.10 for *Example 5.2*. We can observe that for the stopband attenuation of 60 dB, the peak passband ripple amounts to only 1.5×10^{-5} dB. Therefore, there are practically no amplitude distortions in the passband.

The Role of Extra Filter

An effective approach for improving stopband characteristics of IIR decimators and interpolators based on an additional low-order IIR filter was proposed in (Renfors & Saramäki, 1987a&b). The sampling rate conversion system includes an extra filter $G(z)$ as depicted in Figure 5.12. In decimator, filter $G(z)$ is placed at the output and operates at the output sampling rate as shown in Figure 5.12 (a), whereas in the case of interpolator filter $G(z)$ is placed at the input and operates at the input sampling rate, Figure 5.12 (b).

The configurations of Figure 5.12 can be considered as two-stage implementations. The first stage of the decimator consists of the decimation filter $H_1(z)$ followed with a factor-of-M down-sampler, and the second stage is the extra filter $G(z)$. The interpolator in the first stage has the filter $G(z)$, and the second stage consists of a factor-of-L up-sampler followed with the interpolation filter $H_1(z)$. Therefore, the extra filter $G(z)$ in both realizations operates at the lower sampling rate. The single-stage equivalent $H(z)$ is given by

$$H(z) = H_1(z)G(z^M), \text{ and } H(z) = H_1(z)G(z^L) \tag{5.29}$$

for decimator and for interpolator, respectively.

The extra filter $G(z)$, when properly designed, can compensate the undesired peaks in the stopband, and decrease the transition band in the decimation (interpolation) filter $H_1(z)$. This approach is particularly effective in IIR polyphase decimators and interpolators with all-pass subfilters in the polyphase branches. As demonstrated in *Example 5.2*, with the all-pass branches, filters can be designed to satisfy only *Case c* tolerance scheme. The following example illustrates the use of a low-order extra filter in constructing a decimation (interpolation) filter that satisfies *Case a* tolerance scheme.

Figure 5.12. (a) Decimator with the extra filter G(z) at the output. (b) Interpolator with the extra filter G(z) at the input.

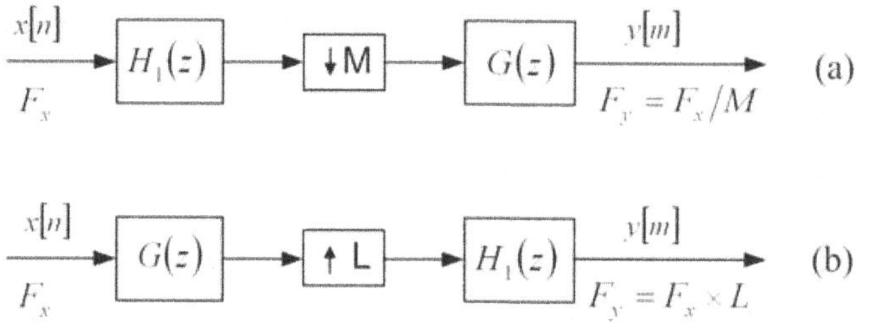

Example 5.3

In this example we examine the frequency response of the sampling rate conversion system of Figure 5.12 and equation (5.29) when the sampling rate conversion factor is 5.

Design specifications:

Decimation filter for $M = 5$ and *Case a* tolerance scheme.
The passband edge frequency $\omega_p = 0.8\pi/5$.
The stopband edge frequency $\omega_s = \pi/5$.
Peak passband ripple $a_p = 0.1$ dB.
Minimal stopband attenuation $a_s = 60$ dB.

Solution:

For the decimation (interpolation) filter $H_1(z)$ we take the polyphase IIR filter of *Example 5.2* whose magnitude response is shown in Figures 5.9 and 5.19. Our goal is to design an extra filter $G(z)$, which will provide the overall magnitude response of the system to satisfy *Case a* tolerance scheme.

Our choice for $G(z)$ is the 5th order elliptic filter with the passband ripple $a_p = 0.1$ dB, the stopband attenuation $a_s = 43$ dB, and passband edge frequency $\omega_p = 0.8\pi$. We compute the filter coefficients using the MATLAB function ellip,

[b,a] = ellip(5,0.1,43,0.8); % Elliptic filter design

The magnitude response of $G(z)$ is given in the first subfigure of Figure 5.13. The second subfigure shows the magnitude response of $H_1(z)$ (thick line) together with the periodic magnitude response of $G(z^5)$ (thin line) . Here, we can observe the role of the extra filter $G(z)$. The periodic characteristic $|G(e^{j5\omega})|$ has the stopbands in the don't-care-band regions and in the transition band of $|H_1(e^{j\omega})|$. The third subfigure presents the overall magnitude response $|H_1(e^{j\omega})|$, which evidently satisfies *Case a* tolerance scheme.

The overall system of *Example 5.3* can be implemented with 16 multiplication constants: 11 coefficients for $H_1(z)$, see *Example 5.2*, and 5 coefficients for $G(z)$. An optimal single- stage decimation (interpolation) FIR filter that meets the specifications of *Example 5.3* should be of the length $N = 142$.

When the IIR sampling rate conversion system of *Example 5.3* is used to convert the sampling frequency of 10000 Hz to the sampling frequency of 2000 Hz, the multiplication rate is 32000 mult/sec. This multiplication rate is considerably lower when compared with the solution based on the single stage FIR filter of the length $N = 142$. For the FIR filter, multiplication rate is 282000 mult/sec when the coefficient symmetry is neglected, and 141000 when exploiting the coefficient symmetry.

In *Example 5.2* and *Example 5.3*, we have demonstrated the high efficiency of IIR polyphase filters. Using the optimization algorithms introduced by Renfors and Saramäki (1987), extremely efficient solutions for high-performance single-stage and multi-stage IIR polyphase filters can be achieved. Besides the algorithms for nonlinear-phase IIR decimators and interpolators, the algorithms for optimal approximately linear phase solutions are also presented in the papers of Renfors and Saramäki (1987).

POLYPHASE IIR STRUCTURE WITH TWO ALL-PASS SUBFILTERS: IIR HALFBAND FILTER

Particularly efficient IIR decimators and interpolators can be reached for the sampling rate conversion factor $M(L) = 2$. It was observed (Valenzuela & Constantinides, 1983), that a sampling-rate conversion system implemented with two polyphase all-pass branches exhibits favorable pass-stopband characteristics. Actually, the polyphase IIR filter constructed with two all-pass subfilters is a well-known *IIR halfband filter* (Ansari, 1983; Valenzuela & Constantinides, 1983; Renfors & Saramäki, 1987; Wegener, 1979; Lutovac, Tošić, & Evans 2000; Gazsi, 1985; Schüssler & Stefen, 1998; Krukowski & Kale, 2003).

The expression for the halfband filter transfer function $H^{HB}(z)$ follows from general equations (5.16) and (5.17) for $M(L) = 2$,

$$H^{HB}(z) = \frac{1}{2}\left(A_0^{HB}\left(z^2\right) + z^{-1}A_1^{HB}\left(z^2\right)\right). \tag{5.30}$$

Hence, an IIR halfband filter consists of two all-pass subfilters denoted by $A_0^{HB}(z)$ and $A_1^{HB}(z)$.

Figure 5.13. Application of extra filter in sampling rate conversion system. From top to the bottom: Extra filter $G(z)$. Decimation filter $H_1(z)$ and periodic extra filter $G(z^5)$. The overall filter $H(z)$.

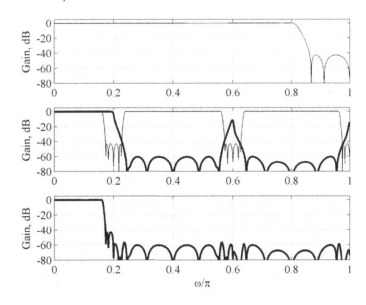

The general magnitude response property for the polyphase IIR filters implemented with all-pass subfilters is given in (5.24). Consequently, the magnitude response property for a halfband filter is characterized by

$$\left| H^{HB}\left(e^{j\omega}\right) \right|^2 + \left| H^{HB}\left(e^{j(\omega+\pi)}\right) \right|^2 = 1. \tag{5.31}$$

When rewriting the above equation in the form

$$\left| H^{HB}\left(e^{j\omega}\right) \right|^2 = 1 - \left| H^{HB}\left(e^{j(\pi-\omega)}\right) \right|^2, \tag{5.32}$$

we observe that the magnitude squared function $|H^{HB}(e^{j\omega})|^2$ is symmetric around $\omega = \pi/2$, i.e.,

$$\omega_s^{HB} = \pi - \omega_p^{HB}, \tag{5.33}$$

where ω_p^{HB} and ω_s^{HB} are passband and stopband edge frequencies, respectively. At the middle of the band, at the frequency $\omega = \pi/2$, the squared magnitude response equals $1/2$,

$$\left| H^{HB}\left(e^{j\omega}\right) \right|^2_{\omega=\pi/2} = \frac{1}{2}. \tag{5.34}$$

Notice that the corresponding gain response at $\omega = \pi/2$ is approximately 3 dB below the maximum value.

It can be observed also that the ripples of the squared magnitude response should be equal in the pass- and stopband. Namely, if the squared magnitude approximates unity with tolerance Δ in the passband region $[0,\omega_p^{HB}]$ with $\omega_p^{HB} < \pi/2$, then it approximates zero with the same tolerance Δ in the stopband $[\pi - \omega_p^{HB}, \pi]$. Consequently, the peak passband ripple and the minimal stopband attenuation are not independent. For the given minimal stopband attenuation of a_s in dB, the peak passband ripple a_p in dB is determined by

$$a_p = -10\log\left(1 + \frac{1}{10^{a_s/10} - 1}\right) \tag{5.35}$$

Table 1 illustrates the practical aspect of equation (5.35). It follows that for the reasonable values of a_s, one obtains very small values for the passband ripple a_p.

Table 5.1. Corresponding values for a_s and a_p for halfband filter transfer function

a_s[dB]	a_p[dB]
10	0.45767491
20	0.04364805
30	0.00434512
40	0.00043432
50	0.00004343
60	0.00000434

Figure 5.14. Implementation structure of IIR halfband filter based on the parallel connection of two all-pass subfilters

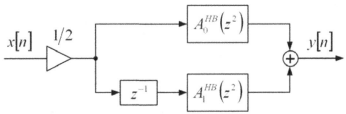

Figure 5.14 displays the implementation structure for a halfband filter based on equation (5.30). This structure is a parallel combination of two all-pass subfilters $A_0^{HB}(z^2)$ and $z^{-1} A_1^{HB}(z^2)$.

For implementing factor-of-two decimation or interpolation, we use efficient commutative structures of Figures 5.6 and 5.7, respectively. The polyphase structure is composed of two all-pass subfilters $A_0^{HB}(z)$ and $A_1^{HB}(z)$ that operate at the lower sampling rate of the system as shown in Figure 5.15.

There exist analytic solutions for the design of an IIR halfband filter transfer function (Valenzuela & Constantinides, 1983; Lutovac, Tošić, & Evans, 2000; Milić & Lutovac, 2002; Milić & Lutovac, 2003; Mitra, 2006). Since the halfband filter frequency response should satisfy symmetry conditions defined in (5.32) and (5.33) the filter design parameters cannot be arbitrarily chosen. For the given filter order N, we are restricted to choose only one of the design parameters $[\omega_p, \omega_s, a_p, a_s]$. If we, for example, choose ω_p, the remaining three parameters $[\omega_s, a_p, a_s]$ should be determined in a manner to provide symmetry conditions (5.32) and (5.33). The design algorithms for IIR halfband filters will be discussed in more detail in Chapter VII, and the MATLAB programs for the design of elliptic halfband filters is given in Appendix A. In this chapter, we show how a halfband filter can be used in constructing factor-of-two decimators and factor-of-two interpolators.

For a given halfband filter transfer function $H^{HB}(z)$ one can use the configurations of Figure 5.15 (a) and (b) to construct a factor-of-two decimator or a factor-of-two interpolator, respectively. When the filter poles are known, it is straightforward to determine the all-pass subfilters $A_0^{HB}(z^2)$ and $A_1^{HB}(z^2)$. For a large class of the halfband filters, the filter order N is an odd number, and the poles of $H^{HB}(z)$ are placed on the imaginary axis of the z-plane, as will be exposed in Chapter VII. One pole is located at the origin producing the term z^{-1} in equation (5.30), and consequently the delay element in Figure 5.14. The remaining poles, which appear in conjugate-complex pairs, are distributed among the all-pass subfilters $A_0^{HB}(z^2)$ and $A_1^{HB}(z^2)$.

A conjugate-complex pole pair forms a second-order all-pass section, which belongs to $A_0^{HB}(z^2)$ or to $A_1^{HB}(z^2)$. Thereby, the all-pass transfer functions $A_0^{HB}(z^2)$ and $A_1^{HB}(z^2)$ are expressible by,

$$A_0^{HB}\left(z^2\right) = \prod_{l=2,4,\ldots}^{(N+1)/2} \frac{\beta_l^{HB} + z^{-2}}{1 + \beta_l^{HB} z^{-2}} \text{ and } A_1^{HB}\left(z^2\right) = \prod_{l=3,5,\ldots}^{(N+1)/2} \frac{\beta_l^{HB} + z^{-2}}{1 + \beta_l^{HB} z^{-2}} \qquad (5.36)$$

where β_l^{HB} is a squared radius of the pole p_l^{HB},

$$\beta_l^{HB} = \left(r_l^{HB}\right)^2 \quad \text{for } l = 2,3,\ldots,(N+1)/2 \quad \text{and} \quad \beta_l^{HB} < \beta_{l+1}^{HB} \text{ for } l = 2,3,\ldots,(N-1)/2 \qquad (5.37)$$

Figure 5.15 Efficient commutative structures for:(a) factor-of-two decimator, (b) factor-of-two inter-polator

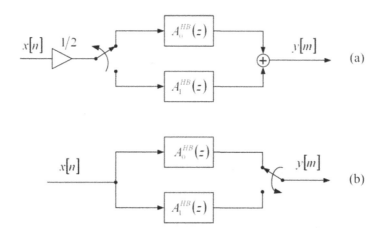

Equation (5.37) indicates how to share the common poles of $H^{HB}(z)$ between the all-pass subfilters in order to provide that the parallel connection of $A_0^{HB}(z^2)$ and $A_1^{HB}(z^2)$ achieves the desired frequency response.

Finally, the all-pass branches $A_0^{HB}(z)$ and $A_1^{HB}(z)$ in decimator and interpolator structures of Figures 5.15 (a) and (b) are obtained when replacing terms z^{-2} with z^{-1} in equations (5.36). That is, the polyphase branches of a factor-of-two decimator/interpolator are composed as a cascade connection of the first-order all-pass sections,

$$A_0^{HB}(z) = \prod_{l=2,4,\ldots}^{(N+1)/2} \frac{\beta_l^{HB} + z^{-1}}{1 + \beta_l^{HB} z^{-1}} \quad \text{and} \quad A_1(z) = \prod_{l=3,5,\ldots}^{(N+1)/2} \frac{\beta_l^{HB} + z^{-1}}{1 + \beta_l^{HB} z^{-1}}.$$ (5.38)

There exist in the literature several implementation structures that implement the first-order all-pass sections with only one multiplier (Mitra & Hirano, 1974; Gazsi, 1985; Ansari & Liu, 1985; Lutovac &Milić, 1997; Lutovac, Tošić & Evans, 2000; Krukowski & Kale, 2003). Observing equation (5.38), we find that for an Nth order halfband filter the total number of sections in both branches is $(N-1)/2$. Thereby, the factor-of-two decimator/interpolator can be implemented with only $(N-1)/2$ multipliers. This leads to extremely efficient solutions as will be demonstrated in the following example.

Example 5.4

In this example, we demonstrate the application of the IIR halfband filter for decimation and interpolation with $M = L = 2$. The 5th order halfband filter is designed using the MATLAB program halfbandiir from Appendix A. The input data are the filter order $N = 5$ and the passband edge frequency located at $\omega_p^{HB} = 0.4\pi$. Hence, we compute the filter parameters,

```
[b,a,z,p,k] = halfbandiir(5,0.4); % IIR halfband filter design
```

The program returns: (i) the coefficients of the filter transfer functions in vectors b and a, (ii) zeros and poles in vectors z and p, respectively, and k is the multiplication constant. The choice $N = 5$ and $\omega_p = 0.4\pi$ determines the remaining filter characteristics. The minimal attenuation in the stopband is 36.2385, and the peak passband ripple amounts to 0.001 dB. According to the halfband filter symmetry condition (5.33), the stopband edge frequency is $\omega_s^{HB} = \pi - \omega_p^{HB} = 0.6\pi$.

With the 5th order halfband filter, the all-pass subfilters $A_0^{HB}(z)$ and $A_1^{HB}(z)$ in efficient configurations of Figures 5.15 (a) and (b) become the first-order all-pass sections. Therefore, from the filter poles computed by halfbandiir,

p = [0.4863i,-0.4863i,0.8453i,-0.8453i] % Halfband filter poles

we compute β_2 and β_3 according to (5.37), and determine the all-pass subfilters $A_0^{HB}(z)$ and $A_1^{HB}(z)$,

$$A_0^{HB} = \frac{\beta_2^{HB} + z^{-1}}{1 + \beta_2^{HB} z^{-1}} \quad \text{and} \quad A_1^{HB} = \frac{\beta_3^{HB} + z^{-1}}{1 + \beta_3^{HB} z^{-1}}. \tag{5.39}$$

Program demo_5_1 given below, performs the decimation and interpolation with the given halfband filter using the efficient configurations of Figures 5.15 (a) and (b). The program computes the coefficients of the all-pass branches according to (5.37), computes the frequency response, generates the input signal, and performs the factor-of-two decimation and interpolation.

```
% Program demo_5_1.m
% Sampling rate conversion with M=L=2.
% 5th order IIR halfband filter
 close all, clear all
% Allpass branches
p = [0.4863i,-0.4863i,0.8453i,-0.8453i];  % Halfband filter poles
beta0 = (abs(p(1)))^2;   beta1 = (abs(p(3)))^2;  % Coefficients of allpass branches
[A0,f] = freqz([beta0,0,1],[1,0,beta0],512,2); [A1,f] = freqz([0,beta1,0,1],[1,0,beta1],512,2);

% Halfband filter frequency response
H = (A0 + A1)/2;  % Computing the IIR halfband filter frequency response
figure (1)
subplot(4,1,1), plot(f,20*log10(abs(H)))
ylabel('Gain, dB'), axis([0,1,-60,2])

% Generating the input signal
F = [0,0.1,0.46,1]; A=[0,1,0,0]; x1 = fir2(256,F,A);
x = x1 +  0.008*cos(2*pi*0.35*(0:256));% Original signal
X = fft(x); % Spectrum of the original signal
subplot(4,1,2), plot((0:127)/128,abs(X(1:128))), ylabel('|X|'), axis([0,1,0,1.2])

% Decimation-by-2
u0 = x(1:2:length(x));
```

```
u1 = [0,x(2:2:length(x)-1)];
y0 = filter([beta0,1],[1,beta0],u0);
y1 = filter([beta1,1],[1,beta1],u1);
ydec = (y0 + y1)/2;
Ydec = fft(ydec);
subplot(4,1,3), plot((0:63)/64,abs(Ydec(1:64))), ylabel('|Y_{dec}|'), axis([0,1,0,0.6])

%Interpolation
y0 = filter([beta0,1],[1,beta0],ydec);
y1 = filter([beta1,1],[1,beta1],ydec);
yy = [y0;y1];
yint = yy(:);
Yint = fft(yint);
subplot(4,1,4), plot((0:127)/128,abs(Yint(1:128)))
xlabel('Normalized frequency \omega/\pi'), ylabel('|Y_{int}|'),  axis([0,1,0,1.2])
```

Figure 5.16 displays the results. We observe that the aliasing in decimation is sufficiently suppressed, and also, the imaging is eliminated in interpolation. It is to be pointed out that a high-performance decimator and interpolator are achieved with the filter whose order is only five, and which is implemented with only two multiplication constants. Moreover, all arithmetic operations are evaluated at the lower sampling rate, see Figures 5.15 (a) and (b).

The above example demonstrates the extremely high efficiency achieved in the factor-of-two decimators and interpolators which are based on the IIR halfband filters. This solution is effective in the single-stage and particularly in the multistage systems. When the sampling rate conversion factor is expressible as a power-of-two, i.e., $M(L) = 2^K$ where K is a positive integer, the multistage implementation is usually based on the halfband filters.

IIR STRUCTURES WITH TWO ALL-PASS SUBFILTERS: APPLICATIONS OF EMQF FILTERS

In the preceding section, we have examined factor-of-two polyphase sampling rate alteration systems, which are based on the parallel connection of two all-pass subfilters. Those systems are shown to be extremely efficient when used in decimation/interpolation. In this section, we consider the application of the implementation structures based on the parallel connection of two all-pass subfilters in the cases when the sampling rate conversion factor is other than two. We use the particular class of the elliptic transfer functions, the *elliptic minimal Q-factors transfer functions,* called also *EMQF* filter, which are shown to be very efficient when implemented as a parallel connection of two all-pass subfilters (Lutovac & Milić, 1997; Milić & Lutovac, 1999; Lutovac, Tošić & Evans, 2000; Milić & Lutovac, 2003). The concept of elliptic minimal Q-factors transfer function was first introduced by Rabrenović and Lutovac (1994) and is related to the class of elliptic filters whose *s*-plane poles achieve minimal Q factors when compared to other elliptic filters. It was shown later that digital filters based on the EMQF transfer functions exhibit attractive properties particularly when implemented as a parallel connection of two all-pass subfilters (Lutovac & Milić, 1997; Milić & Lutovac, 1999; Lutovac, Tošić & Evans,

Figure 5.16. Decimation and interpolation with IIR halfband filter: From top to bottom: filter magnitude response, spectrum of the input signal, spectrum of the decimation signal, and spectrum of the interpolated signal

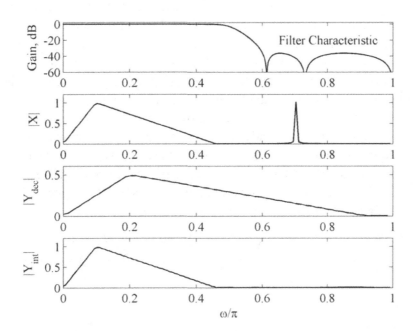

2000; Milić & Lutovac, 2003). It was shown also that the halfband filter belongs to the class of EMQF filters as a special case.

The transfer function of a lowpass EMQF filter $H(z)$ can be expressed as a sum of two all-pass functions $A_0(z)$ and $A_1(z)$,

$$H(z) = \frac{1}{2}\left(A_0(z) + A_1(z)\right).\tag{5.40}$$

and the corresponding implementation structure is depicted in Figure 5.17. Here, $A_0(z)$ is an even-order transfer function and is composed of several second-order all-pass sections, whereas $A_1(z)$ is an odd-order function composed of several second-order all-pass sections and of one first-order all-pass section,

$$A_0(z) = \frac{1}{2}\prod_{l=2,4,\ldots}^{(N+1)/2}\frac{\beta_l + \alpha\left(1 + \beta_l\right)z^{-1} + z^{-2}}{1 + \alpha\left(1 + \beta_l\right)z^{-1} + \beta_l z^{-2}} \qquad A_1(z) = \frac{\alpha_1 + z^{-1}}{1 + \alpha_1 z^{-1}}\prod_{l=3,5,\ldots}^{(N+1)/2}\frac{\beta_l + \alpha\left(1 + \beta_l\right)z^{-1} + z^{-2}}{1 + \alpha\left(1 + \beta_l\right)z^{-1} + \beta_l z^{-2}}$$

$$\tag{5.41a}$$

The useful property of EMQF filters is that in all second-order sections of $A_0(z)$ and $A_1(z)$ the constant α has the same value, which is determined by the position of the filter 3dB cutoff frequency. A constant β_l is a squared radius of the pole p_l, i.e.,

$$\beta_l = (r_l)^2 \quad \text{for } l = 2,3,\ldots,(N+1)/2 \quad \text{and} \quad \beta_l < \beta_{l+1} \quad \text{for } l = 2,3,\ldots,(N-1)/2 \tag{5.41b}$$

The last equation indicates the distribution of the conjugate complex pole pairs among the all-pass subfilters $A_0(z)$ and $A_1(z)$. The pole pair with the smallest radius is allotted to $A_0(z)$, the next one to $A_1(z)$, and the remaining pole pairs are shared between $A_0(z)$ and $A_1(z)$ in the same alternative manner.

It was shown recently (Milić and Saramäki, 2003 a&b; Saramäki & Milić, Submitted for publication 2007) that the constants of an odd-order EMQF filter can be computed from the constants of an odd-order start-up halfband filter using extremely simple formulae. The mathematical background was found in the well-known lowpass-to-lowpass frequency transformation for digital filters introduced by Constantinides (1970). The frequency transformation provides that the filter frequency response is moved along the frequency axis while the passband and stopband characteristics remain unchanged.

In the above mentioned method of Milić and Saramäki, the frequency transformations are applied to the all-pass structure of the halfband filter in order to obtain the constant values in the corresponding sections of the EMQF filter. Actually, the all-pass functions of the halfband filter $A_0^{HB}(z)$ and $z^{-1}A_1^{HB}(z)$ are transformed to the all-pass functions $A_0(z)$ and $A_1(z)$ of the EMQF filter,

$$A_0^{HB}(z^2) = \prod_{l=2,4,\ldots}^{(N+1)/2} \frac{\beta_l^{HB} + z^{-2}}{1 + \beta_l^{HB} z^{-2}} \quad \Rightarrow \quad A_0(z) = \frac{1}{2} \prod_{l=2,4,\ldots}^{(N+1)/2} \frac{\beta_l + \alpha(1+\beta_l)z^{-1} + z^{-2}}{1 + \alpha(1+\beta_l)z^{-1} + \beta_l z^{-2}} \tag{5.42a}$$

$$z^{-1}A_1^{HB}(z) = z^{-1} \prod_{l=3,5,\ldots}^{(N+1)/2} \frac{\beta_l^{HB} + z^{-1}}{1 + \beta_l^{HB} z^{-1}} \quad \Rightarrow \quad A_1(z) = \frac{\alpha_1 + z^{-1}}{1 + \alpha_1 z^{-1}} \prod_{l=3,5,\ldots}^{(N+1)/2} \frac{\beta_l + \alpha(1+\beta_l)z^{-1} + z^{-2}}{1 + \alpha(1+\beta_l)z^{-1} + \beta_l z^{-2}} \tag{5.42b}$$

Applying the lowpass-to-lowpass transformation in (5.42a) and (5.42b) the following simple expressions for computing α, α_1 and β_l are obtained (Milić & Saramäki, 2003 a&b; Saramäki & Milić, Submitted for publication 2007):

$$\alpha = -\cos(\omega_{3dB}) \tag{5.43}$$

$$\alpha_1 = \frac{1}{\alpha}\left(1 - \sqrt{1-\alpha^2}\right) \tag{5.44}$$

$$\beta_l = \left(\beta_l^{HB} + \alpha_1^2\right)/\left(\beta_l^{HB}\alpha_1^2 + 1\right), \quad l = 2,3,\ldots,(N+1)/2. \tag{5.45}$$

The constant α, which is the common constant for all second-order sections is directly determined by the 3 dB cutoff frequency ω_{3dB}, which is chosen by designer. The constant α_1 is related only to the value of α, i.e. depends on the choice of ω_{3dB}. Finally, the constants β_l are related with the constants β_l^{HB} of the

Figure 5.17. Lowpass filter constructed as a parallel connection of two all-pass subfilters

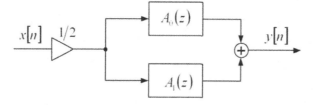

start-up halfband filter. In this way, an EMQF filter is generated from the corresponding halfband filter. The pass- stopband characteristics of both filters are the same, only the 3 dB cutoff frequency is moved from the position $\omega = \pi/2$ to the chosen position $\omega = \omega_{3dB}$. Figure 5.18 displays a family of EMQF filters derived from the 7th order halfband filter (thick line) by using transformation formulae (5.43) - (5.45).

One particular property of EMQF filters has to be pointed out. The value of constant α, which is the same in all second-order sections, is related only with the 3 dB cutoff frequency, see (5.43). Therefore, a very small change in ω_{3dB} gives the opportunity to adjust α to the value suitable for a minimum shift-and-add implementation. That is, with a slight change in ω_{3dB} we provide α to reach one of the forms of the signed digit implementation,

$$\alpha \in \left\{\pm 1/2^m, \pm\left(1 - 1/2^m\right)\right\}, \text{ or} \tag{5.46a}$$

$$\alpha \in \left\{\pm 1/2^m \pm 1/2^p, \pm\left(1 - 1/2^m \pm 1/2^p\right)\right\}, \text{ or} \tag{5.46b}$$

$$\alpha \in \left\{\pm 1/2^m \pm 1/2^p \pm 1/2^q, \pm\left(1 - 1/2^m \pm 1/2^p \pm 1/2^q\right)\right\}, \tag{5.46c}$$

with m, p, q integers, $m < p < q$. Configurations (46a)–(46c) are used here for the sake of simplicity. The use of other configurations (Dempster & Macleod, 1995) increases the possible number of shift-and add combinations.

Presenting α with a small number of shift-and-add operations halves the total number of general multipliers in an Nth order EMQF filter. An interested reader can find details on how to exploit this property in the best manner (Lutovac & Milić, 1997; Milić & Lutovac, 1999; Lutovac, Tošić, & Evans, 2000; Milić & lutovac, 2003).

The passband and the stopband edge frequencies ω_p and ω_s of the resulting EMQF filter can be computed from the passband and stopband edges of the start-up filter, which we denote here as ω_p^{HB} and ω_s^{HB}. The first step is to determined the so-called selectivity factor ξ, which has the same value for the start-up halfband filter and for the resulting EMQF filter. The value of ξ is given by

Figure 5.18. Family of EMQF filters developed from the 7th order halfband filter

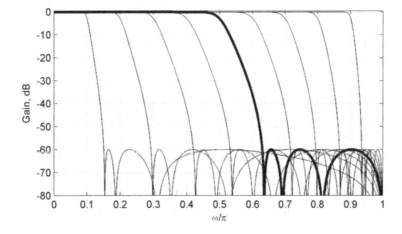

$$\xi = \tan\left(\omega_s^{HB}/2\right)\Big/\tan\left(\omega_p^{HB}/2\right) = \tan\left(\omega_s/2\right)\Big/\tan\left(\omega_p/2\right). \tag{5.47}$$

For the chosen halfband filter, we determine the value of ξ using the values of the corresponding edge frequencies ω_p^{HB} and ω_s^{HB}. In the second step, we compute the passband and the stopband edge frequencies of the EMQF filter whose 3 dB cutoff is located at the desired position ω_{3dB},

$$\omega_p = 2\tan^{-1}\left(\tan\left(\omega_{3dB}/2\right)\Big/\sqrt{\xi}\right) \text{ and } \omega_s = 2\tan^{-1}\left(\sqrt{\xi}\,\tan\left(\omega_{3dB}/2\right)\right). \tag{5.48}$$

There exists a useful relation between ω_{3dB}, ω_p and ω_s,

$$\tan^2\left(\omega_{3dB}/2\right) = \tan\left(\omega_p/2\right)\tan\left(\omega_s/2\right). \tag{5.49}$$

It is of interest for the later use to observe that in the halfband filter case, where $\omega_{3dB} = \pi/2$, equations (5.48) simplify to

$$\omega_p^{HB} = 2\tan^{-1}\left(1/\sqrt{\xi}\right) \text{ and } \omega_s^{HB} = 2\tan^{-1}\left(\sqrt{\xi}\right), \tag{5.50}$$

and accordingly equation (5.49) becomes

$$\tan\left(\omega_p^{HB}/2\right)\tan\left(\omega_s^{HB}/2\right) = 1. \tag{5.51}$$

The design procedure for an EMQF filter can be carried out as follows:

Step 1: Select the filter order N, an odd number.

Step 2: From the specified EMQF filter edge frequencies ω_p and ω_s, determine the selectivity factor ξ according to equation (5.47). In this step, one can slightly modify ω_p or ω_s in order to adjust the constant α to one of the values from the sets given in (5.46a) – (5.46c). See relations (5.43) and (5.47) – (5.51)

Step 3: Use (5.50) to compute the passband edge frequency of the start-up halfband filter, ω_p^{HB}, or the stopband edge frequency ω_s^{HB}. Notice that ω_p^{HB} and ω_s^{HB} should satisfy the symmetry condition (5.33).

Step 4: Run the program halfbandiir to compute the poles of the start-up halfband filter.

Step 5: Compute constants β_l^{HB} according to (5.37), and distribute the constants among the all-pass branches $A_0^{HB}(z)$ and $A_1^{HB}(z)$.

Step 6: Use the transformation formulae (5.43) – (5.45) to compute the constants α, α_1 and β_l of the desired EMQF filter.

Using the 6-step procedure explained above, we simplify the EMQF filter design problem to the halfband filter design problem. Actually, we need only the program halfbandiir to compute halfband filter poles and then, we use extremely simple formulae (5.43) – (5.45) to compute directly the constants of the EMQF filter. In this way, the EMQF filter is generated from the start-up halfband filter.

Using the above procedure, we compute directly the constants of the first-order and the second-order all-pass sections of the all-pass subfilters $A_0(z)$ and $A_1(z)$ as expressed in (5.39). For the implementation, one can choose among several structures developed for the all-pass first and second-order sections

(Ansari & Liu, 1985; Gaszy, 1988; Milić & Lutovac, 1999; Lutovac, Tošić & Evans, 2000).

In the following example, we demonstrate how an EMQF filter improves the overall performance of the multi-stage decimator composed of halfband filters. We consider a factor-of-8 decimator implemented in three stages using IIR halfband filters of *Example 5.4*. A general disadvantage, which arises with halfband filters, is that only *Case b* tolerance scheme can be met. We show in the following example that by introducing an appropriate EMQF filter in the last decimation stage, the overall selectivity of the decimator can be improved to satisfy *Case a* tolerance scheme.

Example 5.5

Consider two solutions for the three-stage decimator with the decimation factor $M = 8$ indicated in Figure 5.19:

(a) Filters $H_1(z)$, $H_2(z)$, and $H_3(z)$ are the 5th order halfband filters of *Example 5.4*.
(b) Filters $H_1(z)$, $H_2(z)$ are the 5th order halfband filters of *Example 5.4*, and $H_3(z)$ is an EMQF filter designed to satisfy *Case a* specifications for $M = 2$.

For the solutions (a) and (b), we compute the gain responses of the corresponding single-stage equivalents, and compare the results. For computing the gain responses, we use the single-stage equivalent for the three-stage decimator of Figure 5.19 whose overall transfer function $H(z)$ is given by

$$H(z) = H_1(z)H_2(z^2)H_3(z^4).$$

(5.49)

Solution (a):
The 5th order halfband filter of *Example 5.4* is used to implement $H_1(z)$, $H_2(z)$, and $H_3(z)$ in the three-stage decimator of Figure 5.19. The passband/stopband edge frequencies are $\omega_p^{HB} = 0.4\pi$, and $\omega_s^{HB} = \pi -\omega_p^{HB} = 0.6\pi$. The gain response of the halfband filter is shown in the top subfigure of Figure 5.20. The dashed line in the third subfigure of Figure 5.20 shows the gain response of the resulting single-stage equivalent. Since the halfband filters in the three stages satisfy *Case b* tolerance scheme, the overall frequency response also satisfies *Case b* tolerance scheme.

Solution (b):
In this solution, the first two stages are identical with the first two stages of *Solution (a)*. Hence, filters $H_1(z)$ and $H_2(z)$ are the 5th order halfband filters with $\omega_p^{HB} = 0.4\pi$ and $\omega_s^{HB} = 0.6\pi$. For the third stage, a new EMQF filter $H_3(z)$ is designed to meet the *Case a* tolerance scheme. We select the passband and stopband edge frequencies of the EMQF filter to satisfy the *Case a* tolerance scheme for $M = 2$. The passband edge frequencies are the same for all filters in the three stage-decimator of Figure 5.19, i.e., $\omega_p = 0.4\pi$. The stopband edge frequency for the *Case a* tolerance scheme for $M = 2$ is $\omega_s = 0.5\pi$. We use the six-step procedure explained above to design the desired EMQF filter.

Figure 5.19. Three-stage decimator, $M = 8$

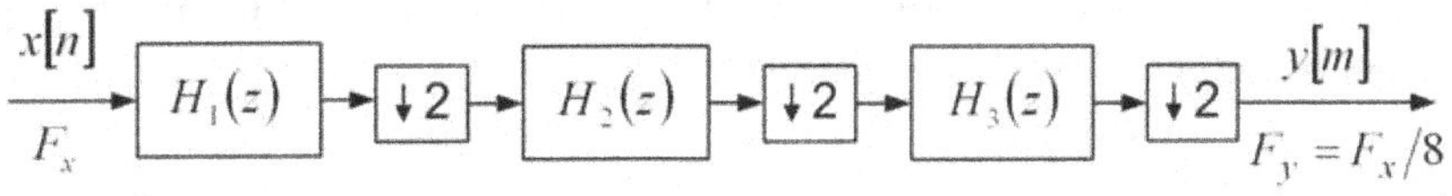

Step 1: We select the filter order $N = 7$. For the *Case a* tolerance scheme, the transition band of $H_3(z)$ should be half of the transition band achieved by $H_1(z)$ and $H_2(z)$. Thereby, we should increase the filter order from $N = 5$ to the first odd integer, $N = 7$.

Step 2: We start from the EMQF filter edge frequencies $\omega_p = 0.4\pi$ and $\omega_s = 0.5\pi$, and using (5.49) we compute the 3-dB cutoff frequency ω_{3dB}, and using (5.43) we compute $\alpha = -0.1584$. This value of α we modify to $\alpha = -(1/8 + 1/32)$, and afterwards, by using (5.48), we re-compute the passband edge frequency to $\omega_p = 0.4013\pi$. Finally, for $\omega_p = 0.4013\pi$ and $\omega_s = 0.5\pi$ we determine the selectivity factor from (5.47) and obtain $\xi = 1.3704$.

Step 3: We use (5.50) to compute passband edge frequency of the 7th order start-up halfband filter and obtain $\omega_p^{HB} = 0.450059\pi$.

Step 4: We compute the parameters of the start-up halfband filter

[b,a,z,p,k] = halfbandiir(7, 0.450059);

Program returns the filer poles in vector p

| 0 | 0 - 0.4359i | 0 + 0.4359i | 0 - 0.7429i | 0 + 0.7429i | 0 - 0.9274i | 0 + 0.9274i |

Step 5: We use (5.37) to compute constants β_i^{HB}, and obtain:

$$\beta_2^{HB} = 0.1900, \quad \beta_3^{HB} = 0.5519, \quad \beta_4^{HB} = 0.8601$$

Notice that β_2^{HB} and β_4^{HB} belong to $A_0(z)$, whereas β_3^{HB} belongs to $A_1(z)$.

Step 6: Using the values of β_i^{HB} computed in *Step 5*, and $\alpha = -(1/8 + 1/32)$ as chosen in *Step 2*, we compute the remaining constants of the EMQF filter, α_1 from (5.44), and β_i from (5.45). The results are the following:

- Subfilter $A_0(z)$ consists of two second-order sections with the common constant $\alpha = -(1/8 + 1/32)$, and $\beta_2 = 0.1960$, $\beta_4 = 0.8617$.
- Subfilter $A_1(z)$ consists of the first order section with $\alpha_1 = -0.0786$, and the second-order section with $\alpha = -(1/8 + 1/32)$ and $\beta_3 = 0.5562$.

The resulting 7th order EMQF filter has only four multiplication constants $\{\alpha_1, \beta_2, \beta_3, \beta_4\}$. The common constant α, which appears in all second-order sections, is implementable with two binary shifts and one adder.

Figure 5.20 displays the following gain responses: (i) 5th order halfband filter which is used for $H_1(z)$, $H_2(z)$, $H_3(z)$ of *Solution (a)* and for $H_1(z)$, $H_2(z)$ of *Solution (b)*. (ii) 7th order EMQF filter used for $H_3(z)$ in *Solution (b)*. (iii) The gain responses of the single stage equivalents of *Solution (a)* plotted with a dashed line, and that of *Solution (b)* plotted with a solid line. We observe from Figure 5.20 that the three-stage decimator implemented with the identical halfband filters achieves only the *Case b* tolerance scheme. But, when replacing in the last stage the halfband filter with an EMQF filter, the three stage decimator achieves to reach the *Case a* tolerance scheme.

Figure 5.21 shows the passband details of the three-stage decimators of *Solution (a)* – dashed line, and that of *Solution (b)* – solid line. Since the very small passband ripple is an inherent property of halfband and of EMQF filters, the overall passband ripples of the three-stage decimators are also very small.

The three-stage decimators presented in *Example 5.5* show that by combining IIR halfband filters and EMQF filters implemented with a parallel connection of two all-pass branches, *Case a* and *Case b* tolerance schemes can be achieved with very simple filters:

- *Solution (a):* Two multiplication constants per stage. The overall three-stage decimator is implemented with only 6 multiplication constants. Each stage can be implemented with the use of efficient configuration of Figure 5.15(a). Solution meets *Case b* tolerance scheme.
- *Solution (b):* In the first two stages, two multiplication constants per stage. The last stage is implemented with 4 multiplication constants. The overall three-stage decimator is implemented with 8 multiplication constants. The first two stages can be implemented using the efficient configuration of Figure 5.15(a), whereas the last stage should be implemented by using configuration of Figure 5.17. Solution meets *Case a* tolerance scheme.

Notice that the *Case a* tolerance scheme is met at the cost of two additional multiplication constants.

In order to illustrate a variety of applications of EMQF filters in sampling rate conversion systems, we show in the following example the three-stage decimator for $M = 30$, which is based on EMQF filters.

Example 5.6

We use EMQF filters to construct the decimator with a high conversion factor $M = 30$. The factor-of-30 decimator is implemented in three stages with conversion factors $M_1 = 5$, $M_2 = 3$, and $M_3 = 2$ as indicated in Figure 22. The decimator is requested to satisfy *Case b* tolerance scheme with the minimal stopband

Figure 5.20. Three-stage decimator of Example 5.5, M = 8. From top to the bottom: Halfband filter gain response. EMQF filter gain response. Gain response of the single stage equivalent: Dashed line is for Solution (a): $H_1(z)$, $H_2(z)$, $H_3(z)$ are identical 5th order halfband filters. Solid line is for Solution (b): $H_1(z)$, $H_2(z)$ are identical 5th order halfband filters, $H_3(z)$ is the 7th order EMQF filter.

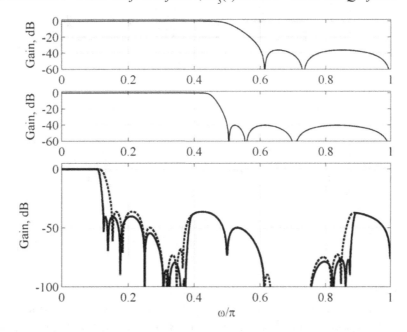

Figure 5.21. The passband details of the three-stage decimator of Example 5.5: Solution (a) – dashed line, Solution (b) – solid line

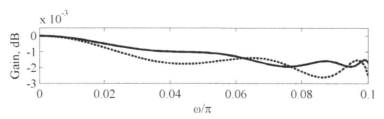

Figure 5.22. Three-stage implementation of factor-of-30 decimator

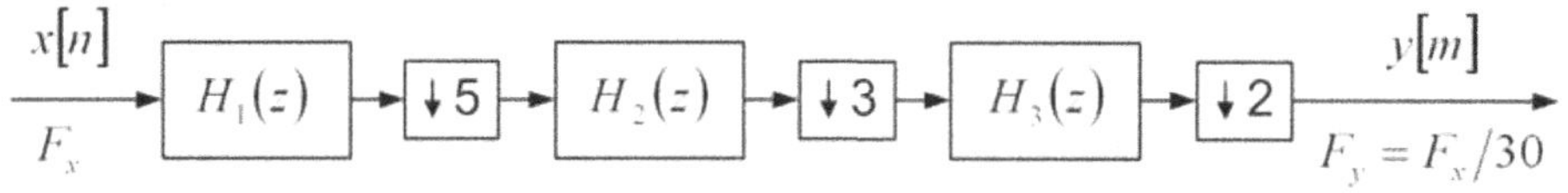

Figure 5.23. Gain characteristics for the three-stage decimator with EMQF filters: Upper subfigure: solid thick line is for $H_1(z)$, solid thin line is for $H_2(z^5)$, dashed thick line is for $H_3(z^{15})$. Second subfigure plots the gain response of $H(z) = H_1(z) H_2(z^5)H_3(z^{15})$.

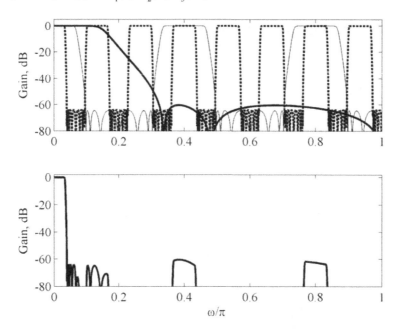

attenuation $a_s = 60$ dB.

Design parameters for filters $H_1(z)$, $H_2(z)$, and $H_3(z)$ are given in Table 5.2. Notice that the minimal stop-band attenuation a_s for all filters is more than 60 dB. The pass-band ripples are very small, approximately 10^{-6}dB. Filters $H_1(z)$ and $H_2(z)$ are the 5th order and the 7th order EMQF filters designed ac-

Table 5.2. EMQF filters in three-stage decimator of Example 5.6

	$H_1(z)$	$H_2(z)$	$H_3(z)$
N	5	7	9
α	$-(1-1/8+1/64)$	$-1/2$	0
ω_p/π	0.0658	0.2172	0.4200
ω_{3dB}/π	0.1502	0.3334	0.5000
ω_s/π	0.3240	0.4800	0.5800
$a_s[dB]$	60.44	64.54	63.81
N° mult.	3	4	4

cording to the six-step procedure explained above. They satisfy *Case b* tolerance scheme for M_1=5 and M_2=3, respectively. Filter $H_3(z)$ is the 9th order halfband filter. In $H_1(z)$ and $H_2(z)$, the values of constant α are adjusted to the minimal number of shift and add operations. It is noticeable that in $H_2(z)$ which is designed for the sampling rate conversion factor M_2 =3, we have $\alpha = -1/2$ that practically means a binary shift.

Table 5.2 illustrates the efficiency of the factor-of-30 decimator based on EMQF filters. The total number of multiplications in the overall filter is only 11. It should be noticed that the EMQF filters $H_1(z)$ and $H_2(z)$ use the configuration of Figure 5.17 for implementation, and therefore they operate at the input sampling rates. The third filter $H_3(z)$ is a halfband filter, which can use the efficient configuration of Figure 5.15 (a) and therefore, $H_3(z)$ can operate at the output sampling rate.

Figure 5.23 shows the resulting gain responses: The upper subfigure plots the gain responses for $H_1(z)$, $H_2(z^5)$, and $H_3(z^{15})$. The second subfigure shows the overall gain response of the single-stage equivalent, $H(z) = H_1(z) H_2(z^5)H_3(z^{15})$.

MATLAB EXERCISES

5.1 Design and implementation of an IIR polyphase interpolator.
(a) Construct the factor-of-5 interpolator by making use of the efficient polyphase configuration of Figure 5.7.
(b) Use the polyphase all-pass branches specified in *Example 5.2*.
(c) Compute and plot the magnitude and phase responses of the interpolation filter.
(d) Generate an input signal of the sampling frequency of 5000 Hz, and perform the factor-of-5 interpolation with the efficient IIR interpolator of Figure 5.7.
(e) Plot the input and output spectra of the signal.
(f) Compute the multiplication rate of the system.

5.2 For the factor-of-5 interpolator designed in the MATLAB Exercise 5.1, design an IIR extra filter to provide the overall system satisfying *Case a* tolerance scheme. Compute the multiplication rate of the system consisting of the extra filter and factor-of-5 interpolator.

5.3 Two-stage decimation with $M = 4$.

(a) Design the 7th order halfband filter with the stopband edge frequency $\omega_s = 0.58\pi$.
(b) Construct the two-stage decimator using the efficient configuration of Figure 5.15.
(c) Compute and plot the magnitude response of the single-stage equivalent.
(d) Generate an input signal of the sampling frequency 20000 Hz, and decimate this signal by 4 using the two-stage decimator designed above.
(e) Compute and plot the input and output spectra of the signal.
(f) Compute the multiplication rate of the system.

5.4 Repeat MATLAB Exercise 5.3 for the two-stage interpolator with $L = 4$.

5.5 Design the 7th order EMQF filter with 3 dB cut-off frequency $\omega_{3dB} = (1/6)\pi$ starting from the 7th order halfband filter from the MATLAB Exercise 5.3. Use the transformation formulae (5.43) – (5.45). Compute the passband and stopband edge frequencies of the resulting EMQF filter. Compute and plot the magnitude responses of the start-up halfband filter and the magnitude response of the EMQF filter.

5.6 Compute the multiplication rate of the three-stage decimator of *Example 5.6* for the input sampling frequency of 2000 Hz.

5.7 The factor-of-30 decimator of *Example 5.6* is implemented in three stages where $M_1 = 5$, $M_2 = 3$, and $M_3 = 2$. The parameters of the EMQF filters are given in Table 5.2. Examine the frequency response of the single-stage equivalents for the following combinations:

(a) $M_1 = 2$, $M_2 = 3$, and $M_3 = 5$
(b) $M_1 = 3$, $M_2 = 5$, and $M_3 = 2$
(c) $M_1 = 3$, $M_2 = 2$, and $M_3 = 5$

Compute and plot the magnitude characteristics of the single-stage equivalents defined above, and compare with the characteristic of Figure 5.23.

REFERENCES

Ansari, R., & Liu, B. (1983). Efficient sampling rate alteration using recursive (IIR) digital filters. *IEEE Transactions on Acoustics, Speech, and Signal Processing, 31*(6), 1366-1373.

Ansari, R., & Liu, B. (1985). A class of low-noise computationally efficient recursive digital filters with applications to sampling rate alternations. *IEEE Transactions on Acoustics, Speech, and Signal Processing, 33*(1), 90-97.

Ansari,R., & Liu,B., (1993). Multirate signal processing. In Sanjit. K. Mitra and James F. Kaiser (ed.), *Handbook for Digital Signal Processing*. New York, NY: John Wiley-Interscience, 981-1084.

Bellanger, M.G., Bonnerot, G., & Coudreuse, M. (1976). Digital filtering by polyphase network: application to sample-rate alteration and filter banks. *IEEE Transactions on Acoustics, Speech, and Signal Processing, 24*(2), 109-114.

Bellanger, M. (2000). *Digital processing of signals: Theory and practice.* 3rd edition. New York, NY: John Wiley.

Constantinides, A. C. (1970). Spectral transformations for digital filters. *IEE Proceedings. 117*(8), 1585-1590.

Crochiere, R.E., & Rabiner, L.R., (1981, March). Interpolation and decimation of digital signals - A Tutorial Review. *Proceedings of the IEEE, 69*(3), 300-331.

Crochiere, R.E., & Rabiner, L.R. (1983). *Multirate digital signal processing.* Englewood Cliffs, NJ: Prentice-Hall.

Dempster, A.G., & Macleod, M.D. (1995). Use of minimum-adder multiplier block in FIR digital filters. *IEEE Trans. Circuits and Systems II: Analog and Digital Signal Processing, 42*(9), 569-577.

Drews, W., & Gazsi, L. (1986). A new design method for polyphase filters using all-pass sections. *IEEE Transactions on Circuits and Systems, 33*(3), 346-348.

Fetweis, A., Levin, H., & Seldmeyer, A. (1974, June). Wave digital lattice filters. *Circuit Theory and Applications, 2*(2), 203-211.

Filter design toolbox for use with MATLAB. User's guide. Version 6. (2006). Natick: MathWorks.

Gazsi, L. (1985). Explicit formulas for lattice wave digital filters. *IEEE Transactions on Circuits and Systems, 32*(1), 68-88.

Harris, F. J., (2004). *Multirate signal processing for communication systems.* Upper Saddle River, NJ: Prentice Hall PTR.

Hentchel, T. (2002). *Sample rate conversion in software configurable radious.* Morwood, MA: Artech House.

Krukowski, A., & Kale I. (2003). *DSP System design: Complexity Reduced IIR Filter Implementation for Practical Applicatios.* Boston: Kluver Academic Publishers.

Lutovac, M. D., & Milić, L. D. (1997). Design of computationally efficient elliptic IIR filters with a reduced number of shift-and add operations in multipliers, *IEEE Transactions on Signal Processing, 45*(10), 2422-2430.

Lutovac, M. D. Tošić, D. V., & Evans, B. L. (2001). *Filter Design for Signal Processing Using MATLAB and Mathematica,* Upper Saddle River, New Jersey, Prentice Hall.

Milić, L. D., & Lutovac, M. D. (1999). Design of multiplierless elliptic IIR filters with a small quantization error, *IEEE Transactions on Signal Processing, 47*(2), 469-479.

Milić, Lj., & Lutovac, M.D. (2002). Efficient multirate filtering. In Gordana Jovanović-Doleček, (ed.), *Multirate Systems: Design & Applications.* Hershey, PA: Idea Group Publishing, 105-142.

Milić, Lj., & Lutovac, M. (2003). Efficient Multirate Filtering using EMQF Subfilters. *Proceedings of th 6ᵗʰ International Conference TELSIKS 2003,* 301-304.

Milić, L. D., & Saramäki, T. (2003a). Three classes of IIR complementary filter pairs with an adjustable crossover frequency. *Proc. IEEE Int. Symp. Circuits Syst. ISCAS 2003, 4*, 145–148.

Milić, L. D., & Saramäki, T. (2003b). Power-complementary IIR filter pairs with an adjustable crossover frequency. *Facta Universitatis, Ser.: Elec. Energ. 16*(3), 295–304.

Mitra, S. K., & Hirano, K. (1974). Digital all-pass networks. *IEEE Trans. Circuits and Systems.* CAS-*21*(5), 688-700.

Mitra, S. K. (2006). *Digital signal processing: A computer based approach.* 3rd edition. New York, NY: The McGraw-Hill Companies, Inc.

Rabrenovic, D., & Lutovac, M. D. (1994). Elliptic filters with minimal Q-factors. *Electronics Letters, 30*(3), 206-207.

Renfors, M., & Saramäki, T. (1987, January). Recursive Nth-band digital filters-part I: design and properties. *IEEE Transactions on Circuits and Systems, 34*(1), 24-39.

Renfors, M., & Saramäki, T. (1987). Recursive n̲th-band digital filters-part II: design of multistage desimators and interpolators. *IEEE Transactions on Circuits and Systems, 34*(1), 40-51.

Surma-Aho, K, & Saramäki, T., (1999). A systematic technique for deigning approximately linear phase recursive digital filters. *IEEE Transactions on Circuits and Systems − II Analog and Digital Signal Processing, 46*(7), 956-963.

Saramäki, T. (1986). Design of optimal multistage IIR and FIR filters for sampling rate alteration. *Proc. 1986 IEEE International Symposium on Circuits and Systems.* pp. 227-230.

Saramäki, T. *Multirate Signal Processing.* (2001). Lecture notes for a graduate course, the Institute of Signal Processing, Tampere University of Technology, Finland.

Saramäki, T., & Milić, L. (Submitted for publication). Tree Classes of Complementary Recursive Filter Pairs with Variable Crossover Frequency. *Submitted to Circuits Systems and Signal Processing,* Birkhäuser.

Schüssler, H. W., & Stefen, P. (1998). Halfband filters and Hilbert transformers. *Circuits Systems Signal Processing, 17*(2), 137–164.

Signal processing toolbox for use with MATLAB. User's guide. Version 6. (2006). Natick: Math-Works.

Valenzuela, R.A., & Constantinides, A.G. (1983). Digital signal processing schemes for efficient interpolation and decimation. *IEEE Proceedings*, Pt. G., *130*(6), 225-235.

Vaidyanathan, P.P. (1990). Multirate digital filters, filter banks, polyphase networks, and applications: A Tutorial. *Proceedings of the IEEE, 78*(1), 56-93.

Vaidyanathan, P.P., (1993). *Multirate systems and filter banks.* Englewood Cliffs, NJ: Prentice Hall.

Wegener, W. (1979). Wave digital directional filters with reduced number of multipliers and adders. *Arch. Elec. Übertragung.* (1979), *33*(6), 239-243.

Chapter VI
Sampling Rate Conversion by a Fractional Factor

INTRODUCTION

We have discussed so far the decimation and interpolation where the sampling rate conversion factor is an integer. However, the need for a non-integer sampling rate conversion appears when the two systems operating at different sampling rates have to be connected, or when there is a need to convert the sampling rate of the recorded data into another sampling rate for further processing or reproduction. Such applications are very common in telecommunications, digital audio, multimedia and others.

In this chapter, we consider the sampling rate conversion by a rational factor, called sometimes a *fractional sampling rate conversion*. We use MATLAB functions from the *Signal Processing* and *Filter Design Toolbox* to demonstrate the fractional sampling rate conversion. We present the technique for constructing efficient fractional sampling rate converters based on FIR filters and the polyphase decomposition. In the sequel, we consider the sampling rate alteration with an arbitrary conversion factor. We present the polynomial-based approximation of the impulse response of a hybrid analog/digital model, and the implementation based on the Farrow structure. We also consider the fractional-delay filter problem. This chapter concludes with MATLAB exercises for individual study.

SAMPLING RATE CONVERSION BY A RATIONAL FACTOR

The change of the sampling frequency by a rational factor L/M, sometimes called the *fractional sampling rate alteration* or *resampling*, can be achieved by increasing the sampling frequency by L first, and then decreasing by M. Hence, the sampling rate conversion by L/M is achieved by a cascading factor-of-L interpolator and a factor-of-M decimator as indicated in Figure 6.1(a). Here, factors L and M are positive relatively prime integers, i.e. there is no common integer between L and M. In the implementation

Figure 6.1. Sampling rate conversion by L/M. (a) General implementation scheme. (b) Efficient implementation

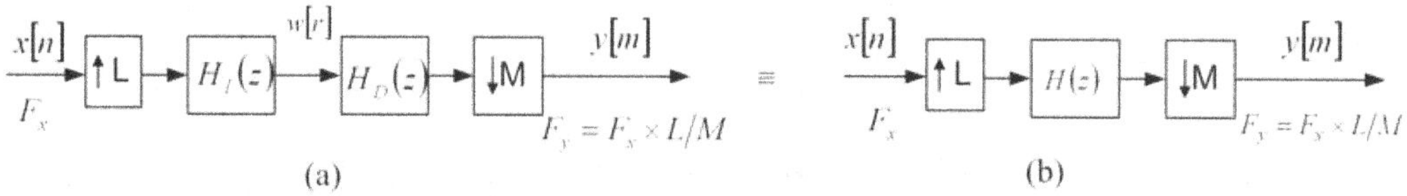

(a)   (b)

scheme of Figure 6.1(a), the original signal $\{x[n]\}$ is up-sampled-by-L and then filtered by the lowpass interpolation filter $H_I(z)$. The interpolated signal $\{w[r]\}$ is filtered with the lowpass antialiasing filter $H_D(z)$, and then down-sampled-by-M. The sampling rate of the output signal $\{y[m]\}$ is L/M times the sampling rate of the original signal $\{x[n]\}$. Since the interpolation filter $H_I(z)$ and the decimation filter $H_D(z)$ operate at the same sampling rate, they can be replaced by the single lowpass filter $H(z)$ as indicated in Figure 6.1 (b). The lowpass filter $H(z)$ should be designed to eliminate imaging caused by the up-sampling, and to avoid aliasing produced in down-sampling. With the properly designed filter $H(z)$, the fractional sampling rate conversion can be implemented by using the computationally efficient structure of Figure 6.1 (b).

The role of filter $H(z)$ in the efficient fractional sampling-rate converter of Figure 6.1(b) is twofold: it acts as the antiimaging filter $H_I(z)$, and also as the antialiasing filter $H_D(z)$. For the adequate removal of images, the stopband edge frequency of the low-pass filter $H(z)$ must be below π/L, and avoiding of aliasing requires the stopband edge below π/M. Therefore, the low-pass filter $H(z)$ in the implementation scheme of Figure 6.1 (b) has the stopband edge frequency at ω_s, which is given by

$$\omega_s = \min\left(\frac{\pi}{L}, \frac{\pi}{M}\right). \tag{6.1}$$

Choosing ω_s according to (6.1) ensures the elimination of imaging which appears in interpolation, and at the same time ensures the suppression of aliasing that may be caused by decimation. Hence, the ideal specifications for the magnitude response of $H(z)$ are given by

$$\left|H\left(e^{j\omega}\right)\right| = \begin{cases} L, & |\omega| \leq \min\left(\dfrac{\pi}{L}, \dfrac{\pi}{M}\right). \\ 0, & \text{otherwise} \end{cases} \tag{6.2}$$

It is important to observe that for a large L or M, filters with very narrow passbands are requested.

In MATLAB, there are several functions that perform the fractional sampling rate alteration according to the implementation scheme of Figure 6.1(b). The simplest for use is the function resample from *Signal Processing Toolbox*. For the given original signal stored in vector x, and the sampling rate conversion factor L/M, the function resample returns the resampled signal y,

```
y = resample(x,L,M);    % Fractional sampling rate alteration by L/M
```

Here, the up-sampling factor L, and the down-sampling factor M are integers. For the description of function resample including the related options, see the *Signal Processing Toolbox User's Guide*. In the following, we illustrate the fractional sampling rate alteration on the example sinusoidal sequence.

Example 6.1

In this example, we generate one period of the sinusoidal sequence, and using resample convert the sampling rate by the alteration factor $L/M = 3/2$. This is performed through the program demo_6_1 given below. Program demo_6_1 generates the example sequence, changes the sampling rate by 3/2, and displays the results.

The sinusoidal sequence is obtained by the uniform sampling of the continuous-time sinusoidal signal $x_c(t)$, which is sampled by the sampling period $T_x = 1/F_x$, where F_x is the sampling frequency, $F_x = 20$ Hz.

Program demo_6_1 generates one period of the discrete sinusoidal signal $\{x[n]\}$ where the value of the nth element is given by,

$$x[n] = x_c\left(t\right)_{t=nT_x} = \sin\left(2\pi nT_x\right), \quad n = 0, 1, \ldots, 20. \tag{6.3}$$

Using resample program changes the sampling frequency by 3/2, and computes the resampled signal $\{y[m]\}$ $= \{y(mT_y)\} = \{\sin(2\pi mT_y)\}$. Program plots the original and the resampled signals on the same graph.

```
% Program demo_6_1.m
% Sampling rate conversion by L/M = 3/2
close all, clear all
Fx = 20;   Tx=1/Fx;              % Original sampling frequency in Hz
tx = 0:Tx:1;                     % Time vector tx
x = 0.9*sin(2*pi*tx);           % Input sequence
y = resample(x,3,2);            % Re-sampling
ty = (0:(length(y)-1))*2*Tx/3;  % New time vector ty
figure (1)
subplot(2,1,1), stem(tx,x,'*')
hold on
stem(ty,y,'-.')
legend('original','resampled'),xlabel('Time'), ylabel('amplitude'), axis([0,1,-1,1])
```

Figure 6.2 shows the original and the resampled signals. Obviously, the continuous-time sinusoidal signal $x_c(t)$ can be reconstructed either from the original, either from the resampled signal. Notice that in the example of Figure 6.2, the resampling process starts at the time instant $t = 0$. Hence, there are no time-offsets between two sequences.

Function resample by default makes use of a FIR filter to perform the sampling rate alteration according to Figure 6.1 (b). The filter design is based on the MATLAB function firls and the Kaiser window.

The MATLAB function upfirdn enables the fractional sampling rate alteration with the particular FIR filter $H(z)$ chosen by the designer. This function belongs to the *Signal Processing Toolbox*, and the MATLAB code for the sampling rate alteration by a rational factor L/M is the following,

```
y = upfirdn(x,h,L,M); % Fractional sampling rate alteration by L/M
```

Figure 6.2. Resampling of the sinusoidal sequence by L/M = 3/2

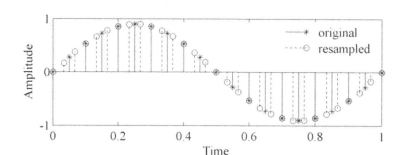

The upfirdn returns the resampled signal y for the following input data: the samples of the input signal stored in vector x, the FIR filter impulse response stored in vector h, up-sampling integer factor L, and the down-sampling integer factor M. Obviously, the FIR filter $H(z)$ should be designed prior to applying upfirdn. In *Signal Processing Toolbox User's Guide*, the reader can find several options for upfirdn that can be of interest in some applications. MATLAB Exercise 6.2 specifies the sampling-rate alteration problem to be solved by using upfirdn.

SPECTRUM OF THE RESAMPLED SIGNAL

Frequency-domain characteristics of up-sampled and down-sampled signals with the integer factors have been considered in Chapter II. In this section, we apply relationships derived in Chapter II to the case of the fractional sampling rate alteration.

Apparently, the antiimaging/antialiasing filter $H(z)$ in the structure of Figure 1(b) approximates with some error the ideal magnitude characteristic specified in (6.2), and consequently, the process of resampling produces distortions in the signal spectrum. Therefore, it is of practical interest to derive the frequency-domain relations for the resampler of Figure 6.1(b).

We derive below the relationship that expresses the spectrum of the resampled signal denoted by $Y(e^{j\omega})$, by means of the spectrum of the original signal $X(e^{j\omega})$, the filter transfer function $H(e^{j\omega})$, and the factors L and M. To achieve this, we derive the corresponding relation in the domain of z-transform, and then replace variable z with $e^{j\omega}$.

Following the structure of Figure 6.1(b), the z-transform of the resampled signal is obtained in three steps:

1. Up-sampling by L: According to equation (2.22), the z-transform of the up-sampled signal is $X(z^L)$.
2. Filtering: Filtering the up-sampled signal by $H(z)$, results in the product $X(z^L) H(z)$.
3. The last step is down-sampling by M. Applying equation (2.17) to the z-transform of the up-sampled and filtered signal gives directly the desired expression for $Y(z)$,

$$Y(z) = \frac{1}{M} \sum_{k=0}^{M-1} X\left(z^{L/M} W_M^{-k}\right) H\left(z^{1/M} W_M^{-k}\right).$$

(6.4)

The replacement $z = e^{j\omega}$ in (6.4) gives the desired spectrum of the resampled signal,

$$Y\left(e^{j\omega}\right) = \frac{1}{M}\sum_{k=0}^{M-1} X\left(e^{j(L\omega - 2\pi k)/M}\right)H\left(e^{j(\omega - 2\pi k)/M}\right). \tag{6.5}$$

When introducing here the ideal characteristic for $H(e^{j\omega})$ as given in (6.2), the spectrum $Y(e^{j\omega})$ becomes,

$$Y\left(e^{j\omega}\right) = \frac{1}{M}X\left(e^{jL\omega/M}\right). \tag{6.6}$$

Therefore, with an ideal filter, we obtain for $Y(e^{j\omega})$ compressed-by-L and stretched-by-M version of the original spectrum $X(e^{j\omega})$, which is additionally scaled by $1/M$.

It is apparent that the overall performance of the resampler depends on the filter characteristics. In the following example, we demonstrate the effects of the stopband characteristic on the spectrum of the resampled signal.

Example 6.2

In this example, we generate the signal $\{x[n]\}$, and convert the sampling rate by the factor $L/M = 5/3$ using two different FIR filters.

For $\{x[n]\}$, we select the signal having the spectrum of a triangular shape. Such a signal can be generated in MATLAB using the function fir2,

```
x = fir2(256,[0,0.8,1], [0,1,0]); % Generating the original signal
```

We design two FIR filters in order to compare the effects of two different stopband attenuations on the spectrum of the resampled signal. Limitations given in (6.2), determine the stopband edge frequency for both filters, $\omega_s = \min(\pi/3,\pi/5) = \pi/5$. For the passband edge, we choose $\omega_p = \pi/6$. The MATLAB function firgr computes the coefficients of two optimal FIR filters for the lengths $N_1 = 81$, and $N_2 = 141$,

```
h1 = firgr(80,[0,1/6,1/5,1],[1,1,0,0]); and h2 = firgr(140,[0,1/6,1/5,1],[1,1,0,0]);
```

The MATLAB function upfirdn converts the sampling rate of the signal stored in the vector x for L = 5, and M = 3,

```
y1 = upfirdn(x,h1,5,3); and y2 = upfirdn(x,h2,5,3);
```

Figure 6.3 displays the magnitude spectrum of the original signal $|X(e^{j\omega})|$, the magnitude responses of the filters $|H_1(e^{j\omega})|$ and $|H_2(e^{j\omega})|$, and the corresponding spectra of the resampled signals $|Y_1(e^{j\omega})|$ and $|Y_2(e^{j\omega})|$.

Figure 6.3 demonstrates significant derogations in the signal spectrum when the stopband attenuation is relatively small. With 35 dB attenuation in the stopband, filter $H_1(z)$ fails to suppress aliasing and to remove images from the spectrum of the resampled signal, see Figures 6.3 (c) and (e). However, Figures 6.3(d) and (f) show that the 55 dB stopband attenuation of $H_2(z)$, eliminates visible effects of the aliasing and imaging from signal spectrum.

Figure 6.3. Resampling with L/M = 5/3: (a) and (b) Magnitude spectrum of the original signal $|X(e^{j\omega})|$. (c) Gain response $20\log|H_1(e^{j\omega})|$. (d) Gain response $20\log|H_2(e^{j\omega})|$. (e) Resampled signal magnitude response$|Y_1(e^{j\omega})|$. (f) Resampled signal magnitude response $|Y_2(e^{j\omega})|$.

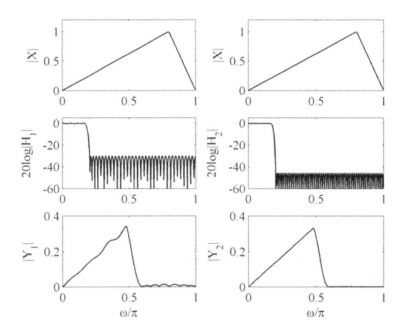

POLYPHASE IMPLEMENTATION OF FRACTIONAL SAMPLING RATE CONVERTERS

For the implementation of the efficient fractional sampling rate converters, the polyphase configurations have been developed which enable the arithmetic operations to be evaluated at the lowest possible sampling frequency. The computationally efficient fractional sampling rate converters based on the polyphase decomposition of FIR filters are described in (Hsiao, 1987; Vaidianathan 1990, 1993), and in (Fliege, 1994). In this approach, the overall resampler structure of Figure 6.1(b) is organized in such a manner that the arithmetic operations in the polyphase components of $H(z)$ are evaluated at the lowest sampling rate. In the following, we present the above mentioned method for the case $L/M < 1$. Using the dual of all the structures, the same derivations can be performed for the case $L/M > 1$.

Let us start from the basic structure of Figure 6.1(b) and consider the polyphase configuration of $H(z)$. Recall that the polyphase decomposition for an FIR filter and the corresponding implementation structures are described in Chapter IV. With the decomposition of $H(z)$ into L polyphase components, $E_0(z)$, $E_1(z)$, ..., $E_{L-1}(z)$, the structure of Figure 6.1(b) obtains the form shown in Figure 6.4. As in the case of integer sampling rate converters considered in Chapter IV, the polyphase structure of Figure 6.4 is convenient to be transformed to a computationally efficient implementation.

In the first step, the up-sampler L is moved to the right into the polyphase branches, and in compliance with the Sixth Identity (see Chapter II, Figure 2.23), the up-samplers-by-L and the polyphase components $E_0(z^L)$, $E_1(z^L)$, ..., $E_{L-1}(z^L)$ exchange their positions. In addition, following the First Identity (Figure 2.17), the down-samplers are moved into the polyphase branches. The modified polyphase realization is given in Figure 6.5.

Figure 6.4. Factor-of-L/M sampling rate converter with the filter implemented in polyphase form

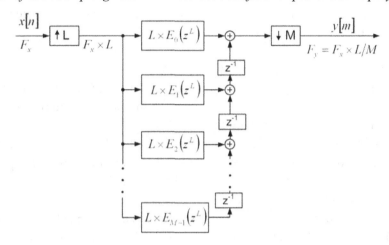

It is evident that the configuration of Figure 6.5 provides already the factor-of-L computational savings, if compared with the initial configuration of Figure 6.4. It will be shown next that with the reorganization of the individual polyphase branches, additional savings can be achieved. Before starting to modify the structure of Figure 6.5, some useful replacements need to be introduced.

The basic idea for the desired modification has been found in number theory. Namely, if two integers L and M are relatively prime, then there exist two integers l_0 and m_0 such that

$$l_0 L - m_0 M = -1. \tag{6.7}$$

Using this relation, the delay element z^{-k} is expressed in the following manner,

$$z^{-k} = z^{k(l_0 L - m_0 M)}. \tag{6.8}$$

Figure 6.5. Polyphase realization of the factor of L/M sampling-rate converter

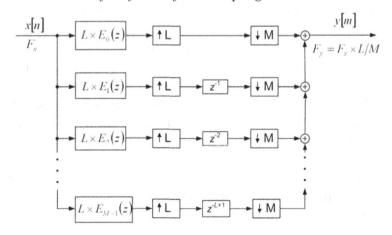

We now select one of the branches from the structure of Figure 6.5, and describe the procedure, which finally results in the resampler of a low computational complexity. The step-by-step procedure is presented in Figures 6.6(a) – (e):

- We start from the kth branch shown in Figure 6.6(a).
- In the next step, depicted in Figure 6.6(b), we introduce relation (6.8) to replace the delay element z^{-k} with the block $z^{k(l_0L-m_0M)}$. This structure enables the following reorganization in the cascade of up-sampler, delay block, and down-sampler.
- Using the Fifth Identity (Figure 2.22), we can put the delay block z^{kl_0} before the up-sampler, and according to the Second Identity (Figure 2.18), we put z^{-km_0} after the down-sampler, see Figure 6.6 (c). Assuming that L and M are relatively prime, the up-sampler and down-sampler can exchange their positions as shown in Figure 6.6(d).
- The part of the structure of Figure 6.6 (d) enclosed in the dashed block is the cascade of the FIR filter $L \times E_k(z)$ followed by the factor-of-M down-sampler. We have seen in Chapter IV that the cascade of an FIR filter and a down-sampler can be efficiently implemented in the polyphase form. Hence, we can decompose the FIR filter $L \times E_k(z)$ into the polyphase components $L \times E_{k0}(z)$, $L \times E_{k1}(z)$, ..., $L \times E_{k(M-1)}(z)$, and apply the configuration of Figure 4.7 (b). In this way, the efficient configuration for the kth branch as shown in Figure 6.6(e) is obtained. In this configuration, the arithmetic operations are to be evaluated at the lowest possible sampling rate.

The resulting polyphase branch of Figure 6.6(e) has a delay block z^{-km_0} on its far right, and the negative delay z^{kl_0} on the left. The delay block z^{-km_0} is to be implemented with a series of delay blocks z^{-m_0}. The negative delay z^{kl_0} indicates that the obtained solution is noncausal. This problem can be solved by introducing the extra delay of $z^{-(L-1)l_0}$ at the input. This extra delay cancels in each branch with the particular delay z^{kl_0}.

Figure 6.7 depicts the overall structure of the sampling rate converter in the case $L/M = 2/3$, which is based on the decompositions explained above. Notice that the delay blocks on the input side and on the output side of the converter (enclosed in the dotted blocks) are both the unit delays z^{-1}. This is because for $L = 2$ and $M = 3$, equation (6.7) is satisfied with $l_0 = 1$, and $m_0 = 1$.

The structure of Figure 6.7 enables a high computational efficiency. If this structure is used to change the sampling rate of the input signal from 48 kHz to 32 kHz, the arithmetic operations are evaluated at the sampling frequency of 16 kHz.

The solution for the rational sampling-rate conversion explained above is derived under the assumption $L/M < 1$. To develop the structure of the sampling rate converters with $L/M > 1$, the duals of the steps of Figures 6.4 – 6.6 have to be performed. To obtain the dual structure, we have to reverse the directions of all branches, change the up-samplers with the down-samplers of the same rate (and vice versa), and exchange inputs and outputs with each other. The resulting structure for $L/M = 3/2$, which is derived from the 2/3-converter of Figure 6.7 is displayed in Figure 6.8. When this structure is used to convert the input rate of 32 kHz to 48 kHz, the arithmetic operations are to be performed at the rate of 16 kHz. Hence, the two dual converter structures of Figures 6.7 and 6.8 achieve the same computational efficiency.

The implementation of efficient sampling-rate converters based on the configurations of Figures 6.7 and 6.8 can be performed in MATLAB. We demonstrate in *Example 6.3* the sampling rate conversion based on the configuration of Figure 6.8

Figure 6.6. Polyphase decomposition of the kth branch

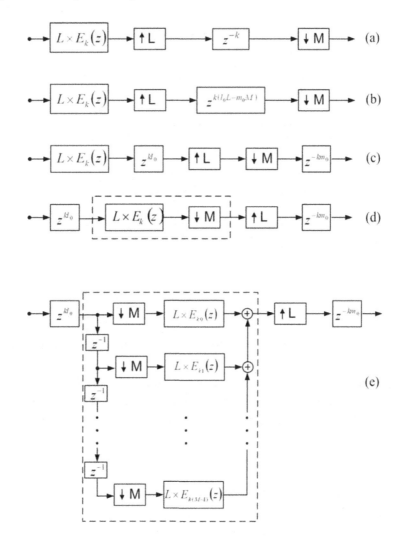

Example 6.3

In this example, we show the MATLAB program demo_6_2, which converts the sampling rate of the original signal from F_x = 32 kHz to F_y = 48 kHz. Here, the sampling rate conversion factor is L/M = 3/2, and the efficient implementation structure of Figure 6.8 is used.

Program demo_6_2 creates 256 samples of the original sequence $\{x[n]\}$ and converts the sampling rate of $\{x[n]\}$ by the conversion factor L/M = 3/2. Program demo_6_2 implements the structure of Figure 6.8.

For the low-pass filter, we choose the stopband edge frequency ω_s = $\pi/3$ to satisfy the condition given in (6.2), for L = 3 and M = 2. The lowpass filter is designed as an optimal FIR filter of the length N = 144.

Figure 6.7. Structure of the efficient sampling-rate converter for L/M = 2/3

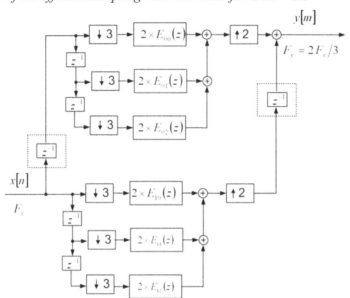

To obtain the input sequence $\{x[n]\}$ with the spectrum of a triangular shape, the MATLAB function fir2 is used.

Program demo_6_2 computes the FIR filter impulse response, creates the input signal $\{x[n]\}$, implements the structure of Figure 6.8, and using this structure computes the output signal $\{y[m]\}$.

Program demo_6_2 returns the following plots: filter gain response, spectrum of the original signal, and spectrum of the resampled signal.

```
% Program demo_6_2.m
% In this program, we create the input sequence x[n] and convert
% the sampling rate of x[n] by the conversion factor L/M = 3/2.
% Program demo_6_2 implements the structure of Figure 6.8.
clear all, close all
% Lowpass filter design
h = firgr(143,[0,0.9/3,1/3,1], [1,1,0,0]); [H,f]=freqz(h,1,512,2);
figure (1)
subplot(3,1,1),  plot(f,20*log10(abs(H)))
xlabel('Normalized Frequency [\omega/\pi]'),  ylabel('Gain [dB]')
axis([0,1,-60,2])

% Input signal x[n]
Fx = 32000; Fy = 48000;
F = [0,2/3+1/5,1]; m = [0,1,0];  % Setting the input parameters for fir2
x = fir2(255,F,m);   % Generating the original signal 'x'
[X,f] = freqz(x,1,512,Fx);  % Spectrum of the original signal
```

Figure 6.8. Structure of the efficient sampling-rate converter for L/M = 3/2. This converter is dual to the one of Figure 6.7.

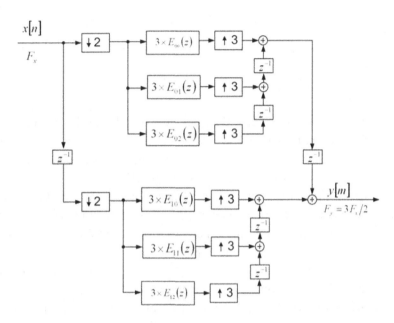

```
subplot(3,1,2),  plot(f/1000,abs(X))
xlabel('Frequency [kHz]'), ylabel('|X|'), axis([0,16,0,1])

% Polyphase decomposition
e0 = h(1:2:length(h)); e1 = h(2:2:length(h));
e00 = 3*e0(1:3:length(e0)); e01 = 3*e0(2:3:length(e0)); e02 = 3*e0(3:3:length(e0));
e10 = 3*e1(1:3:length(e1)); e11 = 3*e1(2:3:length(e1)); e12 = 3*e1(3:3:length(e1));

% Down-sampling and polyphase filtering
u0 = x;
u1 = [0,x(1:length(x)-1)];
u02 = u0(1:2:length(u0));
u12 = u1(1:2:length(u1));
x00 = filter(e00,1,u02);
x01 = filter(e01,1,u02);
x02 = filter(e02,1,u02);
x10 = filter(e10,1,u12);


x11 = filter(e11,1,u12);
x12 = filter(e12,1,u12);

% Up-sampling
yy0 = [x00;x01;x02];
```

```
yy1 = [x10;x11;x12];
y0 = yy0(:);
y1 = yy1(:);
yy = [0,(y0(1:length(y0)-1))'];
y = (yy'+y1); [Y,f] = freqz(y,1,512,Fy); % Output signal
subplot(3,1,3), plot(f/1000,abs(Y))
xlabel('Frequency [kHz]'), ylabel('|Y|'), axis([0,24,0,1.5])
```

Figure 6.9 displays the results. Evidently, the selected FIR filter of the length $N = 144$, and the efficient polyphase configuration of Figure 6.8 provide the resampling nearly without derogations in the signal spectrum. Notice that each of the six polyphase components has 24 coefficients, and that filtering in those polyphase components is evaluated at the rate of 16 kHz.

It is left to the reader to modify program demo_6_2 according to the structure of Figure 6.7, and to perform the sampling rate conversion by the factor $L/M = 2/3$, see MATLAB Exercise 6.3.

We have demonstrated in this section that efficient rational sampling rate conversion is achieved when utilizing an FIR filter implemented in the polyphase form.

It is to be noticed however that the polyphase structures of the fractional sampling rate converters presented above fail to exploit the coefficient symmetry in the case of a linear-phase FIR filter. It is well-known that a linear-phase FIR filter exhibits coefficient symmetry property, $h[n] = h[N-n-1]$, $n = 0, 1, ..., N-1$. Due to this property, the number of multiplication constants can be halved in the direct implementation forms, see Chapter I. In the polyphase structures, except for $L = M = 2$, the particular polyphase subfilters are no more symmetric. Therefore, the polyphase implementation structures become unsuitable for exploiting the coefficient symmetry property of the original overall transfer function, and thus reduce the number of multiplication constants.

Figure 6.9. Sampling-rate conversion with $L/M = 3/2$: Filter characteristic. Spectrum of the original signal. Spectrum of the resampled signal.

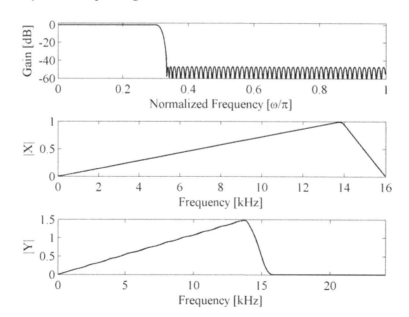

An efficient implementation structure for linear-phase rational sampling rate converters has been proposed recently by Bregović, Saramäki and Lim (2006). This structure has a reduced computational complexity compared to other existing structures for the rational sampling rate converters. Actually, this method is developed by extending the approach given in (Mou, 1996) where the coefficient symmetry property is utilized to reduce the computational complexity in the linear-phase decimators and interpolators.

Obviously, instead of the FIR filter, an infinite impulse response (IIR) filter can undertake the role of antiimaging and antialiasing filtering in the basic structure of Figure 6.1 (b). The use of IIR filters in rational sampling rate converters has received smaller attention in the literature. It is worth mentioning the paper of Russel, 2000 where the rational sampling rate converters based on IIR filters are considered. It has been shown in the paper of Russel that an efficient rational sampling rate converter is obtained when utilizing the well-known method of modifying the IIR filter transfer function as introduced by Bellanger, Bonnerot, and Coudreuse (1976), and by applying the polyphase decomposition to the non-recursive part of the transfer function. With this approach, the considerable gain in efficiency over the direct implementation is obtained.

RATIONAL SAMPLING RATE ALTERATION WITH LARGE CONVERSION FACTORS

Configurations based on polyphase decomposition are convenient in applications where L and M are small numbers. Otherwise, filters of a very high order are requested. To illustrate this, we show an example of sampling rate conversion from 48 kHz to 44.1 kHz, which is usually requested in digital audio.

Example 6.4

Changing the sampling rate from 48 kHz to 44.1 kHz can be achieved with the rational factor $L/M = 147/160$. Hence, the sampling rate conversion factor is the ratio of two large integers. This fact imposes severe requirements on filter specifications. Namely, for avoiding effects of aliasing and imaging, the stopband edge frequency should be located at $\omega_s = \pi/160$, see equation (6.2).

Let us estimate the order of an optimal FIR filter that meets the following specifications:

- Passband: edge frequency $\omega_p = \pi/250$, peak passband ripple $a_p = 0.2$ dB.
- Stopband: edge frequency $\omega_s = \pi/160$, minimal stopband attenuation $a_p = 50$ dB.

Using the MATLAB function firpmord we compute the filter order $N_{ord} = 1995$, i.e. the filter length amounts to $N = 1996$. FIR filters of such complexity are very difficult, if not impossible for implementation. Also, the solution based on the polyphase decomposition discussed in the previous section is impractical because of the number of polyphase branches, which is too large for $L/M = 147/160$.

However, in MATLAB we can perform the sampling rate alteration from 48 kHz to 44.1 kHz using the function upfirdn, already used in *Example 6.3*, or using the object mfilt.firsrc from *Filter Design Toolbox*.

In this example, we generate a simple original signal composed of two sinusoidal sequences,

```
x = sin(2*pi*0.045*n)+0.6*sin(2*pi*0.032*n);
```

To perform the sampling rate conversion, we design first the FIR filter $H(z)$ using m-files firpmord and firpm,

```
L = 147;, M = 160;
ap = 0.2; as = 50;  f = [1/250,1/160];  a = [1 0];
dev = [(10^(ap/20)-1)/(10^(ap/20)+1)  10^(-as/20)];
[n,fo,ao,w]  = firpmord(f,a,dev,2);   h = firpm(n,fo,ao,w);
```

The filter gain response including the passband details is displayed in Figure 6.10.

With the filter coefficients computed above, and stored in vector h, and with conversion factors L=147, M=160, we define object hm, which contains the structure for sampling rate conversion,

```
hm = mfilt.firsrc(L,M,L*h);
```

And finally, we apply function filter to the object hm and signal x to compute the samples of the re-sampled signal,

```
y = filter(hm,x);
```

Figure 6.11 plots the results of the re-sampling process presenting the original signal in the upper subfigure, and the resampled signal in the bottom subfigure. The resampled signal is a delayed and slightly modified replica of the original signal.

The above example illustrates the difficulties in constructing rational sampling rate conversion when L and M are large integers. Apparently, the FIR filter whose response is shown in Figure 6.10 is very difficult if not impossible for implementation. However, there are many applications requiring the sampling rate conversion between arbitrary sampling rates.

Design of a digital sampling rate converter with an arbitrary conversion factor is not simple. In particular, design is quite difficult and expensive when the conversion factor is a ratio of two very large integers or an irrational number. The concept of the sampling rate alteration by an arbitrary factor we consider in the next section.

Figure 6.10. Gain response of the FIR filter used in Example 6.4. Filter length is N = 1996

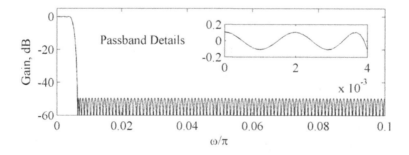

Figure 6.11. Upper subfigure: the original signal. Bottom subfigure: resampled signal

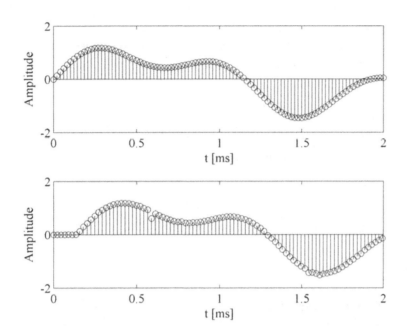

SAMPLING RATE ALTERATION BY AN ARBITRARY FACTOR

In many applications, there is a need to estimate the value of the discrete signal at an arbitrary time instant between two existing samples. This is of particular importance for telecommunications and digital audio. Some of the applications are arbitrary sampling rate conversion, symbol synchronization in digital receivers, echo cancellation in modems, and time-delay estimation.

The new sample value at arbitrary points can be evaluated by utilizing the digital signal processing technique known as *general interpolation filtering*. This filter is sometimes simply called the *interpolator*. The term *fractional-delay (FD) filter* is also used in this context. In this section, we expose an efficient technique based on the polynomial interpolation and the Farrow structure.

Time-Domain Approach to Interpolation

The interpolation by an arbitrary factor can be viewed as the computation of the new sample values at arbitrary points between the existing samples. The concept is indicated in Figure 6.12. The input sequence $\ldots, x((n_l-2)T_x), x((n_l-1)\,T_x), x(n_lT_x), x((n_l+1)T_x), x((n_l+2)\,T_x),\ldots$ is formed of the uniformly spaced samples with the sampling interval T_x. The new sample $y(lT_y)$, also called the *interpolant*, occurs between the samples $x(n_lT_x)$ and $x((n_l+1)T_x)$, at the point $lT_y = n_lT_x + \mu_lT_x$. We call the input sample index n_l the *basepoint index*, and μ_l the *fractional interval*. The fractional interval can take any value in the range $0 \le \mu_l < 1$.

The new sample value $y(lT_y)$ is computed from the existing samples by utilizing some interpolation algorithm. Figure 6.13 indicates the interpolation process. Given the input sequence $\{x[n]\}$, the output sample $y[l]$ is computed when applying the interpolation algorithm for the basepoint n_l and for the fractional interval μ_l. The interpolation algorithm may be defined as a time-varying digital filter with the impulse response $h(n_l,\mu_l)$.

The new sample $y(t_l) = y(lT_y)$ is placed between the input samples $x(n_lT_x)$ and $x((n_l+1)T_x)$ at the time instant t_l, which we express in terms of the input sampling interval T_x, the basepoint index n_l, and the fractional interval μ_l, see Figure 6.12,

$$t_l = \left(n_l + \mu_l\right)T_x. \tag{6.9}$$

For the given time instant t_l, the basepoint index n_l is determined by

$$n_l = \lfloor t_l/T_x \rfloor, \tag{6.10}$$

and the fractional interval,

$$\mu_l = t_l/T_x - \lfloor t_l/T_x \rfloor, \tag{6.11}$$

where $\lfloor x \rfloor$ stands for the integer part of x.

Let us consider the sequence $\{x[n]\}$ as a discrete signal obtained by sampling the bandlimited continuous-time signal $x_c(t)$ at uniformly spaced time instants nT_x,

$$x[n] = x_c\left(nT_x\right). \tag{6.12}$$

An ideal interpolation can be imagined as sampling $x_c(t)$ at the time instant t_l, and thus giving the interpolant $y[l] = x_c(t_l)$. For achieving the supposed ideal case, it is requested to reconstruct ideally the continuous signal $x_c(t)$ from the given samples $\{x[n]\}$, and to resample the reconstructed signal at the time instant t_l. In reality, we generate an approximate waveform $y_c(t)$, and determine the interpolant $y[l]$ by sampling $y_c(t)$ at the time instant t_l,

$$y[l] = y_c\left(t_l\right) = y_c\left((n_l + \mu_l)T_x\right). \tag{6.13}$$

Figure 6.12. Time relations between the input sequence and the interpolated sequence

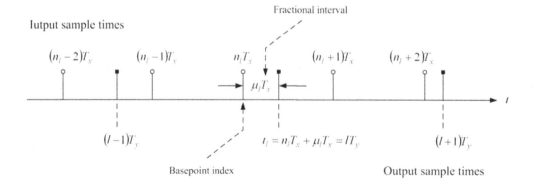

Figure 6.13. Simplified model of general interpolator

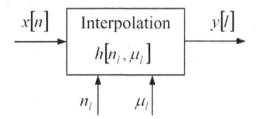

Therefore, the interpolation problem can be considered as a reconstruction problem. In practice, we apply an interpolation algorithm to generate the continuous-time function $y_c(t)$ from a finite number of existing samples $x[n]$. We use an interval of $N = N_2 + N_1 + 1$ consecutive samples surrounding the basepoint sample n_l in a manner $-N_1 + n_l \leq n_l \leq n_l + N_2$.

The commonly used interpolation techniques are based on the polynomial approximations. In this approach, for the given set of the input samples $\{x[-N_1 + n_l], \ldots, x[n_l], \ldots, x[n_l + N_2]\}$, a polynomial approximation $y_c(t)$ to $x_c(t)$ is defined as

$$y_c(t) = \sum_{k=-N_1}^{N_2} P_k(t) x[n_l + k] \tag{6.14}$$

where $P_k(t)$ are polynomials. In most applications, two general classes of polynomials are used: *Lagrange polynomials*, and *B-spline functions*. In this chapter, we will demonstrate the interpolation process based on the Lagrange polynomials.

The Lagrange polynomial $P_k(t)$ is given by

$$P_k(t) = \prod_{\substack{i=-N_1 \\ i \neq k}}^{N_2} \frac{t - t_k}{t_k - t_i}, \; k = -N_1, -N_1 + 1, \ldots, N_2 - 1, N_2. \tag{6.15}$$

Notice that the degree of $P_k(t)$ is $N - 1$ since $N = N_2 + N_1 + 1$. When we choose $N = 4$, the Lagrange polynomials are of the 3-rd order, and this type of approximation is called the cubic Lagrange interpolation.

The attractive property of Lagrange approximation is the exact reconstruction of the input sample values. This is due to the Lagrange polynomial property

$$P_k(t_n) = \begin{cases} 1, & k = n \\ 0, & k \neq n \end{cases}, -N_1 \leq n \leq N_2, \tag{6.16}$$

which, when applied to equation (6.14), gives

$$y_c(nT_x) = x[n], -N_1 \leq n \leq N_2. \tag{6.17}$$

Figure 6.14 illustrates the polynomial interpolation by means of example. Given four input samples $\{x[n_l - 1], x[n_l], x[n_l + 1], x[n_l + 2]\}$ obtained by sampling the continuous-time signal $x_c(t)$ with the sample interval $T_x = 10$ ms, we compute the sample value $y[l] = y(t_l)$ located between $x(n_lT_x)$ and $x(n_lT_x+1)$ at the time instant $t_l = n_lT_x + (2/3)T_x$. Using equation (6.15), we evaluate four third-order Lagrange polynomials. In this case, we have chosen $N_1 = 1$ and $N_2 = 2$, and therefore, the cubic Lagrange polynomials evaluated for $k = -1, 0, 1, 2$ are denoted as $P_{-1}(t)$, $P_0(t)$, $P_1(t)$ and $P_2(t)$. The plots of $P_{-1}(t)$, $P_0(t)$, $P_1(t)$ and $P_2(t)$ are displayed in Figure 6.14. Finally, we utilize expression (6.14) to compute the sample value $y[l]$ $= y(t_l)$ as desired. It can be observed from Figure 6.14 that the resulting interpolant, indicated with the square-ended dashed line, is a very good approximation of the original continuous-time signal $x_c(t)$ for the time instant $t = t_l$.

The example shown in Figure 6.14 demonstrates the process of Lagrange polynomial interpolation. As mentioned earlier in this section, an alternative approach for the polynomial interpolation is to utilize B-spline functions, see (Mitra, 2006). For generating a splin in MATLAB, one can utilize the function spline. There are also several functions that simplify the computations with splins. For example, function interp1 provides interpolation with a cubic spline function. The command

yl = interp1(t,x,tl,'spline');

computes the output sample (or a sequence) yl as the result of the cubic splin interpolation for the inputs: t is the vector of the input sample times (minimum 9); x is the vector of input sample values (of the same length as t); tl is the vector of the output sample times. The interpolation is performed with the cubic spline functions. As an illustrative example, solve MATLAB Exercise 6.4.

The Polynomial Interpolation Based on the Hybrid Analog/Digital Model

As mentioned earlier in this section, the interpolation can be considered as a basically reconstruction problem. We can imagine the process of determining the interpolant value as a two-step procedure. In the first step, the approximating continuous-time signal $y_c(t)$ is reconstructed based on the samples of the existing input sequence $\{x[n]\}$. In the second step, $y_c(t)$ is sampled at desired time instants. The hybrid analog/digital model for signal reconstruction and resampling is depicted in Figure 6.15. The continuous-time signal $y_c(t)$ is reconstructed by using a digital-to-analog converter (DAC) and the reconstruction analog filter $h_c(t)$. The output sample $y[l]$ is obtained by sampling $y_c(t)$ at $t_l = n_l T_x + \mu_l T_x$. As will be shown later in this section, a very efficient interpolation algorithm can be developed by mimicking the hybrid model of Figure 6.15.

Let us consider the processing of the signal in the hybrid model of Figure 6.15. An ideal DAC converts the sequence $\{x[n]\}$ to the sequence $x_s(t)$ consisting of the weighted and shifted continuous-time impulses,

$$x_s(t) = \sum_{n=-\infty}^{\infty} x[n]\delta(t - nT_x). \tag{6.18}$$

The sequence $x_s(t)$ is then filtered using the analog filter $h_c(t)$, and the output continuous-time signal $y_c(t)$ is obtained by the convolution

$$y_c(t) = \sum_{n=-\infty}^{\infty} x[n]h_c(t - nT). \tag{6.19}$$

Finally, the interpolant $y[l]$ is obtained by sampling $y_c(t)$ at the time instant t_l.

In practice, only several samples of the input signal are used for interpolation, and $h_c(t)$ is a fictive linear-phase analog filter whose impulse response is zero outside the interval $-NT_x/2 \leq t \leq NT_x/2 - T_x$.

Figure 6.14. Lagrange interpolation: Lagrange interpolation with cubic polynomials $P_{-1}(t)$, $P_0(t)$, $P_1(t)$ and $P_2(t)$. The circle-ended solid lines indicate the input samples $\{x[n_l - 1], x[n_l], x[n_l + 1], x[n_l + 2]\}$, and square-ended dashed line indicates the interpolant $y[l]$.

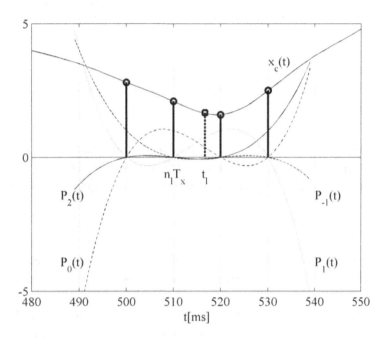

Figure 6.15. Hybrid analog/digital model for interpolation

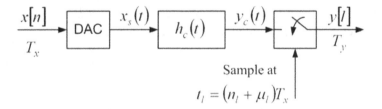

Therefore, the value of $y[l]$ is determined as

$$y[l] = y_c(t_l) = \sum_{k=-N/2}^{N/2-1} x[n_l - k] h_c \left((k + \mu_l) T_x \right). \tag{6.20}$$

It is assumed that the basepoint sample n_l is the central sample of the interval $-NT_x/2 \leq t \leq NT_x/2 - T_x$, where the interval length N is an even integer.

The interpolation system of Figure 6.15 can be implemented digitally when mimicking the impulse response of the fictive analog filter $h_c(t)$ by means of a piecewise polynomial of t. Precisely, $h_c(t)$ is composed as a polynomial of t in each interval: $\{kT_x, (k+1)T_x\}$ for $k = -N/2, -N/2 + 1, \ldots, N/2 - 1$.

To achieve the polynomial interpolation, we express $h_c(t)$ in the form:

$$h_c\big((k+\mu_l)T_x\big) = \sum_{m=0}^{M} c_m(k)\mu_l^{\,m}, \quad \text{for } k = -N/2,\ -N/2+1,\ \dots,\ N/2-1, \tag{6.21}$$

where $c_0(k),\ c_1(k),\ \dots,\ c_M(k)$ are the coefficients, and $M \le N-1$ is the degree of the polynomials.

In equation (6.20), μ_l is a continuous variable, and when μ_l varies from 0 to 1, $h_c(t)$ in each time interval, $\{kT_x,\ (k+1)T_x\}$, takes the form

$$h_c(t) = \sum_{m=0}^{M} c_m(k)\left(\frac{t - kT_x}{T_x}\right)^m, \quad \text{for } k = -N/2,\ -N/2+1,\ \dots,\ N/2-1. \tag{6.22}$$

The polynomial interpolation filter $h_c(t)$ is determined by selecting the degree of the polynomials and the coefficient values $c_0(k),\ c_1(k),\ \dots,\ c_M(k)$. In many applications, the satisfactory results can be achieved with the Lagrange polynomials or B-spline functions. Table 6.1 displays the coefficients for the cubic Lagrange polynomials.

When the polynomial coefficients are known it is straightforward to compute the impulse response $h_c(t)$ for each interval. The frequency-domain characteristics are given by the Fourier transform $H_c(j2\pi F)$. Figure 6.16 depicts the impulse response and the magnitude response for the piecewise polynomial filter $h_c(t)$ whose coefficients are displayed in Table 6.1. The impulse response is symmetric, and therefore, the phase characteristic is linear.

In the polynomial interpolation methods discussed in this section, only time-domain has been considered. No information of the frequency domain characteristics of the signal was taken into account when selecting the design parameters.

The interpolation filter $h_c(t)$ is an approximation of the ideal reconstruction filter, which has been discussed in Chapter I of this book, section Sampling the Continuous-Time Signal. It has to be observed that the signal spectrum at the input to $h_c(t)$ is periodic, and the role of $h_c(t)$ is to eliminate undesired spectral images. The spectrum of the signal at the filter output is given by

$$Y_c(j2\pi F) = H_c(j2\pi F)X\left(e^{j2\pi F/F_x}\right). \tag{6.23}$$

Evidently, the frequency response of the cubic Lagrange filter shown in Figure 6.16 can become unsatisfactory if there are frequency components close to the half of the sampling frequency. The problem with time domain methods is a small number of parameters that can be used in adjusting the filter characteristics.

The frequency domain approach, which provides efficient polynomial filters with desired passband/stopband frequency response have been developed recently (Vesma, 2000; Vesma and Saramäki,

Table 6.1. Lagrange Coefficients for Cubic Interpolation

k	$m = 0$	$m = 1$	$m = 2$	$m = 3$
−2	0	−1/6	0	1/6
−1	0	1	1/2	−1/2
0	1	−1/2	−1	1/2
1	0	−1/3	1/2	−1/6

Figure 6.16. Impulse response and magnitude response of the piecewise Lagrange polynomial filter, N = 4, M = 3

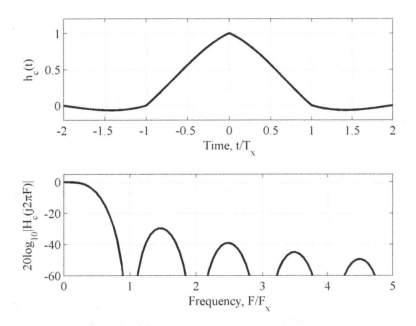

2007). Vesma and Saramäki developed a design method that enables the independent optimization of the passband and of the stopband magnitude characteristics. In this way, the passband and the stopband characteristics can be independently specified. This possibility is particularly useful in decimation, and also in interpolation when the original signal has the spectral components rather close to the half of the sampling rate.

Usually, the number of segments N of the polynomial-based impulse response $h_c(t)$ is an even number. Recently, Babic, Ghadam, and Renfors (2004) have proposed an extension which includes the polynomial filters with the odd number of pieces.

Efficient Implementation of Polynomial Interpolation Filters Using Farrow Structure

The polynomial-based approximation of the analog filter impulse response enables the digital implementation of the analog/digital model of Figure 6.15. By substituting $h_c((k+\mu_l)T_x)$ in equation (6.20) with the relation from (6.21), we obtain the following expression for computing the interpolant $y[l]$:

$$y[l] = \sum_{k=-N/2}^{N/2-1} x[n_l - k]\left(\sum_{m=0}^{M} c_m(k)\mu_l^m\right), \qquad (6.24)$$

which can be rearranged into the form

$$y[l] = \sum_{m=0}^{M} \mu_l^m\left(\sum_{k=-N/2}^{N/2-1} c_m(k)x[n_l - k]\right). \qquad (6.25)$$

To simplify the computation procedure, we can write,

$$y[l] = \sum_{m=0}^{M} v_m[n_l]\mu_l^m, \tag{6.26}$$

where $v_m[n_l]$ is the result of the convolution

$$v_m[n_l] = \sum_{k=-N/2}^{N/2-1} c_m[k]x[n_l - k]. \tag{6.27}$$

Equation (6.27) presents the input/output relation of an FIR filter whose impulse response coefficients are $c_m(-N/2)$, $c_1(-N/2+1)$, ..., $c_m(N/2-1)$. Obviously, the transfer function of that FIR filter, $C_m(z)$, is given by

$$C_m(z) = \sum_{k=-N/2}^{N/2-1} c_m(k)z^{-k}, \quad m = 0, 1, ..., M. \tag{6.28}$$

The coefficients of $C_m(z)$ are fixed numbers independent of μ_l, determined only by the polynomial-based impulse response $h_c(t)$. Equation (6.26) is itself polynomial in μ_l, which is the only variable parameter in the structure.

The corresponding structure that implements equations (6.26) and (6.27) is the Farrow structure, which consists of M constant-coefficient FIR filters in parallel whose outputs are multiplied by the powers of μ_l, and added together as shown in Figure 6.17. Notice that the block diagram of Figure 6.17 implements a causal Farrow filter.

In this section, we have been concentrated on the problem of computing the sample value of a single interpolant located between the known samples. As has been shown, the position of the new interpolant can be arbitrarily chosen in the time interval between two consecutive samples. The only adjustable parameter is the fractional parameter μ_l. We can adapt the changes of μ_l in order to convert the input sampling rate F_x to the desired output sampling rate F_y. Generally, this approach is applicable when the ratio F_y/F_x is a rational, or an irrational number, and is suitable for the sampling rate increase and for the sampling rate decrease.

Using the polynomial-based approach and the Farrow structure, the sampling rate conversion by an arbitrary factor can be implemented efficiently. In the following example, we demonstrate the application of the Farrow structure for converting the sampling rate of the input signal by the factor 16/15.

Example 6.5

In this example, we use program demo_6_3 to generate the input sequence $\{x[n]\}$, and perform factor-of-16/15 sampling-rate conversion. The program simulates the Farrow structure of Figure 6.17 with the cubic Lagrange coefficients from Table 6.1. Program plots the input sequence $\{x[n]\}$ and the resampled output sequence $\{y[l]\}$.

Figure 6.17. Farrow structure for polynomial-based interpolation filters: (a) The overall structure, (b) FIR filter details

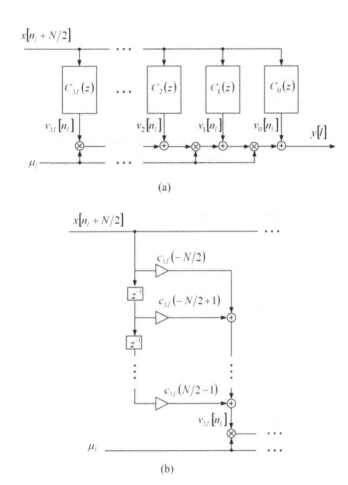

(a)

(b)

```
% Program demo_6_3.m
clear all, close all
% Sampling-rate alteration by the fractional factor R = 16/15

% Generating the input signal x[n]
n = -2:32;
f1 = 0.1; f2 = 0.16;
x = n+2*sin(pi*f1*n)+5*cos(pi*f2*n)+1;
n=0:length(x)-5;
figure (1)
subplot(2,1,1),stem(n,x(3:length(x)-2),'k');
xlabel('Time index [n]'), ylabel('x[n]')
```

```
% Defining fractional intervals for sampling rate alteration by 16/15
R = 16/15;                      %  Re-sampling factor
s = 1:15;
mu = [0,1-(1-1/R)*s];
mu = repmat(mu,1,3);            % Vector of fractional intervals

% Farrow filter, coefficients:
C0 = [1/6,-1/2,1/2,-1/6];       % Vector of Lagrange coefficients for C_0(k)
C1 = [0,1/2,-1,1/2];            % Vector of Lagrange coefficients for C_1(k)
C2 = ([-1/6,1,-1/2,-1/3]);      % Vector of Lagrange coefficients for C_2(k)
C3 = [0,0,1];                   % Vector of Lagrange coefficients for C_3(k)

% Farrow filter, initial states:
xs0 = [0,0,0]; xs1 = [0,0,0]; xs2 = [0,0,0]; xs3 = [0,0];

% Farrow filtering:
l=1;
for n=1:1:31
    xnl = x(n+3);
[v0,xs0] = filter(C0,1,xnl,xs0);
[v1,xs1] = filter(C1,1,xnl,xs1);
[v2,xs2] = filter(C2,1,xnl,xs2);
[v3,xs3] = filter(C3,1,xnl,xs3);
if mu(l)==0
    y(l) = ((v0*mu(l)+v1)*mu(l)+v2)*mu(l)+x(n+2);
    y(l+1) = ((v0*mu(l+1)+v1)*mu(l+1)+v2)*mu(l+1)+x(n+2);
    l = l + 2;
else
    y(l) = ((v0*mu(l)+v1)*mu(l)+v2)*mu(l)+x(n+2);
    l = l + 1;
end
end
figure (1)
subplot(2,1,2)
stem(0:length(y) -1,y,'sk'), xlabel ('Time index [l]'),ylabel('y[l]')
axis([0,33,0,40])
```

Figure 6.18 displays the results of the resampling process.

By means of *Example 6.5*, we have demonstrated the efficiency of the resampler based on the piecewise polynomial analog filter and implemented with the Farrow structure. The constant-coefficient FIR subfilters of the Farrow structure are of a very low order (only 3, in this example), and the fractional interval μ_l is the only variable element which determines the positions of the computed samples $y[l]$.

We have demonstrated here the application of the Farrow structure when increasing the sampling rate. The reader can develop an application for sampling rate decrease by solving MATLAB Exercise 6.6.

The polynomial-based filter with the Farrow structure is a suitable approach for constructing re-samplers when the conversion factor is the ratio of two large numbers, and in the case of an irrational conversion factor. An attractive practical solution in digital audio for converting sampling rate from 44.1 kHz (compact disc) to 48 kHz (digital audio tape) is exposed by Rajamani, Yhean-Sen Lai, and Farrow (2000).

Besides the Farrow structure shown in Figure 6.17, the modified Farrow structures have been developed, see (Vesma, 2000; Hamila, Vesma & Renfors, 2002; Vesma & Saramäki, 2007). The modifications are aimed to provide the coefficient symmetry (anti-symmetry) in the subfilters $C_m(z)$, and to provide flexible solutions for the frequency-domain design approach.

Utilizing the modified Farrow structure, Johansson and Gustafsson (2005) introduced a novel approach for interpolation, decimation and Lth-band filtering. Those filters can be used also for decimation and interpolation by integer factors. This method can be used to simultaneously implement several sampling rate converters at a low cost.

FRACTIONAL-DELAY FILTERS

The need for delaying a discrete-time signal for a fractional delay appears in telecommunications, speech processing, digital audio, modelling the musical instruments, and others. Fractional-delay filters are designed to provide a fractional delay, which can be adjusted to the desired value.

Ideally, the output sequence $\{y[n]\}$ of a fractional-delay filter inputted by the sequence $\{x[n]\}$ is given by

$$y[n] = x[n - D],\tag{6.29}$$

Figure 6.18. Sampling rate conversion by 16/15. Upper subfigure: the input sequence $\{x[n]\}$. Bottom subfigure: the resampled sequence $\{y[l]\}$.

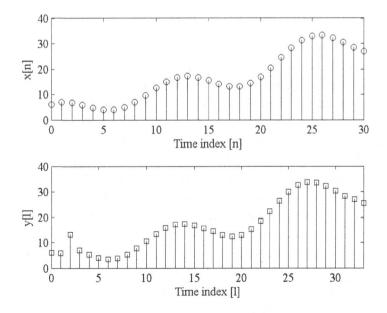

where D is a delay. Equation (6.29) holds for integer values of D only. For the non-integer values of D, equation (6.29) must be approximated in some way. For $D \geq 1$, the total delay D is expressed by two terms,

$$D = N_D + \mu, \tag{6.30}$$

where N_D is the integer part of D and μ is the fractional delay. The integer part is then implemented as a chain of N_D unit delays, and the fractional delay μ is a subject of approximation.

In the frequency domain, an ideal fractional-delay filter can be described as

$$H_{id}^{FD}\left(e^{j\omega}\right) = e^{-j(N_D + \mu)\omega}. \tag{6.31}$$

Thereby, an ideal fractional-delay filter can be considered as a linear-phase all-pass function where

$$\left| H_{id}^{FD}\left(e^{j\omega}\right) \right| = 1, \tag{6.32}$$

$$\phi_{id}^{FD}(\omega) = -\left(N_D + \mu\right)\omega \tag{6.33}$$

If the input signal $\{x[n]\}$ is band limited to the half of the sampling frequency, the delayed signal sample $y(D)$ can be computed by convolving the input signal with the filter impulse response,

$$y(D) = \sum_{n=-\infty}^{\infty} x[n] h_{id}^{FD}\left(n - N_D - \mu\right), \tag{6.34}$$

where $h_{id}^{FD}[n]$ is the time-shifted discrete sinc function,

$$h_{id}^{FD}[n] = \mathrm{sinc}\left(n - N_D - \mu\right) = \frac{\sin\left(\pi\left(n - N_D - \mu\right)\right)}{\pi\left(n - N_D - \mu\right)}, \qquad -\infty < n < \infty. \tag{6.35}$$

The impulse response is, therefore, of an infinite-length and corresponds to the noncausal discrete-time system. Moreover, the filter is not stable since the sequence $\left\{ h_{id}^{FD}[n] \right\}$ is not absolutely summable. To produce a realizable fractional delay a wide range of fractional-delay filters have been developed. The existing design methods can be roughly divided into two main groups: the first group of methods is based on FIR filters designed to approximate a fractional delay, whereas the second group utilizes the IIR approximation approaches based on all-pass filters, see (Laakso & all, 1996; Välimäki & Laakso, 2000).

In the following, we will demonstrate how to perform fractional-delay filtering in MATLAB. We use the maximally-flat FIR approximation of fractional delay, which is based on the so-called Lagrange interpolation considered earlier in this chapter, see subsection Time-Domain Approach to Interpolation. The Lagrange interpolation is a time-domain approach that leads to the polynomial-based filters that can be efficiently implemented by using the Farrow structure. In this section, we use the MATLAB object functions to create the Lagrange fractional-delay filters and to demonstrate operations on sequences.

We expose first the performances of the first-order Lagrange interpolator when applied as a fractional-delay filter. In the sequel, we show the effects of the filter order on the phase-delay and magnitude characteristics of the Lagrange filter. Finally, on the example of a sinusoidal sequence, we demonstrate fractional-delay filtering with a variable delay.

To generate MATLAB design objects for fractional-delay filtering, we use functions farrow.linearfd to create linear fractional-delay Farrow filter, and function fdesign.fracdelay to create the higher order filters. Functions, farrow.linearfd and fdesign.fracdelay, are included in the MATLAB *Filter Design Toolbox*.

The MATALAB program demo_6_4 that follows, generates the object h based on the 1st order Lagrange polynomial, and computes the phase-delay and magnitude responses of the fractional-delay filters with the delays of μ = 0.1, 0.2, 0.3, 0.4, 0.5, 0.6, 0.7, 0.8, and 0.9 samples. In the sequel, demo_6_4 generates the sequence $\{x[n]\}$ composed of two sinusoidal sequences and performs fractional-delay filtering with the fractional delays μ = 0.2, and μ = 0.5.

```
% Program demo_6_4.m
% Generating linear fractional delay filters
clear all, close all
for k = 0:9
    mu = k/10;
    h(k+1)= farrow.linearfd(mu);
end
[phi,f] = phasedelay(h,512,2); % Computing the phase delay
figure (1)
plot(f,phi,'k','LineWidth',1)
xlabel('\omega/\pi'), ylabel('Phase delay (samples)'), grid
[H,f] = freqz(h,512,2);  % Computing the frequency response
figure (2)
plot(f,abs(H),'k','LineWidth',1)
xlabel('\omega/\pi'), ylabel('Magnitude'), axis([0,1,0,1.1]),grid

% Aplying the fractional delay filter
t=0:9;
x=0.5*sin(pi*0.2*t)+(sin(pi*0.1*t));  % Original sequence
y1=filter(h(3),x);                     % 0.2-sample delay
y2=filter(h(6),x);                     % 0.5-sample delay
figure (3)
stem(t,x,'k'); hold on; stem(t-0.2,y1,'kv--','filled')
hold on; stem(t-0.5,y2,'ks-.','filled')
xlabel('n'),ylabel('Amplitude')
legend('\mu=0','\mu=0.2','\mu=0.5'), grid
```

Figures 6.19 – 6.21 display the results. The plots of the phase-delay and magnitude characteristics are shown in Figures 6.19 and 6.20, respectively. It can be observed that the 1st order Lagrange interpolator provides a satisfactory approximation only in the low-frequency range. The behaviour of the magnitude responses is particularly restrictive, see Figure 6.20. In Figure 6.21, we illustrate the fractional delay of the signal when processed by a fractional-delay filter. Since the frequencies of the original signal are located in the low-frequency range (normalized frequencies f_1 = 0.1, f_2 = 0.2), the 1st order polynomial approximation of the fractional delay gives the satisfactory results in this case.

As shown in Figures 6.19 and 6.20, the 1st order Lagrange interpolator can be used only in the low-frequency range. In order to demonstrate how an increase in the filter order contributes to the phase-delay and the magnitude characteristics, the MATLAB program demo_6_5 has been created. This program generates Lagrange fractional-delay filters of the orders $N = 1, 3, 5,$ and 7 with the fixed value of fractional delay of $\mu = 0.4$. Program demo_6_5 computes and plots the phase-delay and the magnitude characteristics for all filters.

```
% Program demo_6_5.m
% Lagrange fractional-delay filters for N=1, N=3, N=5, and N=7.
clear all, close all
mu = 0.4; % Fractional delay
h1 = farrow.linearfd(mu)
fd = fdesign.fracdelay(mu,'N',3)
h3 = design(fd,'lagrange','FilterStructure','fd')
fd = fdesign.fracdelay(mu,'N',5)
h5 = design(fd,'lagrange','FilterStructure','fd')
fd = fdesign.fracdelay(mu,'N',7)
h7 = design(fd,'lagrange','FilterStructure','fd')
[phi1,f] = phasedelay(h1,512,2);
[phi3,f] = phasedelay(h3,512,2);
[phi5,f] = phasedelay(h5,512,2);
[phi7,f] = phasedelay(h7,512,2);
```

Figure 6.19. Phase-delay characteristics of the 1st order fractional-delay filters designed for $\mu = 0.1$, 0.2, 0.3, 0.4, 0.5, 0.6, 0.7, 0.8, and 0.9 samples

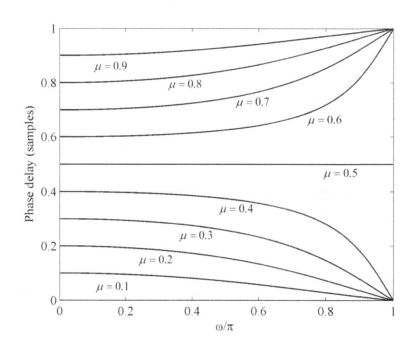

Figure 6.20. Magnitude characteristics of the 1ˢᵗorder fractional-delay filters designed for μ = 0.1, 0.2, 0.3, 0.4, 0.5, 0.6, 0.7, 0.8, and 0.9 samples

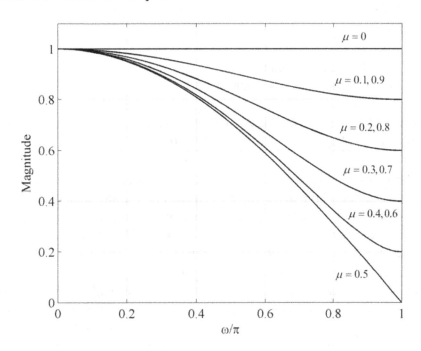

Figure 6.21. Fractional-delay filtering: Original sequence μ = 0, *and delayed sequences for 0.2 (*μ = 0.2) and 0.5 (μ = 0.5) samples

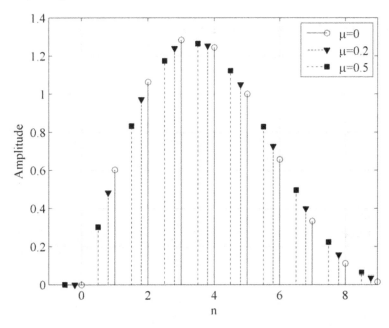

```
[H1,f] = freqz(h1,512,2);
[H3,f] = freqz(h3,512,2);
[H5,f] = freqz(h5,512,2);
[H7,f] = freqz(h7,512,2);
figure (1)
plot(f,phi1,'k',f,phi3,'k',f,phi5,'k',f,phi7,'k','LineWidth',1)
xlabel('\omega/\pi'), ylabel('Phase delay (samples)'), grid
figure (2)
plot(f,abs(H1),'k',f,abs(H3),'k',f,abs(H5),'k',f,abs(H7),'k','LineWidth',1)
xlabel('\omega/\pi'), ylabel('Magnitude'), grid, axis([0,1,0,1.1])
```

The results are displayed in Figures 6.22 and 6.23. From Figure 6.22, we observe that the increase of the filter order contributes a little to the linearity of the phase-delay characteristics. On the other hand, Figure 6.23 shows that the magnitude characteristic improves for the higher values of the filter order N. It is to be expected that an increase in the filter order should increase the overall delay D of the fractional-delay filter. For the Nth order FIR filter, we obtain $D = (N-1)/2 + \mu$, and consequently for $N = 1, 3, 5, 7$ and $\mu = 0.4$, $D = 0.4, 1.4, 2.4,$ and 3.4 as Figure 6.22 illustrates.

The application of Farrow structure in constructing fractional-delay filters provides a suitable mean to implement a variable fractional delay. Consider subsection Efficient Implementation of Polynomial Interpolation Filters Using Farrow Structure of this Chapter. Program demo_6_6 illustrates the processing of a sinusoidal sequence when the fractional delay factor is changed from $\mu = 0.4$ to $\mu = 0.2$. Program demo_6_6 generates the 1st order Lagrange fractional-delay filter, creates 10 samples of the sinusoidal sequence, and performs filtering with the sudden change of the delay factor from $\mu = 0.4$ to $\mu = 0.2$ at the time-index $n = 5$.

```
% Program demo_6_6.m
% Fractional-delay filtering with the change of the delay
clear all, close all
t = 0:9;
x = sin(2*pi*0.1*t);
for i=1:10
    if i <= 5, mu(i)=0.4;
    else
        mu(i)=0.2;
    end
end
h = farrow.linearfd(mu);
h.PersistentMemory = true;
for i=1:10
    h.Fracdelay = mu(i);
    ty(i) = t(i)-mu(i);
    y(i) = filter(h,x(i));
end
```

Figure 6.22. Phase-delay characteristics of the Lagrange fractional-delay filters for N = 1, 3, 5, and μ = 0.4

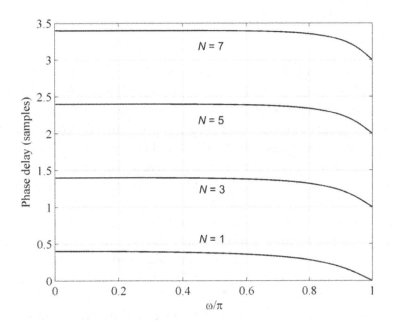

Figure 6.23. Magnitude characteristics of the Lagrange fractional-delay filters for N = 1, 3, 5, 7 and μ = 0.4.

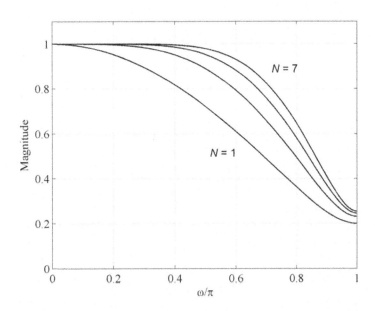

```
figure (6)
stem(t,x,'k'); hold on; stem(ty,y,'ks--')
xlabel('n'),ylabel('Amplitude')
legend('Original Signal','Delayed Signal')
grid
```

The results of program demo_6_6 are shown in Figure 6.24.

The problem of implementing a fractional delay can be solved by utilizing the methods developed for the resampling purposes. The desired solution can be achieved by mimicking the hybrid analog/digital model for interpolation given in Figure 6.15. Following this model, the continuous-time bandlimited signal is reconstructed first, then time-shifted and resampled, see (Vesma, 2000; Vesma & Saramäki; 2007). In this way, the problem of introducing a fractional delay can be considered as the interpolation problem. But most importantly, in fractional delay filtering the input and output sampling rates are the same.

One possibility to design a fractional delay filter is to use the Lagrange approximation and the Farrow structure discussed in the previous sections. Notice that with the Farrow structure the value of the fractional delay can be easily adjusted by changing only the value of the fractional parameter (μ). The interested reader is recommended to construct the Farrow-structure based fractional-delay filter by modifying program demo_6_3, see MATLAB Exercise 6.7.

During the last decade, many attractive methods have been developed for the design and implementation of fractional-delay filters with fixed or with variable coefficients, see (Laakso & all, 1996; Vesma, 2000; Hamila, Vesma, & Renfors, 2002; Johansson & Löwenborg, 2003; Vesma & Saramäki, 2007; Välimäki, & Haghparast, 2007; Dam, Cantoni & Nordholm, 2008; Hermanowitcz, Johansson & Rojewski, 2008), and the references within.

Figure 6.24. Illustration of the sudden change of the fractional delay

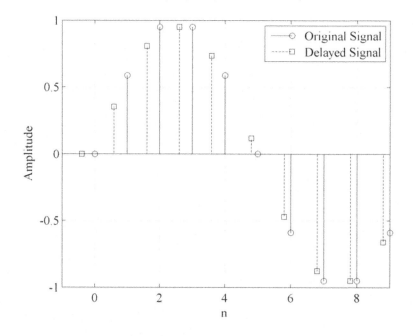

MATLAB EXERCISES

6.1 Modify program demo_6_1 to convert the sampling frequency by the factor 5/4. Plot the samples of input and the samples of the output signals.

6.2 Use the MATLAB function upfirdn to convert the sampling rate by the factor $L/M = 3/4$:
(a)   Design an optimal FIR filter of the length $N = 64$ to approximate the ideal specifications as given in (6.2).
(b)   Generate 512 samples of the input signal $\{x[n]\}$ given by $x[n] = \sin(2\pi \times 0.08 \times n) + \sin(2\pi \times 0.1 \times n) + \sin(2\pi \times 0.3 \times n)$ Compute the samples of the resampled signal $\{y[n]\}$.
(c)   Plot the frequency response of the filter, and the spectra of the signal $\{x[n]\}$, and those of the resampled signal $\{y[n]\}$. Plot the first 50 samples of the original signal, and the first 50 samples of the resampled signal. Comment on the results.

6.3 Modify program demo_6_2 to perform the sampling rate alteration with the fractional factor $L/M = 2/3$. The implementation structure is given in Figure 6.7.

6.4 Use the MATLAB function interp1 with the option **spline** to convert the sampling rate of the signal x=sin(2*pi*2000*t)+0.5*cos(2*pi*1000*t); from 44.1 kHz to 48 kHz.

6.5 Modify program demo_6_3 to perform the interpolation by a fractional factor 14/13 using the cubic Lagrange polynomials and the Farrow structure.

6.6 Based on the program demo_6_3, develop the program which simulates the factor-of-15/16-decimator using the cubic Lagrange polynomials and the Farrow structure.

6.7 Modify program demo_6_3 for fractional delay filtering by means of the Farrow structure. Recall that the input and the output rates in the case of a fractional filter are identical. Generate the first 30 samples of the signal $x[n] = \sin(0.1\pi n)$ and perform the fractional delay filtering for the delay factor $\mu = 0.3$. Plot the input sequence $\{x[n]\}$ and the delayed sequence $\{y[n]\}$.

REFERENCES

Ansari,R., & Liu,B., (1993). Multirate signal processing. In Sanjit. K. Mitra and James F. Kaiser (ed.), *Handbook for Digital Signal Processing*. New York, NY: John Wiley-Interscience, 981-1084.

Babic, D., Ghadam, A. S. H., & Renfors, M. (2004). *Polynomial-based filters with odd number of polynomial segments for interpolation. IEEE Signal Processing Letters*, *11*(2), Part 2, 171-174.

Bellanger, M.G., Bonncrot, G., & Coudrcusc, M. (1976). Digital filtering by polyphase network: appli cation to sample-rate alteration and filter banks. *IEEE Transactions on Acoustics, Speech, and Signal Processing*, *24*(2), 109-114.

Bellanger, M. (2000). *Digital processing of signals: Theory and practice*. 3rd edition. New York, NY: John Wiley.

Bregović, R., Saramäki, T., & Lim, Y. C. (2006). An efficient implementation of linear-phase FIR filters for a rational sampling rate conversion. *Proc. IEEE Int. Symp. Circuits and Systems, ISCAS 2006*, 5395-5398.

Crochiere, R.E., & Rabiner, L.R., (1981, March). Interpolation and decimation of digital signals - A Tutorial Review. *Proceedings of the IEEE, 69*(3), 300-331.

Crochiere, R.E., & Rabiner, L.R. (1983). *Multirate digital signal processing*. Englewood Cliffs, NJ: Prentice-Hall.

Dan, H.H., E., Cantoni, A., & Nordholm, S. (2008). Variable digital filter with group delay flatness specification or phase constraints. *IEEE Trans. On Circuits and Systems – II: Express Briefs, 455*(5), 442-446.

Erup, L., Gardner, F. M., & Harris, R. A. (1993). Interpolation in digital modems – Part II: implementation and performance. *IEEE Transactions on Communications, 41*(6), 998-1008.

Farrow, C. W. (1988). A continuously variable digital filter element. *Proc. IEEE Int. Symp. Circuits and Systems, ISCAS*, 2641- 2645.

Filter design toolbox for use with MATLAB. User's guide. Version 6. (2006). Natick: MathWorks.

Fliege, N. J. (1994). *Multirate digital signal processing*. New York, NY: John Wiley.

Hamila, R., Vesma, J., & Renfors, M. (2002). Polynomial-based maximum-likelihood technique for synchronization in digital receivers. *IEEE Trans. On Circuits and Systems – II: Analog and Digital Signal Processing, 49*(8), 567-576.

Harris, F. J., (2004). *Multirate signal processing for communication systems*. Upper Saddle River, NJ: Prentice Hall PTR.

Hentchel, T. (2002). *Sample rate conversion in software configurable radios*. Morwood, MA: Artech House.

Hermanowicz, E., Johansson, H., & Rojewski, M. (2008). A fractionally delaying complex Hilbert transform filter. *IEEE Trans. On Circuits and Systems – II: Express Briefs, 455*(5), 452-456.

Hsiao, C.C. (1987). Polyphase Filter Matrix for Rational Sampling Rate Conversion. *Proc. . of IEEE Int. Conf. on Acoustics, Speech, and Signal Processing, ICASSP-87*. 2173-2176.

Johansson, H., & Gustafsson, O. (2005). Linear-phase FIR interpolation, decimation, and Mth-band filters utilizing Farrow structure. *IEEE Transactions on Circuits and Systems I: Regular Papers, 52*(10), 2197-2207.

Johansson, H., & Löwenborg, P. (2003). On the design of adjustable fractional delay FIR filters. *IEEE Transactions on Circuits and Systems II: Analog and Digital Signal Processing, 50*(4), 164-169.

Laakso, T. T., Välimäki, V., Karjalainen, M., & Laine, U. K. (1996). Splitting the unit delay – tools for fractional delay filter design. *IEEE Signal Processing Magazine, 13*(1), 30-60.

Mitra, S. K. (2006). *Digital signal processing: A computer based approach.* 3rd edition. New York, NY: The McGraw-Hill Companies, Inc.

Mou, Z. J. (1996, October). Symmetry exploitation in digital interpolators/decimators. *IEEE Trans. on Signal Processing, 44*(10), 2611-2615.

Proakis J. G., & Manolakis D.G. (1996). *Digital signal processing: Principles, algorithms, and applications.* London: Prentice Hall.

Rajamani, K., Yhean-Sen Lai & Farrow, C. W. (2000). An efficient algorithm for sample rate conversion from CD to DAT. *IEEE Signal Processing Letters, 7*(10), 288-290.

Ramstad, T. A. (1984). Digital methods for conversion between arbitrary sampling frequencies. *IEEE Trans. Acoust. Speech, Signal Processing.* ASSP-32(3). 577-591.

Russel, A. I., (2000). Efficient Rational Sampling Rate Alteration Using IIR Filters. *IEEE Signal processing Letters, 7*(1), 6-7.

Saramäki, T. *Multirate Signal Processing.* (2001). Lecture notes for a graduate course, the Institute of Signal Processing, Tampere University of Technology.

Signal processing toolbox for use with MATLAB. User's guide. Version 6. (2006). Natick: Math-Works.

Vaidyanathan, P.P., (1990). Multirate digital filters, filter banks, polyphase networks, and applications: A Tutorial. *Proceedings of the IEEE, 78*(1), 56-93.

Vaidyanathan, P.P., (1993). *Multirate systems and filter banks.* Englewood Cliffs, NJ: Prentice Hall.

Välimäki, V., & Laakso, T. T. (2000). Principles of fractional delay filters. *Proc. of IEEE Int. Conf. on Acoustics, Speech, and Signal Processing - ICASSP'00,* 693-696.

Välimäki, V., & Haghparast, A. (2007). Fractional delay filter design based on truncated Lagrange interpolation. *Signal processing Letters, 14*(11), 816-819.

Vesma, J. (2000). A frequency-domain approach to polynomial-based interpolation and the Farrow structure. *IEEE Trans. On Circuits and Systems – II: Analog and Digital Signal Processing, 47*(3), 206-209.

Vesma, J. & Saramäki, T. (2007). Polynomial-based interpolation filters—Part I: Filter synthesis. *Circuits, Systems & Signal Processing, 26*(2), 115 – 146.

Chapter VII
Lth–Band Digital Filters

INTRODUCTION

Digital Lth-band FIR and IIR filters are the special classes of digital filters, which are of particular interest both in single-rate and multirate signal processing. The common characteristic of Lth-band lowpass filters is that the 6 dB (or 3 dB) cutoff angular frequency is located at π/L, and the transition band is approximately symmetric around this frequency. In time domain, the impulse response of an Lth-band digital filter has zero valued samples at the multiples of L samples counted away from the central sample to the right and left directions. Actually, an Lth-band filter has the zero crossings at the regular distance of L samples thus satisfying the so-called *zero intersymbol interference property*. Sometimes the Lth-band filters are called the *Nyquist filters*.

The important benefit in applying Lth band FIR and IIR filters is the efficient implementation, particularly in the case $L = 2$ when every second coefficient in the transfer function is zero valued.

Due to the zero intersymbol interference property, the Lth-band filters are very important for digital communication transmission systems. Another application is the construction of Hilbert transformers, which are used to generate the analytical signals. The Lth-band filters are also used as prototypes in constructing critically sampled multichannel filter banks. They are very popular in the sampling rate alteration systems as well, where they are used as decimation and interpolation filters in single-stage and multistage systems.

This chapter starts with the linear-phase Lth-band FIR filters. We introduce the main definitions and present by means of examples the efficient polyphase implementation of the Lth-band FIR filters. We discuss the properties of the separable (factorizable) linear-phase FIR filter transfer function, and construct the minimum-phase and the maximum-phase FIR transfer functions. In sequel, we present the design and efficient implementation of the halfband FIR filters ($L = 2$). The class of IIR Lth-band and halfband filters is presented next. Particular attention is addressed to the design and implementation of IIR halfband filters. Chapter concludes with several MATLAB exercises for self study.

*L*TH-BAND LINEAR-PHASE FIR FILTERS: DEFINITIONS AND PROPERTIES

In this section, we consider the basic properties of the linear-phase *L*th-band FIR filters. The filter transfer function *H*(*z*) of such a filter can be expressed in the form

$$H(z) = \sum_{n=0}^{2K} h[n] z^{-n},$$ (7.1)

where, obviously, the filter length *N* is an odd number,

$$N = 2K + 1.$$ (7.2)

Since the filter is of a linear phase, the impulse response coefficients are symmetric,

$$h[2K - n] = h[n] \quad \text{for } n = 0, 1, \ldots, 2K.$$ (7.3)

The frequency response of a linear-phase filter is expressible in the form

$$H(e^{j\omega}) = e^{-jK\omega} H(\omega),$$ (7.4)

where *H*(ω) is the zero-phase frequency response given by

$$H(\omega) = 1/L + 2\sum_{n=1}^{K} h[K - n] \cos(\omega n).$$ (7.5)

The filter *H*(*z*) is an *L*th-band filter if the impulse response coefficients satisfy the following conditions

$$h[K] = 1/L, \, h[K \pm rL] = 0 \quad \text{for } r = 1, 2, \ldots, \lfloor K/L \rfloor,$$ (7.6)

where $\lfloor x \rfloor$ stands for the integer part of *x*. Figure 7.1(a) illustrates the above conditions for the case *K* = 10 and *L* = 4. Here, the value of the central coefficient *h*[10] is exactly 1/*L* = 1/4, and the zero crossings occur at *n* = 10 ± 4, and *n* = 10 ± 8.

Equation (7.6) defines the time-domain conditions for the *L*th-band filter. It was proved by Mintzer (1982) that filters satisfying the time-domain conditions defined by (7.6) satisfy also the following condition in the frequency domain

$$\sum_{r=0}^{L-1} H(\omega + 2\pi r/L) = 1.$$ (7.7)

Apparently, this frequency-domain condition restricts the passband of the lowpass filter to be smaller than π/*L*. The filter transition band is symmetric around the angular frequency $\omega_c = \pi/L$, and the zero-phase frequency response has the value 0.5 at that frequency, see Figure 7.1(b).

The passband and the stopband edge frequencies, ω_p and ω_s, are symmetric around $\omega_c = \pi/L$, and usually are expressed in terms of the so-called *roll-off factor* ρ,

$$\omega_p = (1-\rho)\pi/L, \quad \text{and} \quad \omega_s = (1+\rho)\pi/L, \tag{7.8}$$

where ρ is a number in the range $0 < \rho < 1$. Obviously, the overall filter transition bandwidth, $\omega_s - \omega_p$, amounts to

$$\omega_s - \omega_p = 2\pi\,\rho/L. \tag{7.9}$$

It is also important to point out that the peak passband ripple δ_p is related with the peak stopband ripple δ_s (Mintzer, 1982; Saramäki, 1993, 1998), i.e.,

$$\delta_p = (L-1)\delta_s. \tag{7.10}$$

Usually, the passband ripple δ_p is considerably smaller than this upper-limit value.

In MATLAB, function firnyquist from the *Filter Design Toolbox* designs a lowpass linear-phase *L*th-band FIR filter with an equiripple magnitude characteristic. With the following code, we compute the impulse response coefficients

```
h = firnyquist(Nord,L,ro);
```

Program firnyquist, for the given filter order Nord, factor L, and the roll-off factor ro, returns in the vector h the impulse response coefficients of the linear-phase *L*th-band filter. The filter magnitude response exhibits equiripple characteristic in the pass and stopbands. Several options are available with firnyquist, and some of them will be used later on in this chapter.

The illustrative example plotted in Figure 7.1 was designed by means of firnyquist with the following parameters

```
h = firnyquist(20,4,0.2); % Computing the Nyquist filter coefficients
```

*L*th-band filters exhibit a very attractive property when they are used in interpolation. Namely, when the interpolation-by-*L* is performed with an *L*th-band filter, the original values of the input samples appear at the output without any distortion at the regular time intervals of *L* samples. The in-between *L*−1 samples are determined by interpolation. This becomes evident when observing the interpolation process in detail.

It is well known that the interpolation- by-*L* consists of two operations: up-sampling-by-*L*, and the lowpass filtering. Hence, the samples $y[m]$ appearing at the interpolator's output, are to be computed by convolving the up-sampled signal $\{x_u[m]\}$ with the lowpass filter impulse response $\{h[n]\}$,

$$y[m] = L\sum_{k=0}^{2K} h[k]\,x_u[m-k]. \tag{7.11}$$

Figure 7.1. Illustration of the Lth-band FIR filter properties, example K = 10, L=4: (a) Impulse response; (b) Magnitude response

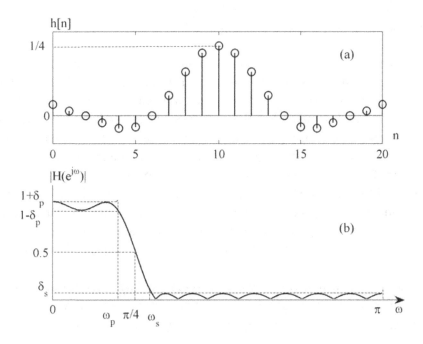

Obviously, the sequence $\{x_u[m]\}$ contains $L-1$ zero-valued samples between each two nonzero samples, and the nonzero samples in $\{x_u[m]\}$ are those of the original signal $\{x[n]\}$.

The ability of an Lth-band filter to preserve the original sample values of $\{x[n]\}$ is due to the favourable coincidence that the zero-crossings in the impulse response $\{h[n]\}$, and the nonzero sample values in $\{x_u[m]\}$ appear at the same regular time intervals. This is demonstrated visually in the following example.

Example 7.1

Let us consider a factor-of-3 interpolator constructed with the third-band filter of the length $N = 21$ ($K = 10$). The m-file firnyquist is used to compute the impulse response coefficients,

```
h = firnyquist(20,3,0.4); % Computing the Nyquist filter coefficients
```

The resulting impulse response is shown in Figure 7.2(a). Notice that the central sample, marked with the arrow, is of the value $h[10] = 1/3$, whereas the zero crossings are located at $k = 10 \pm 3, 10 \pm 6, 10 \pm 9$.

The interpolation is demonstrated on the signal $\{x[n]\}$ composed of two sinusoidal sequences,

```
x = 0.6*cos(2*pi*1.2*n/16)+0.8*sin(2*pi*0.7*n/16);
```

In interpolation, the up-sampling precedes filtering, and here we use function upsample to perform the up-sampling operation and obtain the sequence $\{x_u[m]\}$,

xu = upsample(x,3); % Upsampling

Finally, we can demonstrate what happens when computing the convolution (7.11) of the Lth-band filter impulse response with an up-sampled-by-L sequence. In Figures 7.2(b) and 7.2(c), we show sequence $\{x_u[m-k]\}$ for two distinct values of m (m_0 and m_1). The sequence in Figure 7.2(b) illustrates the computation of the output sample when the nonzero sample in $\{x_u[m-k]\}$ (marked with the arrow) coincides with the central sample of $\{h[k]\}$. In that case, the positions of the remaining nonzero samples of $\{x_u[m_0-k]\}$ coincide exactly with the zero crossings of the filter impulse response. According to this, the convolution sum (7.11) has only one nonzero term, and therefore the sample value at the output is identical with that at the input. Figure 7.2(c) shows the case when the position of the zero-valued sample of $\{x_u[m-k]\}$ coincides with the central sample of the impulse response. In that case, the positions of the nonzero samples of $\{x_u[m_1-k]\}$ coincide with the nonzero samples of the filter impulse response, and consequently, the convolution sum (7.11) computes the output sample $y[n]$ as a nonzero interpolated value.

Figure 7.3 shows the overall result of the above interpolation. The sample values of the original signal are preserved as indicated with the thick lines. The in-between samples indicated with thin lines are the interpolated values. Evidently, there is a delay of 10 samples since the filter length is $N = 21$.

Figure 7.2. Interpolation with the Lth-band filter with L=3: (a) Impulse response of the interpolation filter. (b) Sequence $\{x_u[m_0-k]\}$, $m_0 = 16$. (c) Sequence $\{x_u[m_1-k]\}$, $m_1 = 17$.

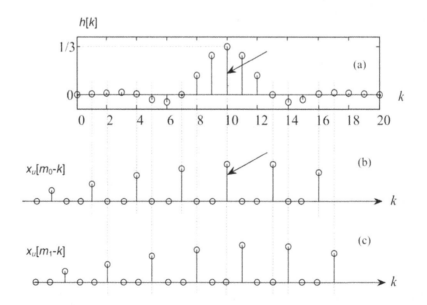

Figure 7.3. Interpolated signal: Thick lines with circles denote the samples that are identical with the input samples. Thin lines with stars denote the interpolated sample values.

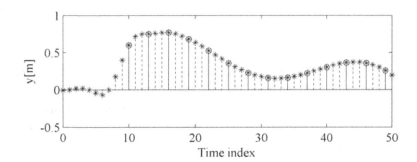

POLYPHASE IMPLEMENTATION OF FIR *L*TH-BAND FILTERS

The impulse response of an *L*th-band filter contains many zero-valued samples providing the possibility of computationally efficient implementation. This is particularly the case when the *L*th-band filter is implemented as a parallel connection of *L* polyphase subfilters. In that case, one out of *L* polyphase subfilters reduces to a constant. Without loss of generality, we use the example third-band filter of the length $N = 15$ to present the efficiency of the polyphase implementation.

The polyphase implementation of FIR filters and their application in decimators and interpolators has been given in Chapter IV. Here, we apply the polyphase decomposition described in Chapter IV to express the third-band filter transfer function $H(z)$ in terms of three polyphase components $E_0(z)$, $E_1(z)$, and $E_2(z)$.

Since $H(z)$ is the third-band filter of the length $N = 15$, the central coefficient is determined as $h[7] = 1/3$, and accordingly, the zero-valued samples are $h[7\pm 3]$ and $h[7\pm 6]$. With this assumption, we write $H(z)$ in the following form

$$H(z) = h[0] + h[2]z^{-2} + h[3]z^{-3} + h[5]z^{-5} + h[6]z^{-6} + (1/3)z^{-7}$$
$$+ h[8]z^{-8} + h[9]z^{-9} + h[11]z^{-11} + h[12]z^{-12} + h[14]z^{-14} \qquad . \tag{7.12}$$

Following the polyphase decomposition described in Chapter IV, we extract three polyphase components,

$$E_0(z) = h[0] + h[3]z^{-1} + h[6]z^{-2} + h[9]z^{-3} + h[12]z^{-4} \tag{7.13a}$$

$$E_1(z) = (1/3)z^{-2} \tag{7.13b}$$

$$E_2(z) = h[2] + h[5]z^{-1} + h[8]z^{-2} + h[11]z^{-3} + h[14]z^{-4}. \tag{7.13c}$$

It is apparent that $E_1(z)$ collects the central term and all zero-valued coefficients of $H(z)$. In this way, subfilter $E_1(z)$ has only one multiplication constant which equals 1/3. Finally, the polyphase form of $H(z)$ is represented by

$$H(z) = E_0\left(z^3\right) + z^{-1}(1/3)z^{-6} + z^{-2}E_2\left(z^3\right). \tag{7.14}$$

Evidently, the polyphase Lth-band filters lead to considerable computational savings when used in decimators and interpolators. Figure 7.4 illustrates the efficient configuration of the factor-of-3 interpolator. This example shows that applying the third-band filter, the total number of multiplication constants reduces by one third. The reader is recommended to consider the decimation with an Lth-band filter by solving MATLAB Exercise 7.3.

SEPARABLE LINEAR-PHASE LTH-BAND FIR FILTERS, MINIMUM-PHASE AND MAXIMUM-PHASE TRANSFER FUNCTIONS

In many applications, such as telecommunications, and signal analysis and synthesis, it is desired to construct a *separable Lth-band filter* transfer function $H(z)$, which is factorizable in two so-called *spectral factors* $H_1(z)$ and $H_2(z)$,

$$H(z) = H_1(z)H_2(z) \tag{7.15}$$

Here, $H_1(z)$ and $H_2(z)$ have the same zero-phase frequency responses, and their impulse responses are time-reversed versions of each other. Therefore, the transfer functions $H_1(z)$ and $H_2(z)$ are related by

$$H_2(z) = z^{-(N-1)/2}H_1\left(z^{-1}\right) \tag{7.16}$$

where N is the impulse response length of $H(z)$. The order of $H_1(z)$ and $H_2(z)$ is the half order of $H(z)$. The filters $H_1(z)$ and $H_2(z)$ are sometimes also called *half Nyquist filters*.

For $H(z)$ being separable (factorizable) in a manner explained above it is required that its zero-phase frequency response is nonnegative. Let us illustrate the properties of the separable Lth-band filters by

Figure 7.4. Efficient factor-of-3 interpolator implemented with the third-band filter

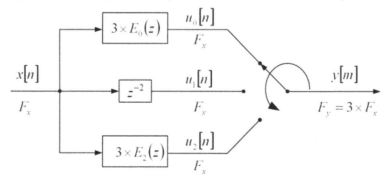

means of example. The MATLAB function firnyquist already used earlier in this chapter computes the coefficients for this class of filters. Let us design a filter of the length $N = 34$, $L = 4$, and $\rho = 0.2$,

h = firnyquist(34,4,0.2,'nonnegative'); % Computing the coefficients of the factorizable Nyquist filter

The filter characteristics are displayed in Figure 7.5. Evidently, the zero-phase frequency response $H(\omega)$ is nonnegative, and the transfer function zeros on the unit circle are double zeros. The remaining zeros are symmetric in respect to the unit circle.

The factorization of $H(z)$ into the spectral factors $H_1(z)$ and $H_2(z)$ is carried out by selecting a set of zeros belonging to $H_1(z)$, or that belonging to $H_2(z)$. Usually, it is demanded to separate $H(z)$ into a minimum-phase and a maximum-phase spectral factors. In that case, $H(z)$ is expressible as the product of the minimum-phase and the maximum phase filters $H_{min}(z)$ and $H_{max}(z)$, respectively,

$$H(z) = H_{min}(z)H_{max}(z) \tag{7.17}$$

The minimum-phase factor $H_{min}(z)$ collects the inside unite circle zeros, and one of each double zeros from the unit circle. The maximum-phase factor $H_{max}(z)$ collects the outside unit circle zeros, and one of each double zeros from the unit circle. As already defined for the spectral factors, equation (7.15), $H_{min}(z)$ and $H_{max}(z)$ have the same zero-phase frequency responses, and their impulse responses are time-reversed versions of each other.

The direct design of the minimum-phase transfer function $H_{min}(z)$ can be carried out by making use of firnyquist with the corresponding option. Let us select $H_{min}(z)$ of the order $N_{ord} = 17$, factor $L = 4$, and roll-of factor $\rho = 0.2$. The following code computes the impulse response $\{h_{min}[n]\}$,

hmin = firnyquist(17,4,0.2,'minphase') % Computing the coefficients of the minimum-phase FIR filter

Figure 7.5. Separable 4th-band filter

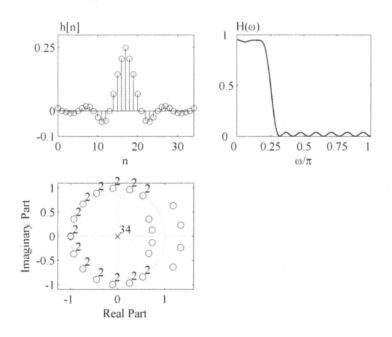

The impulse response of the maximum-phase spectral factor $\{h_{max}[n]\}$ being time-reversed to $\{h_{min}[n]\}$ can be obtained by inverting $\{h_{min}[n]\}$,

hmax = fliplr(hmin); or equivalently, hmax = hmin(end:-1:1);

The minimum-phase and the maximum-phase filters meet the condition,

h = conv(hmin,hmax);

where the vector h contains the impulse response of the linear-phase Lth-band Nyquist filter $H(z)$ whose order is twice the order of $H_{min}(z)$ $[H_{max}(z)]$.

Figure 7.6 illustrates the impulse responses and zero-pole plots for $H_{min}(z)$ and $H_{max}(z)$. It is left to the reader to compute and plot magnitude, phase, and group delay responses of $H_{min}(z)$ and $H_{max}(z)$, see MATLAB Exercise 7.4.

Design of $H_{min}(z)$ or $H_{max}(z)$ by selecting a half of the zeros of the separable linear-phase filter $H(z)$ may become computationally impractical. Namely, the computation of the double-zeros placed on the unit circle may produce an error to the measure that the computed zeros are not the double-zeros. The advanced algorithms have been developed, which provide better accuracy if compared with the widely used computer programs for optimal filter design (Saramäki, 1993; Orchard & Wilson, 2003).

Figure 7.6. Minimum-phase and maximum-phase spectral factors: impulse responses and zero-pole plots

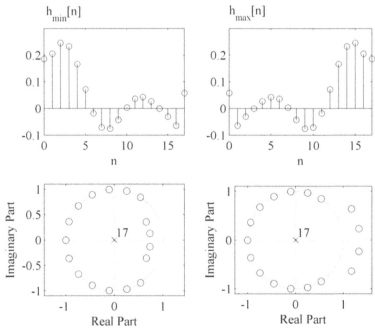

HALFBAND FIR FILTERS

A *halfband filter* is an *L*th-band filter with *L* = 2, and consequently, the halfband filter divides the baseband of the signal into two equal subbands. In the linear-phase halfband filter, half of the constants are zero-valued making the implementation very attractive.

Linear-Phase Halfband Filters

The transfer function of a linear-phase FIR halfband filter is given in the form

$$H(z) = \sum_{n=0}^{2K} h[n] z^{-n}, \tag{7.18}$$

where K is odd, and coefficients are symmetric in respect to the central coefficient $h[K]$,

$$h[2K - n] = h[n], \qquad \text{for } n = 0, 1, \ldots, 2K. \tag{7.19}$$

The filter length N is an odd number,

$$N = 2K + 1, \quad K = 1, 3, 5, \ldots \tag{7.20}$$

The time-domain conditions defined by (7.6) for an *L*th-band filter, in the case of a halfband filter become

$$h[K] = 1/2, \, h[K \pm 2r] = 0 \text{ for } r = 1, 2, \ldots, \lfloor K/2 \rfloor. \tag{7.21}$$

This equation says that the odd-indexed coefficients in $\{h[n]\}$ are zero-valued except for the central coefficient $h[K]$, which is equal to 1/2. Since K is an odd number, the impulse response begins and terminates with nonzero samples, $h[0] \neq 0$, $h[2K] \neq 0$. Figure 7.7 illustrates the impulse response of a typical linear-phase FIR halfband filter. The impulse response shown in Figure 7.7 has the length of $N = 11$ samples, giving $K = (N-1)/2 = 5$. Since K should be odd, the filter length can be augmented (diminished) by the multiple of four samples. Hence, we can choose $N = 7, 11, 15, 19, \ldots$

In the frequency domain, the zero-phase frequency response of a halfband filter exhibits symmetry property

$$H(\omega) + H(\pi - \omega) = 1. \tag{7.22}$$

Notice that the above property follows from the general *L*th-band filter frequency-domain condition (7.7) when reduced to $L = 2$.

The frequency-domain condition (7.22) implies the symmetry of passband and stopband characteristics in respect to the middle of the baseband, $\omega_c = \pi/2$. The passband and stopband edges are symmetric

$$\omega_s = \pi - \omega_p, \tag{7.23}$$

and peak passband and stopband ripples are equal,

$$\delta_p = \delta_s = \delta. \tag{7.24}$$

In the passband, $H(\omega)$ approximates unity within the tolerances $[1-\delta, 1+\delta]$, and in the stopband $H(\omega)$ approximates zero within the tolerance $[-\delta, \delta]$, as Figure 7.7 illustrates. The value of $H(\omega)$ at $\omega_c = \pi/2$ is exactly 0.5,

$$H\left(\frac{\pi}{2}\right) = 0.5. \tag{7.25}$$

An equiripple halfband filter can be designed by using the Parks-McClellan-Rabiner algorithm, but for a large N this numerical procedure produces an error in the values of coefficients. A convenient method, named *halfband filter trick*, is described in Vaidyanathan and Nguen (1987), and in Saramäki (1993). This method decreases the computational requirements by a factor of two. In the first step, a modified filter of about half the required length is designed using the Parks-McClellan-Rabiner program. In the second step, the desired halfband filter is created by simply inserting zeros for the odd-index coefficients, scaling the even-index coefficients by 0.5, and adding the central coefficient of 0.5. In this way, a sufficient coefficient accuracy is achieved, and the design time is considerably decreased.

The function firhalfband from the *Filter Design Toolbox* provides several options for FIR halfband filter design in MATLAB. Using firhalfband the coefficients can be computed for the halfband filter with equiripple characteristics in the pass and stopbands, and also for the halfband filter with impulse response truncated by the specified window.

For the equiripple design, we can specify two parameters. We may choose to specify the filter order N_{ord} and the passband edge frequency $f_p = \omega_p/\pi$,

h = firhalfband(Nord,fp);

The program returns the filter coefficients in vector h. Notice that the passband and the stopband edge frequencies are symmetric, see (7.23). The example displayed in Figure 7.7 has been designed using the following code

h = firhalfband(10,0.42); % Computing the coefficients of linear-phase halfband filter

Alternatively, we can specify the passband edge frequency f_p and the peak ripple δ, and search for the minimum filter order that meets the requirements. In that case, we use the option,

h = firhalfband('minorder',fp,delta);

Program computes the coefficients of the equiripple halfband filter, which with the minimum order attains the passband edge frequency f_p and the peak ripple δ denoted here by delta.

When it is desired to design a halfband filter using the truncated impulse response window-method, we use the option,

h = firhalfband(Nord,win)

Figure 7.7. Linear-phase FIR halfband filter: impulse response and zero-phase frequency response

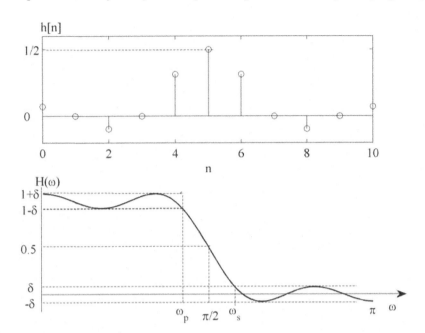

Here, we specify the filter order and the desired window denoted above by win.

We demonstrate the application of firhalfband in FIR linear-phase halfband filter design by means of two examples.

Example 7.2

In this example, we design a minimum order equiripple halfband filter to satisfy the following specifications:

The minimal stopband attenuation $a_s = 40$ dB (maximal stopband gain, -40 dB)
The stopband edge frequency $\omega_s = 0.6\pi$.
The filter design can be carried out with the following code:

```
as = 40;                                 % Specifying the minimal stopband attenuation
delta = 10^(-as/20);                     % Computation of the peak ripple
fs = 0.60;                               % Specifying the stopband edge frequency
fp = 1-fs;                               % Computation of the passband edge frequency
h = firhalfband('minorder',fp,delta);    % Computation of filter coefficients
```

Here, we have determined first the peak ripple δ, and the passband edge frequency f_p. Afterwards, the m-file firhalfband with the option minorder is used to compute the filter coefficients. The requested specifications are met with the filter length $N = 23$. Figure 7.8 shows the filter impulse response and the gain characteristic. As expected, all odd-indexed coefficients of the impulse response are zero-valued, and the central coefficient is $h[11] = 0.5$. In the frequency domain, the stopband attenuation exceeds the requested 40 dB. This is so because the filter order is restricted to integers. Program firhalfband deter-

Figure 7.8. Impulse response and gain characteristic for the halfband filter of Example 7.2

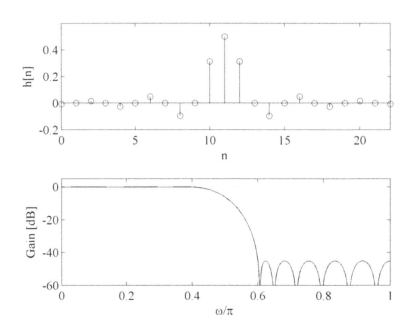

mines the minimal filter order, an integer number, which guarantees the peak ripple smaller or equal to the specified value.

Example 7.3

In this example, we design a halfband filter using the truncated impulse response window-method. We take for the filter length $N = 23$ ($N_{ord} = 22$), and the Hamming window. The filter length is that of *Example 7.2.*

The filter design can be performed as follows:

```
Nord = 22;                  % Specifying the filter order
win = hamming(23);          % Specifying window
h = firhalfband(Nord,win);  % Computation of filter coefficients
```

The filter characteristics in time and frequency domain are displayed in Figure 7.9. It is to be noticed that the impulse response coefficients behave strictly according to the halfband filter conditions given in (7.22).

Efficient Implementation of Linear-Phase Halfband Filters

An important advantage of a linear-phase halfband filter is the efficient implementation, which follows from two favourable properties of the filter impulse response:

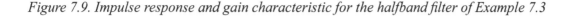

Figure 7.9. Impulse response and gain characteristic for the halfband filter of Example 7.3

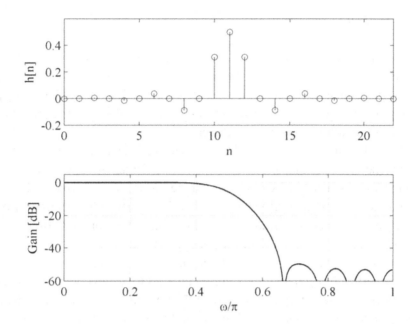

- The number of nonzero-valued coefficients is nearly half of the filter length.
- The nonzero coefficients exhibit symmetry property.

Let us demonstrate the efficiency of implementation by means of a linear-phase halfband filter of the length $N = 11$. Taking into account that every second coefficient is zero-valued except for the central coefficient whose value is 0.5, the filter transfer function $H(z)$ can be written in the form,

$$H(z) = h[0] + h[2]z^{-2} + h[4]z^{-4} + 0.5z^{-5} + h[6]z^{-6} + h[8]z^{-8} + h[10]z^{-10}. \qquad (7.26)$$

Introducing the coefficient symmetry property in (7.26), the transfer function of the linear-phase half-band filter becomes

$$H(z) = h[0] + h[2]z^{-2} + h[4]z^{-4} + 0.5z^{-5} + h[4]z^{-6} + h[2]z^{-8} + h[0]z^{-10}. \qquad (7.27)$$

It is convenient to represent $H(z)$ in terms of two polyphase components (subfilters) $E_0(z)$ and $E_1(z)$,

$$H(z) = E_0(z) + z^{-1}E_1(z), \qquad (7.28)$$

where

$$E_0(z) = h[0] + h[2]z^{-1} + h[4]z^{-2} + h[4]z^{-3} + h[2]z^{-4} + h[0]z^{-5}, \qquad (7.29a)$$

$$E_1(z) = 0.5z^{-2}. \qquad (7.29b)$$

The coefficients in $E_0(z)$ are symmetric, and according to this one writes

$$E_0(z) = h[0]\left(1 + z^{-5}\right) + h[2]\left(z^{-1} + z^{-4}\right) + h[4]\left(z^{-2} + z^{-3}\right). \tag{7.29c}$$

We can see that only three multiplication constants are needed to implement the linear-phase FIR halfband filter of the length $N = 11$. The value of central coefficient is 0.5 and therefore, this coefficient can be implemented with a binary shift. It is important to point out that the subfilter $E_0(z)$ preserves the coefficient symmetry of the original linear-phase halfband filter transfer function. This property makes the linear-phase FIR halfband filters extremely efficient.

Linear-phase FIR halfband filters when applied in factor-of-2 decimators and in factor-of-2 interpolators lead to very efficient realizations. The efficiency of implementation, which can be achieved, is illustrated here by means of a factor-of-2 decimator and the linear-phase halfband filter of $N = 11$.

Efficient polyphase configuration of FIR decimators and interpolators has been described in Chapter IV. In Figure 7.10 (a), the efficient configuration of a factor-of-2 decimator consisting of two polyphase subfilters $E_0(z)$ and $E_1(z)$ is shown. Since the linear-phase halfband filter of $N = 11$ is to be considered, we implement subfilters $E_0(z)$ and $E_1(z)$ according to equations (7.29a – c).

The favourable properties of $E_0(z)$ and $E_1(z)$ can be utilized to achieve very efficient implementation: the subfilter $E_1(z)$ has only one term, see equation (7.29b), and $E_0(z)$ itself is of a linear phase with coefficient symmetry property, see equation (7.29c). Exploiting these favourable properties of $E_0(z)$ and $E_1(z)$, we arrive to the efficient implementation scheme for the overall decimator, which is shown in Figure 7.10(b). The total number of multiplication constants is only three. Moreover, all the multiplications are to be evaluated at the rate of the low-rate (output) signal. Notice that the delay chain in Figure 7.10(b) has only 5 elements. This is so because the memory-saving solution is used that reduces the total number of delays to $(N-1)/2$.

In the example that follows, we simulate in MATLAB decimation-by-2 based on the efficient implementation structure of Figure 7.10 (b). Since the filter of $N = 11$ has poor pass stopband characteristics,

Figure 7.10. Efficient implementation of the linear-phase FIR halfband filter. (a) Polyphase configuration. (b) Implementation scheme for N = 11.

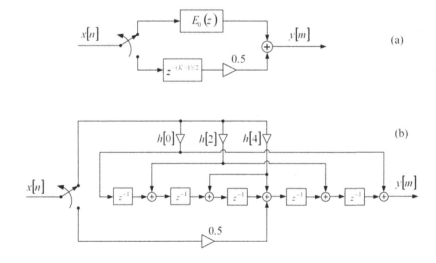

we extend in this example the structure of Figure 7.10 (b) to the filter length of $N = 23$. Actually, we perform the decimation-by-2 using the equiripple halfband filter of *Example 7.1.*

Example 7.4

In this example, we present program demo_7_1 which simulates in MATLAB the factor-of-2 decimation. Program demo_7_1 performs

- Design of the linear-phase FIR halfband filter of $N = 23$ (the same as in *Example 7.1*).
- Generates 257 samples of the original signal $\{x[n]\}$.
- Implements the decimation-by-2 according to the efficient scheme presented in Figure 7.10 (b). The implementation structure of Figure 7.10 (b) is extended since the filter length of this example is $N = 23$ (instead of 11). The delay chain has 11 elements, and the number of multiplication constants is 6.
- Computes and plots the spectrum of the original signal and that of the decimated signal.

```
% Program demo_7_1.m
% Decimatin-by-2 with linear-phase halfband FIR filter
% Efficient polyphase configuration with memory-saving structure

clear all, close all
h = firhalfband(22,0.4); % FIR filter design

% Generation of the input signal
F = [0,0.05,0.45,1]; A = [0,1,0.1,0.05];
x1 = fir2(256,F,A); x2 = sin(2*pi*(0:256)*0.4)/150;
x = x1 + x2;
[X,f] = freqz(x,1,512,2);  % Spectrum of the original signal
figure (1)
subplot(2,1,1)
plot(f,abs(X(1:512))), xlabel('Normalized frequency \omega/\pi'),ylabel('|X(e^{j\omega})|')
axis([0,1,0,1.1])

% Polyphase decomposition
e0 = h(1:2:(length(h)-1)/2);

% Setting the initial states
xk = zeros(size(1:(length(h)+1)/2));

% Decimation
rot = 0; % Initial swithch position
y = [];
for n=1:length(x)
xn = x(n);
```

```
if rot==0 ss=1;
  pro = xn*e0;
  xk = [pro+xk(1:length(xk)/2),fliplr(pro) + xk((length(xk)/2+1):length(xk))];
  yn = xk(length(xk));
  y = [yn,y];
  xk = [0,xk(1:length(xk)-1)];
else
end
if rot==1 ss=0;
  xk(length(xk)/2+1) = 0.5*xn + xk(length(xk)/2+1);
else
end
rot=ss;
end


Y = freqz(y,1,512,2); % Spectrum of the decimated signal
subplot(2,1,2)
plot(f,abs(Y(1:512))), xlabel('Normalized frequency \omega/\pi'),ylabel('|Y(e^{j\omega})|')
axis([0,1,0,0.6])
```

Figure 7.11 displays the magnitude spectrum of the original signal $|X(e^{j\omega})|$ and the magnitude spectrum of the decimated signal $|Y(e^{j\omega})|$. Notice that the filter characteristics are shown in Figure 7.8.

This example shows the advantage of using linear-phase FIR halfband filters in sampling rate conversion. Very efficient solutions are achieved in multistage sampling rate converters when the conversion factor is expressible as a power-of-two, i.e., $M(L) = 2^p$, p is a positive integer. In that case, each of p stages converts the sampling rate by 2 using the efficient halfband filter.

In this subsection, we demonstrated a solution for the efficient decimation with the linear-phase halfband filter. An interested reader can develop the efficient interpolator structure by solving MATLAB Exercise 7.6.

Minimum-Phase and Maximum-Phase FIR Halfband Filters

The minimum-phase and maximum-phase Lth-band filters have been considered earlier in this chapter. As in the case of the Lth-band filters, the halfband minimum-phase and the maximum-phase filters, $H_{min}(z)$ and $H_{max}(z)$, respectively, are the spectral factors of the linear-phase separable (factorizable) halfband filter $H(z)$,

$$H(z) = H_{min}(z)H_{max}(z) \tag{7.30}$$

The zero-phase frequency response of $H(z)$ is nonnegative, and the zeros of $H(z)$ occurring on the unit circle are double-zeros. The filter order should be even and twice an odd number. The central coefficient of the impulse response has the value 0.5, and the remaining odd-indexed coefficients are zero-valued. Therefore, $H(z)$ is a halfband Nyquist filter with nonnegative zero-phase frequency response. Sometimes this filter is called a *valid halfband filter* (Fliege, 1994). An efficient design procedure exists for

Figure 7.11. Magnitude spectra: Original signal – upper subfigure. Decimated signal – bottom subfigure.

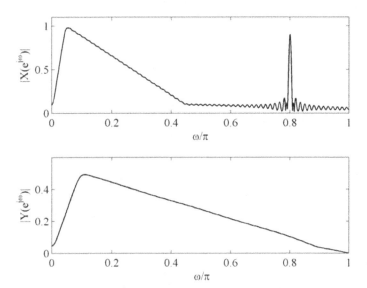

$H(z)$ based on the Remez exchange algorithm and the above mentioned filter properties (Fliege, 1994; Saramäki, 1993, 2001).

Minimum-phase and maximum-phase filters $H_{min}(z)$ and $H_{max}(z)$ share the transfer function zeros of $H(z)$ in such a manner that $H_{min}(z)$ picks up the zeros from the inside of the unit circle and one from each double-zero from the unit circle, whereas $H_{max}(z)$ picks up the outside unit circle zeros and one from each double-zero from the unit circle. The order of $H_{min}(z)$ and $H_{max}(z)$ is odd and half the order of $H(z)$. The impulse responses of $H_{min}(z)$ and $H_{max}(z)$ are reversed to each other.

Figure 7.12 illustrates the impulse responses and the pole-zero plots of the minimum-phase and the maximum-phase filters.

In the frequency domain, minimum-phase and maximum-phase halfband filters exhibit the power-symmetry properties. The term power-symmetry means that the squared magnitude response, $|H(e^{j\omega})|^2$, has equal pass and stopband ripples.

The passband and the stopband edge frequencies ω_p and ω_s are symmetric in the respect of the central frequency $\omega_c = \pi/2$, and the frequency-symmetry property is expressed as

$$\omega_p + \omega_s = \pi \tag{7.31}$$

The magnitude response oscillates between $[1-\delta_p, 1]$ in the range $[0, \omega_p]$, and between $[0, \delta_s]$ in the range $[\omega_s, \pi]$. The peak passband and stopband ripples δ_p and δ_s are related by

$$\delta_p = 1 - \sqrt{1-\delta_s^2} \tag{7.32}$$

thus ensuring that the passband and the stopband ripples of $|H(e^{j\omega})|^2$ are equal.

When the minimal stopband attenuation and the peak passband ripple are to be expressed in dB, their relation is determined as

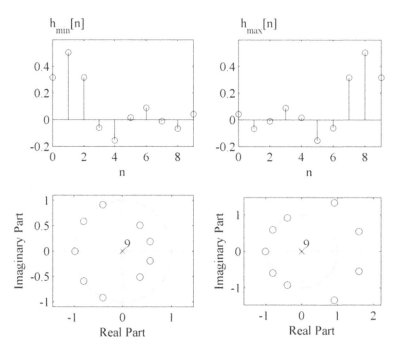

Figure 7.12. Impulse responses and pole-zero plots for minimum-phase and maximum-phase FIR half-band filters

$$a_p = 10\log_{10}\left(1+\frac{1}{10^{a_s/10}-1}\right) \qquad (7.33)$$

The magnitude response at the middle of the baseband, at $\omega_c = \pi/2$, amounts to $1/\sqrt{2}$, i.e. the level of the magnitude response at this frequency is approximately 3 dB below the maximum. The magnitude response for both minimum-phase and maximum-phase filters is identical, and a typical characteristic is shown in Figure 7.13.

The minimum-phase halfband filter can be designed by using the MATLAB function firhalfband with the option minphase. The filter order should be an odd number. One writes,

hmin = firhalfband(Nord,fp,'minphase');

The impulse response of the maximum-phase filter is obtained by inverting the impulse response of the minimum-phase filter,

hmax = fliplr(hmin); or equivalently, hmax = bmin(end:-1:1).

The minimum-phase and the maximum-phase filters meet the condition,

h = conv(hmin,hmax);

Figure 7.13. Typical magnitude characteristic of a halfband minimum-phase (maximum-phase) FIR filter

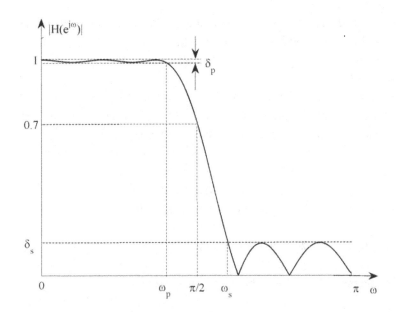

where the vector h contains the impulse response of the linear-phase nonnegative halfband Nyquist filter $H(z)$ whose order is twice the order of $H_{min}(z)$ $[H_{max}(z)]$.

Notice that minimum-phase and maximum-phase halfband filters are used in multirate systems for constructing filter banks with perfect reconstruction.

It is recommended to the interested reader to investigate properties of the minimum-phase and maximum-phase halfband filters by solving MATLAB Exercise 7.7.

FIR Filters with Maximally-Flat Magnitude Response

In this Chapter, we have considered the Lth-band and halfband FIR filters whose magnitude responses approximate unity in the passband and zero in the stopband in a minimax sense. Hence, we have considered the magnitude responses exhibiting the equiripple passband/stopband characteristics. This type of frequency response provides an economic solution when the maximal selectivity for the given filter order is desired.

In applications where the signal waveform has to be preserved, filters with smooth magnitude responses are preferable. This class of FIR filters was introduced first by Hermann (1971) and is usually called *maximally-flat FIR filters*. Filters with maximally-flat magnitude responses are important for digital signal processing particularly for constructing wavelet filter banks.

The transfer function of a maximally-flat lowpass filter has the multiple zero at the point $z = -1$, thus providing the flatness of the filter magnitude response at $\omega = \pi$. The number of multiple zeros at $z = -1$ is the order of flatness at $\omega = \pi$, which is related with the filter regularity order. In this subsection, we illustrate applications of some MATLAB functions dedicated to maximally-flat FIR filter design. For the mathematical background and for the design and implementation methods of this class of FIR filters see (Hermann, 1971; Selesnick and Burrus, 1996 and 1998; Saramäki, 1993; Samadi and Nishikara, 2007) including the references within.

The MATLAB *Signal Processing Toolbox* contains the function maxflat which can be used for designing linear-phase FIR filters with maximally-flat magnitude response. Using the function maxflat we can design a generalized Butterworth maximally-flat linear-phase FIR filter introduced by Selesnick and Burrus (1996, 1998).

In the following, we illustrate the application of maxflat for FIR filter design. We define the filter order N_{ord} and the normalized cutoff frequency denoted as Wn. Selecting the option 'sym', we choose the FIR filter with symmetric coefficients. Hence, we use the following code,

```
Nord = 12                    % Filter order
Wn = 0.5;                    % Normalized cutoff frequency
h = maxflat(Nord,'sym',Wn)   % Designing maximally flat FIR filter
figure (1)
subplot(2,2,1), impz(h,1)    % Impulse response
subplot(2,2,2), zplane(h,1)   % Pole-zero plot
subplot(2,1,2), zerophase(h,1) % Zero-phase frequency response
axis([0,1,0,1.1])
```

Figure 7.14 displays the performances of the resulting 12th-order linear-phase maximally flat FIR filter: impulse response, pole-zero plot, and zero-phase frequency response. The multiplicity of the transfer function zero at $z = -1$ is four. The remaining zeros meet the mirror image symmetry in the respect to the unit circle.

Figure 7.14. Impulse response, pole-zero plot, and zero-phase frequency response of the 12th order linear-phase maximally-flat FIR filter from maxflat.m

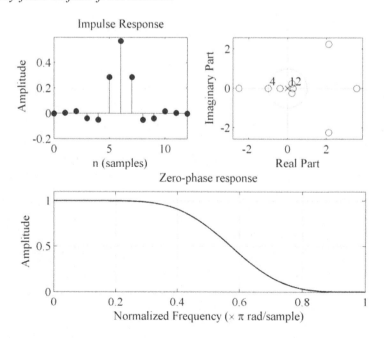

Very important maximally-flat halfband FIR filters are given in the MATLAB *Wavelet Toolbox* since they are used in constructing wavelet filter banks. The class of Daubechies wavelet filters, for example, is very popular for various wavelet applications. In this subsection, we show the design and performances of the lowpass Daubechies filter db5. A Doubechies filters are denoted by dbN, where N = 1, 2, 3 …, is the number of the transfer function zeros at $z = -1$. The filter has also N–1 zeros inside the unit circle. Accordingly, the Doubechies filter is a minimum-phase FIR filter, and the transfer function order is 2N–1. The function dbwavf from the *Wavelet Toolbox* can be used for computing the filter coefficients. With the following code, we demonstrate the design and performances of the filter db5.

```
wname = 'db5';                  % Set Daubechies wavelet name.
h = dbwavf(wname)               % Filter design
figure (1)
subplot(2,2,1), impz(h,1)
subplot(2,2,2), zplane(h,1)
subplot(2,1,2), zerophase(h,1)
axis([0,1,0,1.1])
```

Figure 7.15 displays the performances of the resulting db5 filter: impulse response, pole-zero plot, and zero-phase frequency response. The designed db5 filter has 5 zeros at $z = -1$, and 4 zeros inside the unit circle. The filter order is 9; see the number of poles at the origin. The resulting filter is a minimum-phase FIR filter with the maximally-flat magnitude response.

The application of the Daubechies filters for constructing the orthogonal filter banks with perfect-reconstruction properties will be demonstrated in Chapter XII of this book.

*L*TH-BAND IIR FILTERS

An infinite impulse response *L*th-band filter is characterised by the following properties:

- The passband edge frequency must satisfy $\omega_p < \pi/L$.
- The stopband edge frequency ω_s is located at $\omega_s = 2\pi/L - \omega_p$.
- The magnitude frequency response satisfies

$$\sum_{r=0}^{L-1} \left| H\left(e^{j(\omega + 2\pi r/L)} \right) \right|^2 = 1 \qquad (7.34)$$

- The filter transfer function can be represented in the polyphase form,

$$H(z) = \frac{1}{L} \sum_{r=0}^{L-1} z^{-r} A_r\left(z^L \right). \qquad (7.35)$$

where $A_r(z)$, $r = 0, 1, …, L - 1$, are stable allpass transfer functions.

The design and properties of this filter class are presented in the two papers of Renfors and Saramäki (1987). It was proved that the conditions stated above pose some constraints on the frequency characteristics of the filter. For $L > 2$, the filter frequency response is subject to the following constraints:

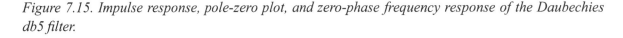

Figure 7.15. Impulse response, pole-zero plot, and zero-phase frequency response of the Daubechies db5 filter.

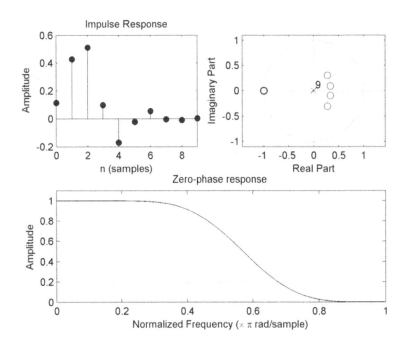

- The filter magnitude response has always peaks around the frequencies $(2k + 1)\pi/L$ for $k = 1, 2, \ldots$
- The filter has multiple stopbands of width $2\omega_p$ centred at the frequencies $k2\pi/L$ for $k = 1, 2, \ldots$
- The filter can satisfy *Case c* specifications, see Figure 3.5(c).

In the case $L = 2$, the filter satisfies the *Case b* specifications of Figure 3.5 (b). This filter is recognized under the name IIR halfband filter, which we consider in the next subsection.

In Chapter V, we have demonstrated the efficient polyphase IIR decimator and interpolator structures where the polyphase branches are allpass subfilters. Actually, IIR polyphase filters in Chapter V exhibit properties of the *L*th-band IIR filters. *Example 5.2* illustrates the above mentioned *L*th-band IIR filter properties.

An *L*th-band IIR filter can be designed for a nonlinear-phase, or for an approximately linear-phase characteristic. The design algorithms for both types are given in the papers of Renfors and Saramäki (1987).

HALFBAND IIR FILTERS

IIR halfband filters are very attractive because of very efficient implementations that can be achieved with this filter class. The implementations of decimators and interpolators based on the IIR halfband filters have been considered in Chapter V. Here, we concentrate on the general IIR halfband filter properties and on the transfer function design.

IIR Halfband Filter Properties

The magnitude response characteristics of an IIR halfband filter are those of a minimum-phase FIR halfband filter defined by (7.31) and (7.32). Therefore, an IIR halfband filter is power-symmetric around the central frequency $\omega_c = \pi/2$. It satisfies the frequency symmetry condition

$$\omega_p + \omega_s = \pi,$$
(7.36)

and according to the power-symmetry property, the passband and the stopband ripples are related by

$$\delta_p = 1 - \sqrt{1 - \delta_s^2} \quad \text{or} \quad a_p = 10\log_{10}\left(1 + \frac{1}{10^{a_s/10} - 1}\right).$$
(7.37)

In the middle of the band, at the frequency of $\omega_c = \pi/2$, the filter magnitude equals $1/\sqrt{2}$ corresponding to the attenuation of approximately 3 dB. A typical magnitude response of an IIR halfband filter is plotted in Figure 7.16.

The poles of an odd-order minimum-phase halfband filter are placed on the imaginary axis of the z-plane (Mitra, 2006; Schüssler and Steffen, 1998 and 2001), and the filter transfer function is expressible as the sum of two real all-pass functions $A_0(z)$ and $A_1(z)$,

$$H(z) = \frac{1}{2}\left(A_0(z) + z^{-1}A_1(z)\right)$$
(7.38)

where

Figure 7.16. Typical magnitude response of an IIR halfband filter

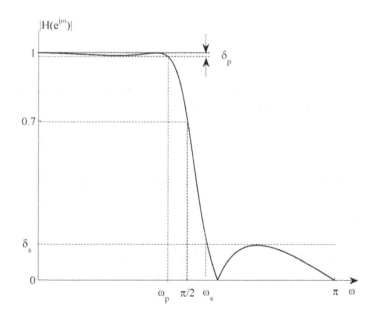

$$A_0(z) = \prod_{l=2,4,\ldots}^{(N+1)/2} \frac{\beta_l + z^{-2}}{1 + \beta_l z^{-2}} \quad \text{and} \quad A_1(z) = z^{-1} \prod_{l=3,5,\ldots}^{(N+1)/2} \frac{\beta_l + z^{-2}}{1 + \beta_l z^{-2}}. \tag{7.39}$$

The term z^{-1} in (7.39) represents the halfband filter first-order section, which implements the pole placed at the origin. Constants β_l represent the conjugate-complex pole pairs, and are computed as the squared modulus of the filter poles,

$$\beta_l = (r_l)^2, \quad \text{with} \quad \beta_l < \beta_{l+1}, \quad l = 2, 3, \ldots, (N+1)/2. \tag{7.40}$$

The most efficient implementation for the given pass- stopband specifications is achieved with the minimum-phase IIR halfband filter implemented with two all-pass subfilters according to (7.38) – (7.40).

IIR Halfband Filter Design in MATLAB

The IIR halfband filter symmetry conditions (7.36) and (7.37) can be met with Butterworth and elliptic filters, whereas Chebyshev and inverse Chebyshev filters cannot produce symmetric magnitude responses. In the next subsection we show the use of MATLAB in designing Butterworth and elliptic halfband filters.

Butterworth Halfband Filters

Digital lowpass Butterworth filters are characterized by a magnitude response that is maximally flat at the angular frequencies $\omega = 0$ and at $\omega = \pi$, and decreases monotonically from the maximum $|H(e^0)|$ = 1 to $|H(e^{j\pi})| = 0$. At the cutoff frequency ω_c, the magnitude response equals $1/\sqrt{2}$.

A Butterworth filter satisfies symmetry conditions (7.36) and (7.37) when choosing the cutoff frequency $\omega_c = 0.5\pi$.

The MATLAB function butter can be used to design Butterworth halfband filter. The Nth-order halfband filter with the cutoff frequency $\omega_c = 0.5\pi$ is obtained with the following code

```
[b,a] = butter(N,0.5);
```

Here, the row vector b returns the numerator coefficients, and the row vector a returns the denominator coefficients. For computing poles and zeros, one writes,

```
[z,p,k] = butter(N,0.5);
```

where the column vector z contains zeros, and the column vector p contains poles.

Elliptic Halfband Filters

Elliptic filters exhibit an equiripple magnitude characteristic in both the pass- and stopbands. In general, elliptic filters meet requested specifications with the lowest order of any filter type.

An elliptic filter is specified by passband and the stopband edge frequencies ω_p and ω_s, maximal attenuation in the passband a_p, and the minimal stopband attenuation a_s. The minimal filter order N is to be determined that guarantees the filter performance satisfying the design parameters. The filter order is an integer, and therefore, some design margin always exists between the specifications and the actual filter characteristics. This design margin can be distributed among the filter parameters in various ways (Lutovac, Tošić & Evans, 2000).

An elliptic halfband filter should exactly satisfy the symmetry conditions given by (7.36) and (7.37), and therefore, the standard computer programs, such as ellip in MATLAB, cannot be directly used for the halfband filter design. An adjustment of the input parameters to meet exactly the design constraints (7.36) and (7.37) is needed before using ellip (Milić & Lutovac, 2002). However, for the higher filter orders, ellip can return an error.

In (Milić & Lutovac, 2002 and 2003), the closed form expressions are developed for computing poles of an elliptic halfband filter for the given edge frequencies and the filter order N. The design procedure is the following:

1. Determine the selectivity factor ξ as a function of the passband or stopband edge frequency

$$\xi = \frac{1}{\tan^2\left(\omega_p/2\right)} \quad \text{or} \quad \xi = \tan^2\left(\omega_s/2\right) \tag{7.41}$$

2. Use the Jacoby elliptic sn function and complete elliptic integral of the first kind to compute the parameters x_l (Lutovac, Tošić & Evans, 2000),

$$x_l = \text{sn}\left(\frac{2l-1}{n}K_J\left(\frac{1}{\xi}\right),\left(\frac{1}{\xi}\right)\right), \quad l = 1,\dots,N. \tag{7.42}$$

For determining Jacobi elliptic sn function use ellipj in MATLAB,

xl = ellipj(((2*l-1)/N+1)*ellipke(1/ksi^2),1/ksi^2)

where the function ellipke computes the complete elliptic integral of the first kind, $K_J(1/\xi)$. Notice that the function ellipke uses the squared modulus $1/\xi^2$.

3. The squared pole magnitudes β_l are then computed from the expression

$$\beta_l = \frac{\xi + x_l^2 - \sqrt{(1-x_l^2)(\xi^2-x_l^2)}}{\xi + x_l^2 + \sqrt{(1-x_l^2)(\xi^2-x_l^2)}}, \quad l = 1,\dots,N. \tag{7.43}$$

4. The filter poles are purely imaginary, and are determined from β_l,

$$p_l = j\sqrt{\beta_l}, \quad p_{l+1} = -j\sqrt{\beta_l} \tag{7.44}$$

Based on the above procedure, the MATLAB program halfbandiir is developed, which is given in Appendix A. The input parameters for halfbandiir are the filter order N and the passband edge frequency $f_p < 0.5$. One writes

[b,a,z,p,k] = halfbandiir(N,fp);

The row vectors b and a return the numerator and denominator coefficients of the filter transfer function. The column vectors z and p return zeros and poles, and k is the gain.

When halfbandiir is used with the option minorder, the input parameters are the passband edge frequency f_p and the passband ripple δ_p,

[b,a,z,p,k] = halfbandiir('minorder', fp,deltap);

With this option program finds the minimal filter order N according to the input parameters and the corresponding stopband conditions computed from (7.36) and (7.37).

In the example that follows, we illustrate an elliptic halfband filter design with specified stopband characteristics.

Example 7.5

Design an elliptic halfband filter that is specified with the stopband edge frequency of $\omega_s = 0.57\pi$, and the minimal stopband attenuation $a_s = 45$ dB.

The design procedure is performed in three steps:

Step 1: Adjustment of the design parameters,

as = 45; ap = 10*log10(1+1/(10^(as/10)-1));
fs = 0.57; fp = 1-fs;

Step 2: Computation of the minimal filter order,

[N,fp] = ellipord(fp,fs,ap,as);

The ellipord returns $N = 7$, and the passband edge frequency $f_p = 0.43$, which satisfies the symmetry condition, $f_p = 1 - f_s = 1 - 0.57$.

Step 3: Computation of the halfband filter parameters

[b,a,z,p,k] = halfbandiir(N,fp);

The gain response of the resulting halfband filter is shown in Figure 7.17. The minimal stopband attenuation exceeds the requested value of 45 dB since N is an integer.

Figure 7.18 displays the zeros and poles of the 7th-order halfband filter with the indication how the poles are shared between the all-pass subfilters $A_0(z)$ and $A_1(z)$. When the distribution of the poles is determined, it is straightforward to compose $A_0(z)$ and $A_1(z)$ according to (7.39) and (7.40).

Regular Filters

Poles of the minimum-phase IIR halfband filter are placed on the imaginary axis of the z-plane whereas the transfer function zeros are located on the unit circle. When the filter order N is an odd number, at least one of the transfer function zeros is placed at $z = -1$.

Figure 7.17. Gain response of the 7th-order elliptic halfband filter of Example 7.5

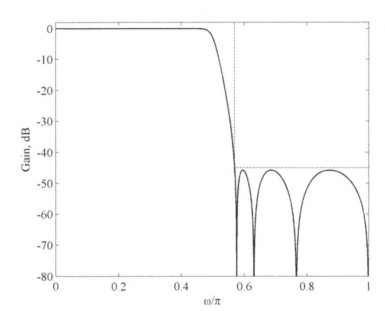

Generally, the transfer function of an odd-order low-pass halfband filter $H(z)$ can be expressed in the form

$$H(z) = z^{-1} \frac{C_0 \left(1 + z^{-1}\right)^K \prod_{l=1}^{\frac{N-K}{2}} \left(1 - a_l z^{-1} + z^{-2}\right)}{\prod_{l=1}^{\frac{N-1}{2}} \left(1 + \beta_l z^{-2}\right)}, \tag{7.45}$$

where N is the filter order, and K denotes the number of zeros at the point $z = -1$, N, K odd numbers.

The number of zeros at $z = -1 = e^{j\pi}$ is of particular importance to wavelets, since it corresponds to the number of vanishing moments of the wavelet and scaling basis functions. So, K zeros at $z = e^{j\pi}$ means K vanishing moments that are related to the filter regularity.

In the case of a Butterworth filter, all z-plane zeros are placed at $z = -1$, i.e., at that point the multiplicity of the transfer function zero is equal to the filter order N. An elliptic filter has a single zero at $z = -1$. Figure 7.19 illustrates the positions of the z-plane poles and zeros for the Butterworth and elliptic filters for $N = 5$. Hence, in the Butterworth filter case the regularity is maximal since $K = N$, and in the elliptic filter case, the regularity is minimal, $K = 1$.

With the multiplicity of zeros $K = N$ at the point $z = -1$, the Butterworth filter exhibits the maximally flat magnitude response at $\omega = \pi$. The elliptic filter is regular at $\omega = \pi$, but the multiplicity of zeros is minimal since $K = 1$. Between those two boundary solutions, $(N-3)/2$ intermediate transfer functions having different regularities can be generated. An Nth order halfband filter with selected $K \leq N$ can be designed to satisfy conditions specified by (7.36) and (7.37).

The design procedure derived by Zhang and Yoshikava (1999) can be used to generate the family of Nth order IIR filters having the desired number of zeros at $z = -1$, $1 \leq K \leq N$, K and N odd numbers.

Figure 7.18. Pole-zero plot for the example 7th-order elliptic halfband filter

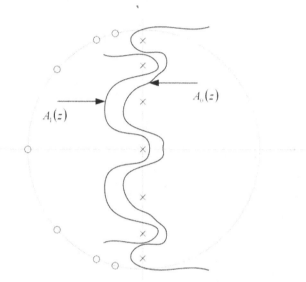

Figure 7.19. Zero-pole plot of the 5th-order Butterworth and the 5th-order elliptic filter

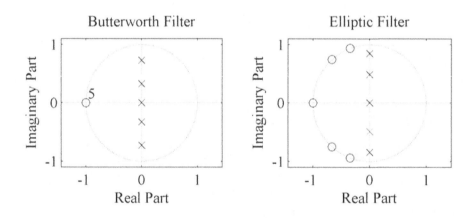

The procedure is performed on the basis of an eigenvalue solution and the Remez exchange algorithm. For $K < N$, the remaining degrees of freedom are used to provide equal ripple pass/stop-band characteristic with a minimal transition bandwidth. Figure 7.20 plots the example gain responses of 5th-order transfer functions: a Butterworth filter with $K=5$, an intermediate filter with $K=3$, and an elliptic filter with $K=1$.

Figure 7.21 illustrates the pole-zero plot for the example 5th-order intermediate filter: poles are placed on the imaginary axis, one pair of zeros is on the unit circle, while the location of the third order zero is the point $z = -1$.

As Figure 7.20 demonstrates, the higher filter regularity is achieved at the cost of decreased selectivity. Intermediate solutions between Butterworth and elliptic filters may be used as a compromise between the regularity and selectivity.

Figure 7.20. The 5th-order halfband filters with different regularities: elliptic filter K = 1, intermediate filter K = 3, and Butterworth filter K = 5

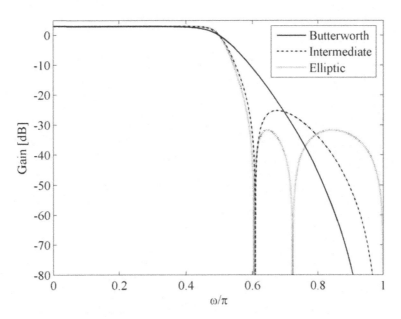

IIR HALFBAND FILTERS WITH APPROXIMATELY LINEAR PHASE

The main disadvantage of minimum-phase IIR filters is their very nonlinear phase response. The simultaneous optimization of the phase and amplitude responses of an IIR filter is a very difficult problem because the phase linearity and amplitude selectivity are conflicting requirements. An approximately linear-phase IIR filter is of interest when the delay of the liner-phase FIR counterpart cannot be tolerated. Solutions for the approximately linear phase IIR filters with a requested amplitude response have been published by several authors (Lawson & Wicks, 1992; Gerken, Schüssler & Steffen, 1995; Schüssler and Steffen, 2001; Surma-Aho and Saramäki, 1999).

A solution for the approximately linear-phase IIR halfband filter can be achieved when representing the transfer function $H(z)$ as a sum of two subfilters, one of them a pure delay term,

$$H(z) = \frac{1}{2}\left(A_0(z) + z^{-K_1}\right),\tag{7.46}$$

whereas the subfilter $A_0(z)$ is an all-pass expressible as a product of the second-order and the fourth-order sections,

$$A_0(z) = \prod_{k=1}^{K_{0,1}} \frac{a_k + z^{-2}}{1 + a_k z^{-2}} \prod_{k=1}^{K_{0,2}} \frac{c_k + b_k z^{-2} + z^{-4}}{1 + b_k z^{-2} + c_k z^{-4}}.\tag{7.47}$$

In an iterative procedure, the coefficients of $A_0(z)$ are modified gradually step-by-step in order to achieve the best approximation for the following criteria,

$$e^{j\omega K_1} A_0\left(e^{j\omega}\right) = \begin{cases} 1, & 0 \leq |\omega| \leq \omega_p \\ -1, & \pi - \omega_p \leq \omega \leq \pi \end{cases}.\tag{7.48}$$

Figure 7.21. Pole-zero plot for the intermediate filter, K = 3

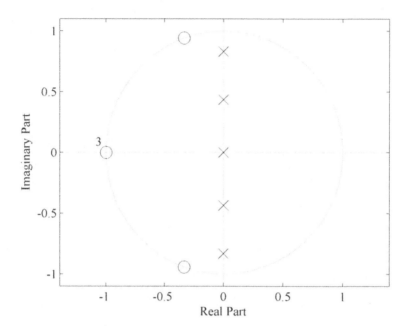

Here, the phase response of the all-pass function $A_0(z)$ is forced to approximate the straight line of the slope $-K_1$. The iterative procedure optimizes the coefficient values $\{a_k, b_k, c_k\}$ in order to achieve the approximation of (7.48) in an equiripple or in a maximally-flat manner. With this approach, one approximates simultaneously the amplitude and phase responses in the passband. From the IIR halfband filter symmetry condition (7.37), which relates the passband and stopband ripples, the stopband attenuation is consequently obtained.

One observes that the poles of $H(z)$ should be located either on the imaginary axis, or they are located symmetrically with respect to the imaginary axis, see equation (7.47). The example that follows illustrates the typical properties of an approximately linear-phase IIR halfband filter.

Example 7.6

The approximately linear-phase IIR halfband filter is specified by the following parameters:

Given the passband/stopband edge frequencies: $\omega_p = 0.4\pi$, $\omega_s = \pi - \omega_s = 0.6\pi$.

The all-pass branch $A_0(z)$ is of degree $K_0 = 10$, and degree of the delay branch is $K_1 = 9$.

The free MATLAB programs written by Schüssler and Steffen (2001) can be used to design an approximately linear-phase IIR halfband filter specified above.

Figure 7.22 displays the resulting filter characteristics: gain response, pole-zero plot, phase response and the group delay. In the transition band only, one observes the significant phase nonlinearity and peak in the group delay characteristic. The passband ripple is very small

The allpass branch $A_0(z)$ being of the 10th-order, consists of one second-order section and two fourth-order sections, where every second coefficient is zero valued, see equation (7.47). Therefore, only 5 multiplication constants are needed to implement the all-pass branch $A_0(z)$. Since the delay branch is a pure delay term, the overall filter can be implemented with only 5 multiplication constants. The delay of the filter is 9 samples.

The linear-phase FIR halfband filter of the length $N = 27$ is requested to achieve the stoppband attenuation and the transition band of the IIR filter from Figure 7.20. This filter introduces a delay of $(N–1)/2 = 13$ samples instead of 9 in the IIR filter case. When exploiting the coefficient symmetry, 7 multiplication constants are needed to implement an FIR halfband filter of the length $N = 27$, whereas the IIR halfband filter of *Example 7.5* requires 5 constants.

It is recommended to the reader interested in the design of approximately linear-phase IIR halfband filters to consider the references mentioned in the first paragraph of this subsection. Notice that the paper by Schüssler and Steffen (2001) is accompanied with free MATLAB programs for approximate linear-phase halfband filter design available at

http://www-nt.e-technik.uni-erlangen.de/~hws/programs/halfbandfilters/

An alternative solution for the approximately linear phase IIR halfband filter may be achieved by using the noncausal implementation of a minimum phase elliptic filter (Lutovac and Milić, 2000). When representing the transfer function of the noncausal IIR filter $H_n(z)$ in the form,

$$H_n(z) = \frac{1}{2}\left(1 + z^{-1} A_0\left(z^{-1}\right) A_1(z)\right),$$ (7.49)

where $A_0(z)$ and $A_1(z)$ are the all-passes of a minimum-phase IIR filter $H(z)$, as given in (7.38). The magnitude response of $H_n(z)$ is identical to that of $H(z)$. Moreover, if $H(z)$ is an elliptic filter, the phase response of $H_n(z)$ exhibits the equiripple magnitude and phase characteristic in the passband.

The causal implementation of $H_n(z)$ suitable for real-time implementation is shown in (Milić & Lutovac, 2000). The solution is based on the use of double filtering with the block processing technique according to the method of Powel and Chau (1991).

Figure 7.22. Characteristics of the approximately linear- phase IIR halfband filter

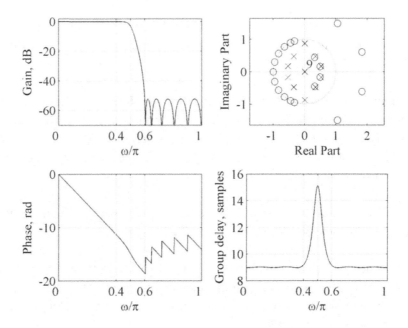

MATLAB EXERCISES

7.1 Use the MATLAB function firnyquist to design linear-phase FIR Lth-band filters of the length N =31, with $L = 3$ and with the roll-off factors: ρ = 0.2, 0.4, and 0.6. Plot the impulse responses and the magnitude responses for all designs.

7.2 Use the Lth-band filters from the MATLAB Exercise 7.1 for interpolation. Generate the original signal $\{x[n]\}$ according to your own choice, and then interpolate this signal by-the-factor-of- 3. Perform the interpolation using three distinct interpolation filters from Exercise 7.1. Compare and comment on the results.

7.3 Design a linear-phase Lth-band FIR filter with the following parameters: $N = 29$, $L =4$, ρ = 0.2. Construct the polyphase factor-of-4 decimator using the efficient configuration of Figure 4.8. Which one of four polyphase subfilters reduces to the constant? Generate a signal to your own choice and decimate this signal by $M = 4$.
Plot the impulse response and the magnitude response of the filter. Plot the signal spectrum before and after decimation.

7.4 Design the minim-phase and the maximum-phase Lth-band FIR filters with the following design parameters: N_{ord} = 17, $L = 4$, and ρ = 0.2. Plot magnitude, phase, and group delay responses for both filters. Compose the linear-phase filter by convolving the impulse responses of the minimum-phase and the maximum phase filters. Plot the impulse response of the linear-phase filter and verify the zero-crossing positions. Comment on the results.

7.5 Design an equiripple linear-phase halfband filter of the length $N = 31$, and the passband edge frequency at ω_p = 0.45π. Compute and plot the impulse response, magnitude response and pole-zero locations in the z-plane.

7.6 Develop the efficient interpolator structure based on the linear-phase FIR halfband filter. Modify program demo_7_1 to perform the decimation-by-2. For the input signal use the decimated signal obtained in demo_7_1.

7.7 Design the minimum-phase and maximum phase halfband FIR filters of the order N_{ord} = 25. Compute and plot the magnitude response and the zero-pole plot for both filters. Compute the impulse response of the linear-phase Nyquist filter by convolving the impulse responses of the minimum-phase and the maximum-phase filters. Compute and plot the magnitude response and the zero-pole plot of the linear-phase Nyquist filter. Comment on the results.

7.8 Use the MATLAB function dbwavf from the *Wavelet Toolbox* to design Daubeschies filter db7. Plot the impulse response, zero-pole plot, magnitude and phase responses.

7.9 Design the 9th-order Butterworth halfband filter. Compute and plot the magnitude response and display the pole-zero plot of the filter. Consider the implementation structure based on the parallel connection of two allpass subfilters $A_0(z)$ and $A_1(z)$ according to (7.38) – (7.40). Determine the constants in the second-order sections of $A_0(z)$ and $A_1(z)$.

7.9 Use the program hallfbandiir given in Appendix A to design the 9th-order elliptic halfband filter. Compute and plot the magnitude response and display the pole-zero plot of the filter. Consider the implementation structure based on the parallel connection of two allpass subfilters $A_0(z)$ and $A_1(z)$ according to (7.38) – (7.40). Determine the constants in the second-order sections of $A_0(z)$ and $A_1(z)$.

REFERENCES

Ansari,R., & Liu,B., (1993). Multirate signal processing. In Sanjit K. Mitra and James F. Kaiser (ed.), *Handbook for Digital Signal Processing*. New York, NY: John Wiley-Interscience, 981-1084.

Crochiere, R.E., & Rabiner, L.R., (1981, March). Interpolation and decimation of digital signals - A Tutorial Review. *Proceedings of the IEEE, 69*(3), 300-331.

Crochiere, R.E., & Rabiner, L.R. (1983). *Multirate digital signal processing*. Englewood Cliffs, NJ: Prentice-Hall.

Filter design toolbox for use with MATLAB. User's guide. Version 6. (2006). Natick: MathWorks.

Fliege, N. J. (1994). *Multirate digital signal processing*. New York, NY: John Wiley & Sons.

Gerken, M., Schüssler, H.W, & Steffen, P. (1995). On the design of recursive digital filters consisting of parallel connection of all-pass sections and delay elements. *AEÜ*, 49, 1-11.

Harris, F. J. (2004). *Multirate signal processing for communication systems*. Upper Saddle River, NJ: Prentice Hall PTR.

Hermann, O. (1971). On the approximation problem in nonrecursive digital filter design. *IEEE Transactions on Circuit Theory*, CT-*18*(3), 411-413.

Lawson, S. S., & Wicks, T. (1992, October). Design of efficient digital filters satisfying arbitrary loss and delay specifications. *Proc. Inst. Elect. Eng., Part G, Circuits, Devices and Systems, 139*(5), 611-620.

Lutovac, M. and Milić, L. (1996). Elliptic halfband filters. *Facta Universitatis, Series Electronics and Energetics, 9*(1), 43-59.

Lutovac, M. D., & Milić, L. D. (2000). Approximate linear phase multiplierless IIR halfband filter. *IEEE Signal Processing Letters, 7*(3), 52-53.

Lutovac, M. D., Tošić, D. V., & Evans, B. L. (2001). *Filter Design for Signal Processing Using MATLAB and Mathematica*, Upper Saddle River, NJ, Prentice Hall.

McClellan, J. H., Parks, T. W., & Rabiner, L. R. (1973). A computer program for designing optimum FIR linear-phase digital filters, *IEEE Transactions on Audio Electroacoustics, 21*(6), 506-526.

Milić, Lj., & Lutovac, M.D. (2002). Efficient multirate filtering. In Gordana Jovanović-Doleček, (ed.), *Multirate Systems: Design & Applications*, Hershey, PA: Idea Group Publishing, 105-142.

Milić, L. D., & Lutovac, M.D. (2003). Efficient algorithm for the design of high-speed elliptic IIR filters. *AEÜ Int. J. Electron. Commun, 57*(4), 255-262.

Mintzer, F. (1982). On halfband, third-band, and nth-band FIR filters and their design. *IEEE Transactions on Acoustics, Speech, and Signal Processing, 30*(5), 734-738.

Mitra, S. K. (2006). *Digital signal processing: A computer based approach.* 3rd edition. New York, NY: The McGraw-Hill Companies, Inc.

Orchard, H. J., & Willson, A. N. (2003). On the computation of a minimum- phase spectral factor. *IEEE Trans. Circuits and Systems-I: Fundamental Theory and Application, 50*(3), 365–375.

Powell, S., & Chau, M. (1991). A technique for realizing linear phase. *IIR filters. IEEE Transactions on Signal Processing, 39*(11), 2425-2435.

Proakis J. G., & Manolakis D.G. (1996). *Digital signal processing: Principles, algorithms, and applications.* London: Prentice Hall.

Renfors, M., & Saramäki, T. (1987). Recursive Nth-band digital filters-part I: design and properties. *IEEE Transactions on Circuits and Systems, 34*(1), 24-39.

Renfors, M., & Saramäki, T. (1987). Recursive nth-band digital filters-part II: design of multistage desimators and interpolators. *IEEE Transactions on Circuits and Systems, 34*(1), 40-51.

Samadi, S., & Nishikara, A. (2007). The word of flatness. *IEEE Circuits and Systems Magazine, 7*(3), 38-44.

Saramäki, T., & Neuvo, Y. (1987). A class of FIR (Nth-band) Nyquist filters with zero intersymbol interference. *IEEE Transactions on Circuits and Systems, 34*(10), 1182-1190.

Saramäki, T. (1993). Finite impulse response filter design., Chapter 4 in *Handbook for Digital Signal Processing.* Edited by S. K. Mitra and J. F. Kaiser, New York, NY: John Wiley Interscience, 155 – 277.

Saramäki, T. *Multirate Signal Processing.* (2001). Lecture notes for a graduate course, the Institute of Signal Processing, Tampere University of Technology, Finland.

Saramäki, T., & Renfors, M. (1998). *N*th band filter design. *Proc. of the IX European Signal Processing Conference, EUSIPCO.* Rhodes, Greece, 1943-1947.

Signal processing toolbox for use with MATLAB. User's guide. Version 6. (2006). Natick: MathWorks.

Schüssler, H. W., & Stefen, P. (1998). Halfband filters and Hilbert transformers. *Circuits Systems Signal Processing, 17*(2), 137–164.

Schüssler, H.W., & Steffen, P. (2001). Recursive halfband-filters. *AEÜ Int. J. Electron. Commun, 55*(6), 377-388.

Selesnick, I.W., & Burrus, C.S. (1996). Generalized digital Butterworth filter design. *Proceedings of the IEEE Int. Conf. Acoust., Speech, Signal Processing,* 1367-1370 vol. 3.

Selesnick, I.W., & Burrus, C.S. (1998). Maximally flat lowpass FIR filters with reduces delay. *IEEE Transactions on Circuits and Systems-II Analog and Digital Signal Processing, 45*(1), 53-68.

Surma-Aho, K, & Saramäki, T., (1999). A systematic technique for designing approximately linear phase recursive digital filters. *IEEE Transactions on Circuits and Systems – II Analog and Digital Signal Processing, 46*(7), 956-963.

Vaidyanathan, P.P., & Nguyen, T.Q. (1987, March). A trick for the design of FIR halfband FIR filters. *IEEE Transactions on Circuits and Systems, 34*(3), 297-300.

Vaidyanathan, P.P., (1990). Multirate digital filters, filter banks, polyphase networks, and applications: A Tutorial. *Proceedings of the IEEE, 78*(1), 56-93.

Vaidyanathan, P.P., (1993). *Multirate systems and filter banks*. Englewood Cliffs, NJ: Prentice Hall.

Wegener, W. (1979). Wave digital directional filters with reduced number of multipliers and adders. *Arch. Elec. Übertragung, 33*(6), 239-243.

Zhang, X. & Yoshikava, T. (1999). Design of orthonormal wavelet filter banks using allpass filters. *Elsevier Science Signal Processing, 78*(1), 91-100.

Chapter VIII
Complementary Filter Pairs

INTRODUCTION

Digital filters with complementary characteristics find many applications in practice. In this chapter, we concentrate on the properties and construction of complementary filters and filter pairs.

An important application of complementary property is deriving a new transfer function from the existing one. A highpass filter can be obtained as a complement of the lowpass filter, and also a bandstop filter can be considered as a complement of the bandpass filter.

Complementary lowpass/highpass and bandpass/bandstop filter pairs are popular because of very attractive implementations. Namely, the two complementary filters in the pair are implemented at the cost of a single one.

Complementary filter pairs, usually lowpass/highpass filter pairs, are widely used whenever there is a need to split the signal into two adjacent subbands and reconstruct it after some processing performed in the subbands. They are used as basic building blocks in constructing analysis and synthesis multi-channel filter banks. Moreover, the complementary filter pairs are used in constructing low sensitivity complex filtering structures. In some applications, such as signal analysis, the complementary filter pairs are used to separate a signal into two bands, and the filtered signals are processed without need to reconstruct it. Another application is digital audio where the signal is separated into two (three) bands resulting in the signals that are feed inside two (three) loudspeakers.

In this chapter, at the beginning we introduce the basic definitions of the complementary properties that will be used through the chapter. We use then the complementary properties to construct FIR and IIR highpass filters from the existing lowpass filters. In the sequel, we consider the analysis and synthesis filter pairs. We present the design and efficient implementations of FIR and IIR complementary filter pairs. Chapter concludes with MATLAB Exercises for individual study.

DEFINITIONS OF COMPLEMENTARY DIGITAL FILTER PAIRS

This section summarizes the definitions of the five complementary relations in a manner similar to those described in (Mitra, 2006) and in (Vaidyanathan, 1993). In addition, the definitions of the angular crossover frequencies for complementary filter pairs are given for later use.

For the definition purposes, consider a digital filter pair, denoted by $[H(z), H_C(z)]$, where $H(z)$ and $H_C(z)$ are the transfer functions of these filters.

Delay-Complementary Filter Pairs

A filter pair $[H(z), H_C(z)]$ is said to be *delay-complementary*, or *strictly-complementary* if the transfer functions $H(z)$ and $H_C(z)$ add up to a delay

$$H(z) + H_C(z) = Kz^{-n_0}.$$ (8.1)

Here, $K > 0$ is a constant, typically $K = 1$; and n_0 is a nonnegative integer.

All-Pass Complementary Filter Pairs

A filter pair $[H(z), H_C(z)]$ is an *all-pass-complementary* filter pair if the sum of $H(z)$ and $H_C(z)$ satisfies

$$H(z) + H_C(z) = A(z).$$ (8.2)

where $A(z)$ is an all-pass transfer function.

Power-Complementary Filter Pairs

A filter pair $[H(z), H_C(z)]$ is a *power-complementary* filter pair if the sum of the squares of their magnitude responses satisfies

$$\left| H\left(e^{j\omega}\right) \right|^2 + \left| H_C\left(e^{j\omega}\right) \right|^2 = K$$ (8.3)

where $K > 0$ is a constant, typically $K = 1$. For this pair, the angular frequency $\omega = \omega_c$, where

$$\left| H\left(e^{j\omega_c}\right) \right|^2 = \left| H_C\left(e^{j\omega_c}\right) \right|^2 = K/2$$ (8.4)

is the *crossover angular frequency*. At this angular frequency, the gain responses of both filters are approximately 3 dB below their maximum values.

Magnitude-Complementary Filter Pairs

A filter pair $[H(z), H_C(z)]$ is *magnitude-complementary* if the sum of their magnitude responses satisfies

$$\left|H\left(e^{j\omega}\right)\right| + \left|H_C\left(e^{j\omega}\right)\right| = K, \tag{8.5}$$

where $K > 0$ is a constant, usually $K = 1$. For this pair, the angular frequency $\omega = \omega_c$, where

$$\left|H\left(e^{j\omega_c}\right)\right| = \left|H_C\left(e^{j\omega_c}\right)\right| = K/2 \tag{8.6}$$

is the *crossover angular frequency*. At this angular frequency, the gain responses of both filters are approximately 6 dB below their maximum values.

Double-Complementary Filter Pairs

A filter pair satisfying two complementary properties is known as a *double-complementary* filter pair. There exist solutions for FIR filters satisfying delay complementary property and magnitude-complementary property, and solutions for IIR filters that are all-pass complementary and power-complementary, or all-pass complementary and magnitude-complementary filter pairs.

CONSTRUCTING HIGHPASS FIR AND IIR FILTERS

Generating Linear-Phase Highpass FIR Filter

Let $H_{LP}(z)$ be a linear-phase lowpass FIR filter of the length N, where N is an odd number. Applying the delay-complementary property defined in (8.1) for $K = 1$, we directly obtain the transfer function of the complementary highpass filter $H_{HP}(z)$,

$$H_{HP}\left(z\right) = z^{-(N-1)/2} - H_{LP}\left(z\right) \tag{8.7}$$

The corresponding implementation structure is shown in Figure 8.1.

There are several options in MATLAB how to perform the complementing operation (8.7), and thus generate the highpass filter $H_{HP}(z)$ from a given linear-phase filter $H_{LP}(z)$. Following the structure of Figure 8.1, one can write,

```
Nord = 24;                                        % Filter order, an even number
hlp = firgr(24,[0 .2 .25 1],[1 1 0 0],[1 1.5]);   % Lowpass filter design
D = Nord/2;                                        % Propagation delay
hhp = zeros(size(1:length(hlp)));  hhp(D+1) = 1;   % Setting the delay path
hhp = hhp - hlp;                                   % Complementary operation
```

Alternatively, one can use the MATLAB function firlp2hp to compute coefficients of the complementary highpass filter,

```
hhp = firlp2hp(hlp,'wide');
```

Figure 8.1. Implementation structure for highpass complementary filter

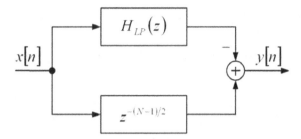

Figure 8.2 shows the impulse responses and the zero-phase frequency responses for the initial lowpass filter and for the corresponding complementary highpass filter computed above. Notice that the plots in Figure 8.2 are accomplished with the use of the MATLAB functions impz for the impulse response, and zerophase for the zero-phase frequency response.

Besides the lowpaass/highpass transformation described above, the structure of equation (8.7) and Figure 8.1 can be applied for other types of transfer functions. Using the same delay-complementary property, a linear-phase stopband filter can be derived from the linear-phase bandpass filter.

Generating Highpass IIR Filter

A highpass IIR filter can be directly derived from a lowpass IIR filter when the transfer function of the lowpass filter is expressible as a sum of two stable all-pass transfer functions.

Figure 8.2. Impulse responses and zero-phase frequency responses for the initial lowpass filter and the complementary highpass filter: Left side – lowpass filter. Right side – highpass filter

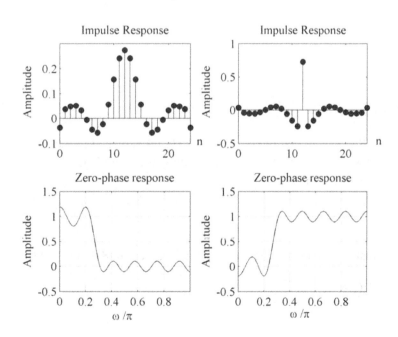

Let us express the transfer function of the lowpass filter $H_{LP}(z)$ in the form

$$H_{LP}(z) = \frac{1}{2}\left(A_0(z) + A_1(z)\right),$$
(8.8)

where $A_0(z)$ and $A_1(z)$ are stable all-pass functions. Since $A_0(z)$ and $A_1(z)$ are two all-passes, their frequency responses $A_0(e^{j\omega})$ and $A_1(e^{j\omega})$ are given by

$$A_0\left(e^{j\omega}\right) = e^{j\phi_0(\omega)} \quad \text{and} \quad A_1\left(e^{j\omega}\right) = e^{j\phi_1(\omega)},$$
(8.9)

where $\varphi_0(\omega)$ and $\varphi_1(\omega)$ are the phase responses of $A_0(z)$ and $A_1(z)$, respectively. Following (8.8) and (8.9), the frequency response of the lowpass filter $H_{LP}(e^{j\omega})$ is expressible as

$$H_{LP}\left(e^{j\omega}\right) = \frac{1}{2}\left(e^{j\phi_0(\omega)} + e^{j\phi_1(\omega)}\right)$$
(8.10)

and consequently, $H_{LP}(e^{j\omega})$ becomes the product of a real cosine function and the complex exponential factor,

$$H_{LP}\left(e^{j\omega}\right) = \cos\left(\left(\phi_0(\omega) - \phi_1(\omega)\right)/2\right)e^{j\left(\phi_0(\omega)+\phi_1(\omega)\right)/2}.$$
(8.11)

The cosine function represents the amplitude response of the filter, and the exponential factor represents the phase. The phase difference, $\phi_0(\omega) - \phi_1(\omega)$ is the essential function which determines the behaviour of the filter amplitude response. When in some range of frequencies the phase difference $\phi_0(\omega) - \phi_1(\omega)$ approximates zero, the cosine function in (8.11) approximates unity, and thereby, the filter exhibits a passband characteristic in that range of frequencies. On the contrary, the stopband occurs in the frequency range where the phase difference $\phi_0(\omega) - \phi_1(\omega)$ approximates π.

From the above discussion and from equations (8.10) and (8.11), one observes that the complementary highpass filter is obtained when the summation in (8.10) is replaced with the subtraction,

$$H_{HP}\left(e^{j\omega}\right) = \frac{1}{2}\left(e^{j\phi_0(\omega)} - e^{j\phi_1(\omega)}\right),$$
(8.12)

resulting in

$$H_{HP}\left(e^{j\omega}\right) = \sin\left(\left(\phi_0(\omega) - \phi_1(\omega)\right)/2\right)e^{j\left(\phi_0(\omega)+\phi_1(\omega)\right)/2}.$$
(8.13)

Equations (8.11) and (8.13) show that the passband of $H_{HP}(z)$ is placed in the stopband range of $H_{LP}(z)$ and vice versa. Therefore, the transfer function of the highpass filter $H_{HP}(z)$ is expressible as a difference of the all-pass functions $A_0(z)$ and $A_1(z)$,

$$H_{HP}(z) = \frac{1}{2}\left(A_0(z) - A_1(z)\right)$$
(8.14)

By adding up the transfer functions $H_{LP}(z)$ and $H_{HP}(z)$, as given in (8.8) and (8.14), one obtains

$$H_{LP}(z) + H_{HP}(z) = A_0(z).$$
(8.15)

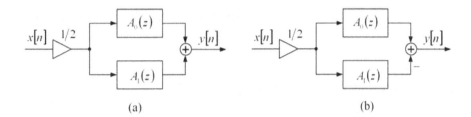

Figure 8.3. All-pass implementation of lowpass and higpass IIR filters. (a) Lowpass filter. (b) Highpass filter.

Comparing (8.15) with the definition of the all-pass complementary property given in (8.2), we verify that the filters $H_{LP}(z)$ and $H_{HP}(z)$ when implemented as a sum and difference of two all-pass subfilters are all-pass complementary transfer functions. The implementation structures for $H_{LP}(z)$ and $H_{HP}(z)$ are indicated in Figure 8.3.

The frequency responses of $H_{LP}(z)$ and $H_{HP}(z)$ satisfy also the power-complementary property defined in (8.3). This can be seen when introducing expressions (8.11) and (8.13) for $H_{LP}(e^{j\omega})$ and $H_{HP}(e^{j\omega})$ in the definition formula (8.3). With this replacement, we apparently obtain that the squared magnitude responses of $H_{LP}(z)$ and $H_{HP}(z)$ add-up to unity, i.e.

$$\left| H_{LP}\left(e^{j\omega}\right)\right|^2 + \left| H_{HP}\left(e^{j\omega}\right)\right|^2 = 1. \tag{8.16}$$

Therefore, we say that IIR filters $H_{LP}(z)$ and $H_{HP}(z)$ are double-complementary filters.

In MATLAB, we can simply generate the transfer function of a highpass filter, which is complementary to the given lowpass filter with the use of the function iirpowcomp from the Filter Design Toolbox. In the MATLAB code that follows, we first generate the 7th order lowpass Chebyshev filter $H_{LP}(z) = B_{LP}(z)/A_{LP}(z)$ specified by the cutoff frequency at $\omega_p = 0.4\pi$ and the passband ripple $a_p = 0.5$ dB,

```
[Blp,Alp] = cheby1(7,.5,.4);
```

In the next step, we use iirpowcomp to compute the coefficients of the power-complementary highpass filter $H_{HP}(z) = B_{HP}(z)/A_{HP}(z)$.

```
[Bhp,Ahp] = iirpowcomp(Blp,Alp);
```

Alternatively, the generation of $H_{HP}(z)$ can be accomplished in a straightforward manner by making use of the all-pass representations of $H_{LP}(z)$ and $H_{HP}(z)$ as given in (8.8) and (8.14), respectively. What is needed here is to provide the decomposition of the transfer function $H_{LP}(z)$ into two all-pass functions according to (8.8). The MATLAB function tf2ca performs this operation. With the code

```
[D0,D1] = tf2ca(Blp,Alp);
```

we obtain denominator coefficients of $A_0(z)$ and $A_1(z)$ in vectors D0 and D1, respectively.

Since the nominators' coefficients of $A_0(z)$ and $A_1(z)$ are obtained by reversing D0 and D1, the desired transfer function of the highpass filter $H_{HP}(z)$ follows directly from (8.14).

For the 7th order Chebyshev filter specified above, Figure 8.4 displays the impulse responses and the squared magnitude responses for $H_{LP}(z)$ and $H_{HP}(z)$. Since the initial lowpass filter $H_{LP}(z)$ is a Chebyshev filter, the complementary highpass filter $H_{HP}(z)$ is the inverse Chebyshev filter. The power-complementary property is particularly visible when comparing the passband ripple of the lowpass filter with the stopband ripple of the highpass filter.

ANALYSIS AND SYNTHESIS FILTER PAIRS

An *analysis filter pair* is defined as a single-input/two-output device consisting of two filters [$H_0(z)$, $H_1(z)$] connected in parallel as indicated in Figure 8.5 (a). Usually, the filter pair [$H_0(z)$, $H_1(z)$] is a lowpass/highpass filter pair, which satisfies some of the complementary properties. The analysis filter pair is used to separate the input signal $x[n]$ into two channel signals $x_0[n]$ and $x_1[n]$ for some processing purposes, such as the subband coding, and for many other signal processing applications.

On the contrary, the role of a synthesis filter pair is to combine two distinct signals into a single composite signal. The *synthesis filter pair* is a two-input/single-output device as shown in Figure 8.5 (b).

The filter pairs are usually called *two-channel filter banks*, or *two-band filter banks*.

In the analysis bank, the input signal $x[n]$ is separated into two channel signals $x_0[n]$ and $x_1[n]$ by the use of the filter pair [$H_0(z)$, $H_1(z)$], Figure 8.5(a). The z-transforms of the two channel signals are expressible as,

$$X_k(z) = H_k(z)X(z), \quad k = 0, 1.$$ (8.17)

Figure 8.4. Impulse responses and squared magnitude responses of the lowpass filter and complementary highpass filter. Left side – lowpass filter. Right side – highpass filter.

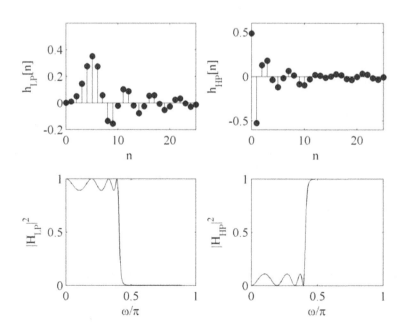

Figure 8.5. (a) Analysis filter pair. (b) Synthesis filter pair

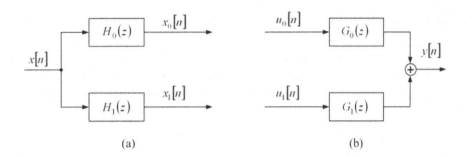

(a)   (b)

In the synthesis bank, the two channel signals $u_0[n]$ and $u_1[n]$ are filtered with the filter pair $[G_0(z)$, $G_1(z)]$, and then added together to give the output signal $y[n]$, Figure 8.5(b). The z-transform of the output signal is given by

$$Y(z) = G_0(z)U_0(z) + G_1(z)U_1(z).$$ (8.19)

In the majority of applications, $H_0(z)$ and $H_1(z)$ $[G_0(z)$ and $G_1(z)]$ are the lowpass and highpass halfband filters dividing the baseband of the signal into two equal subbands. Thereby, on the analysis side the lowpass and highpass signals $x_0[n]$ and $x_1[n]$ can be downsampled-by-2 and each of the down-sampled signals can be processed at the sampling rate which is a half of the input sampling rate. At the synthesis side, the two input signals $w_0[n]$ and $w_1[n]$ can be upsampled-by-2, filtered and added together to compose the output signal. Figure 8.6 shows the analysis and synthesis filter pairs with the down-sampling and up-sampling operations.

Down-sampling of the lowpass and highpass signals produces aliasing. The effects of aliasing produced in the analysis bank, can be eliminated in the synthesis bank. The elimination of aliasing is achieved by the proper choice of the transfer functions $H_0(z)$, $H_1(z)$, $G_0(z)$ and $G_1(z)$. In Chapter XII of this book, the performances of two-channel filter banks and their role in signal decomposition and reconstruction will be discussed in more details.

Figure 8.6. (a) Analysis filter pair with down-sampling. (b) Synthesis filter pair with up-sampling.

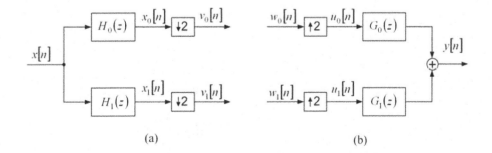

(a)   (b)

FIR COMPLEMENTARY FILTER PAIRS

Delay-Complementary FIR Filter Pairs

In the subsection entitled Generating Linear-Phase Highpass FIR Filter, we have shown how a linear-phase FIR highpass filter $H_{HP}(z)$ is generated by applying the delay-complementary property to the existing linear-phase lowpass filter $H_{LP}(z)$. Hence, those two filters make the delay-complementary filter pair satisfying

$$H_{HP}(z) + H_{LP}(z) = z^{-(N-1)/2}.$$ (8.20)

Here, the filter length N should be an odd number. Due to the complementary property, the filter pair $[H_{LP}(z), H_{HP}(z)]$ can be efficiently implemented at the cost of a single filter as shown in Figure 8.7(a).

Figure 8.7 (b) shows the polyphase implementation structure of a lowpass/highpass complementary filter pair when $[H_{LP}(z), H_{HP}(z)]$ is an FIR halfband filter pair. In the halfband filter case, all odd-indexed coefficients except for the central coefficient are zero valued, see Chapter VII, subsection entitled Efficient Implementation of Linear-Phase Halfband Filters. The polyphase component $E_0(z)$ in Figure 8.7(b) contains the nonzero coefficients of $H_{LP}(z)$ except for the central coefficient, whereas the cascade implementing $0.5z^{-(N-1)/2}$ presents the polyphase component $E_1(z)$.

Figure 8.8 shows the example characteristics of the delay-complementary filter pair where $[H_{HP}(z), H_{LP}(z)]$ are halfband filters. For illustration purposes, we choose the filters of a small length, $N = 15$, to enhance the passband and stopband ripples of the halfband complementary filters. The lowpass filter of Figure 8.8 is obtained with the use of the function firhalfband with the following arguments,

```
hlp = firhalfband(14,0.4);  % Computing the halfband filter coefficients
```

Figure 8.7. Efficient implementation of delay-complementary lowpass/highpass FIR filter pair $[H_{LP}(z), H_{HP}(z)]$. (a) Realization structure that follows directly from equation (8.20). (b) Efficient realization structure for halfband filter pair.

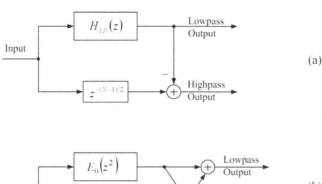

and highpass filter follows simply by using

hhp = firlp2hp(hlp,'wide');

The impulse responses and the zero-phase frequency responses for the delay-complementary filter pair are given in Figure 8.8. One can observe that the sum of the zero-phase frequency responses is unity for all frequencies.

Recall that the delay-complementary filter pairs are also called the *strictly-complementary filter pairs*. They are used as building blocks in constructing complex filters with very narrow transition bands when the solutions based on a single filter are of a very high order and therefore, very difficult, if not impossible, to implement. Chapter 9 considers the application of multirate techniques and complementary filter pairs in constructing filters with narrow transition bands. Moreover, the delay-complementary (strictly-complementary) filter pairs are desirable components in various signal processing applications. One interesting example is the digital television receiver, see (Mitra, 2006).

Delay-Complementary and Magnitude-Complementary FIR Filter Pairs

The delay-complementary lowpass/highpass filter pair can exhibit also the magnitude-complementary properties defined by equations (8.5) and (8.6) when the transfer functions $H_{LP}(z)$ and $H_{HP}(z)$ are designed to meet the following criteria

$$1-\delta \leq \left|H_{LP}\left(e^{j\omega}\right)\right| \leq 1 \quad \text{and} \quad \left|H_{HP}\left(e^{j\omega}\right)\right| \leq \delta \quad \text{for} \quad \omega \in \left[0,\omega_p\right] \tag{8.21a}$$

Figure 8.8. Impulse responses and the zero-phase frequency responses of the delay-complementary halfband filter pair

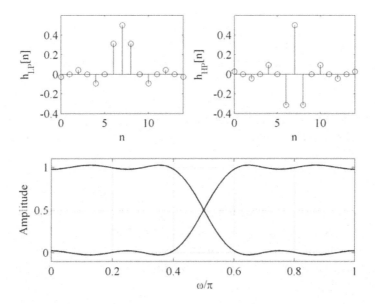

$$1-\delta \leq \left| H_{HP}\left(e^{j\omega}\right)\right| \leq 1 \quad \text{and} \quad \left| H_{LP}\left(e^{j\omega}\right)\right| \leq \delta \quad \text{for} \quad \omega \in \left[\omega_s, \pi\right] \tag{8.21b}$$

$$\left| H_{LP}\left(e^{j\omega}\right)\right| = \left| H_{HP}\left(e^{j\omega}\right)\right| = \frac{1}{2} \quad \text{for} \quad \omega = \omega_c \tag{8.21 c}$$

$$\delta = 10^{-a_s/20} \tag{8.21d}$$

where ω_p and ω_s the are passband/stopband edge frequencies, ω_c is the crossover frequency, and a_s denotes minimal attenuation in the stopband.

The criteria stated above can be met with the transfer functions which exhibit a nonnegative zero-phase frequency response. For those transfer functions, the zero-phase frequency response is nonnegative for all frequencies, and therefore, the magnitude response is identical with the zero-phase frequency response.

Due to the above mentioned properties, the transfer functions of magnitude-complementary FIR filters belong to the class of separable (factorizable) transfer functions, which are expressible as a product of a minimum-phase and a maximum-phase spectral factors.

The design of a filter pair $[H_{LP}(z)\ H_{HP}(z)]$ satisfying criteria (8.21a – d) can be carried out by a slight modification in the design procedure based on the general McClellan, Parks and Rabiner algorithm (1973). The McClellan, Parks and Rabiner algorithm returns the filter whose zero-phase frequency response oscillates between $1-\delta$ and $1+\delta$ in the passband, and between $-\delta$ and $+\delta$ in the stopband, see for example Figure 8.9. However, the criteria given in (8.21a – d) can be met with the zero-phase frequency response that oscillates between $1-\delta$ and unity in the passband, and between zero and $+\delta$ in the stopband. This is achieved with the following three-step design procedure:

Step 1: Compute the modified passband/stopband ripple δ_{ELP},

$$\delta_{ELP} = \frac{\delta}{2\left(1-\delta\right)}. \tag{8.22}$$

Step 2: Use the McClellan, Parks and Rabiner algorithm to design an even-order intermediate optimal FIR filter $E_{LP}(z)$ for the passsband/stopband ripple δ_{ELP} determined in *Step 1*, and for the desired crossover frequency ω_c.

Step 3: Modify the transfer function of the intermediate filter $E_{LP}(z)$ from *Step 2* to determine the transfer function of the desired lowpass filter $H_{LP}(z)$,

$$H_{LP}\left(z\right) = \left(1-\delta\right)E_{LP}\left(z\right) + \left(\delta/2\right)z^{-(N-1)/2}. \tag{8.23}$$

Step 4: Use equation (8.20) to generate the highpass filter $H_{HP}(z)$, which is complementary to the filter $H_{LP}(z)$. Since the magnitude response of $H_{LP}(z)$ satisfies the criteria given in (8.21a – d) for a lowpass filter, the filter $H_{HP}(z)$ satisfies those criteria for a highpass filter. Therefore, the resulting filter pair $[H_{LP}(z), H_{HP}(z)]$ is both magnitude complementary and delay-complementary filter pair, i.e. $[H_{LP}(z), H_{HP}(z)]$ is a double-complementary filter pair.

The example that follows demonstrates MATLAB application of the above procedure. We show here the design and properties of the double-complementary filter pair where the lowpass and higpass filters in the pair are halfband FIR filters. We demonstrate the design procedure based on the existing MATLAB function firceqrip for FIR filter design.

Example 8.1

Design the magnitude-complementary and delay-complementary halfband FIR filter pair to meet the specifications given below.

Specifications:
Filter order N_{ord} = 18, the passband/stopband ripple δ = 0.04, the crossover frequency ω_c = 0.5π.

Solution:

```
Nord = 18;  delta = 0.04; fc = 0.5            % Setting the input parameters
delp = delta/(2*(1 - delta));                 % Adjusting the ripple value, Step 1
elp = firceqrip(Nord,fc,[delp,delp]);         % Design of the intermediate filter Elp(z), Step 2
hlp = (1 - delta)*elp;                         % Modifying the intermediate filter Elp(z), Step 3
hlp(Nord/2 + 1) = hlp(Nord/2+1)+delta/2;       % Modifying the intermediate filter Elp(z), Step 3
hhp=zeros(size(1:length(hlp)));                % Zero settings
hhp(10) = 1;  hhp = hhp - hlp;
```

The resulting filters are shown in Figure 8.9 where the impulse response and the pole-zero locations of the lowpass filter are displayed, and also the magnitude responses of the lowpass/ higpass filter pair $[H_{LP}(z), H_{HP}(z)]$. This figure illustrates the basic properties of the filter pair: The impulse response is symmetric since the filters are the linear-phase filters. The zeros occurring on the unit circle are of a double multiplicity, whereas the remaining zeros satisfy the reciprocal symmetry. The magnitude responses for both filters oscillate between 1–δ and unity in the passband, and between zero and +δ in the stopband. Since both filters are halfband filters, the crossover frequency is located at the middle of the baseband at ω_c = 0.5π, and the magnitudes at this frequency are exactly 0.5. The sum of the magnitude responses amounts to one for all frequencies.

The above example demonstrates the design and properties of the magnitude-complementary filter pair for the case when both filters in the pair are halfband filters, i.e. the crossover frequency is located exactly at the middle of the baseband. For an arbitrary crossover frequency, the design procedure is the same as the procedure shown in *Example 8.1*. The only difference is in the value of the crossover frequency. Instead of specifying ω_c = 0.5π, one specifies the desired value for ω_c. The reader can verify the design procedure explained above by solving MATLAB Exercise 8.2.

Recall that the transfer functions with a nonnegative zero-phase frequency response, we have considered earlier in the case of *L*th-band filters, see Chapter VII, particularly Figure 7.7. In Chapter VII, we have shown how the MATLAB function firnyquist generates a linear phase *L*th-band filter exhibiting the nonnegative zero-phase frequency response.

The magnitude-complementary/delay-complementary filter pairs are used in digital audio as a suitable solution for the loudspeaker crossover networks, see (Regalia & Mitra, 1987) and (Salom, Todorović & Milić, 2007).

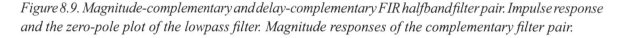

Figure 8.9. Magnitude-complementary and delay-complementary FIR halfband filter pair. Impulse response and the zero-pole plot of the lowpass filter. Magnitude responses of the complementary filter pair.

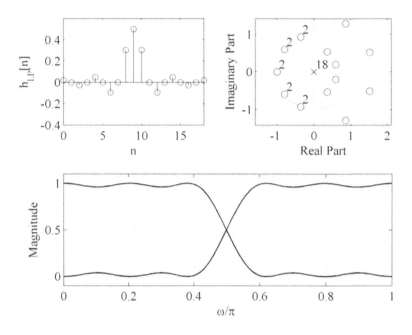

Power-Complementary FIR Filter Pairs

The power-complementary property defined in (8.3) is satisfied when the sum of squared magnitude responses of the lowpass and highpass filters of the pair $[H_{LP}(z), H_{HP}(z)]$ is equal to unity for all frequencies. The filter pair $[H_{LP}(z), H_{HP}(z)]$ achieves the power-complementary property when the magnitude responses of $H_{LP}(z)$ and $H_{HP}(z)$ meet the following criteria

$$\sqrt{1-\delta^2} \leq \left|H_{LP}\left(e^{j\omega}\right)\right| \leq 1 \quad \text{and} \quad \left|H_{HP}\left(e^{j\omega}\right)\right| \leq \delta \quad \text{for} \quad \omega \in \left[0, \omega_p\right] \tag{8.24a}$$

$$\sqrt{1-\delta^2} \leq \left|H_{HP}\left(e^{j\omega}\right)\right| \leq 1 \quad \text{and} \quad \left|H_{LP}\left(e^{j\omega}\right)\right| \leq \delta \quad \text{for} \quad \omega \in \left[\omega_s, \pi\right] \tag{8.24b}$$

$$\left|H_{LP}\left(e^{j\omega}\right)\right| = \left|H_{HP}\left(e^{j\omega}\right)\right| = \frac{\sqrt{2}}{2} \quad \text{for} \quad \omega = \omega_c \tag{8.24c}$$

$$\delta = 10^{-a_s/20}. \tag{8.24d}$$

where ω_p and ω_s are passband/stopband edge frequencies, ω_c is the crossover frequency, and a_s denotes minimal attenuation in the stopband.

The design criteria stated above can be met with the minimum-phase and maximum-phase FIR transfer functions.

In the most of applications, a power-complementary filter pair $[H_{LP}(z), H_{HP}(z)]$ is a halfband filter pair satisfying the so called *power-complementary conjugate quadrature filters'* conditions. Filters $H_{LP}(z)$ and $H_{HP}(z)$ of the length N are said to be conjugate symmetric if their transfer functions are related by

$$H_{HP}(z) = z^{-(N-1)} H_{LP}(-z^{-1}). \tag{8.25}$$

Such filter pairs play a very important role in filter banks as will be shown in Chapter XII.

When $H_{LP}(z)$ is a minimum-phase low-pass filter, $H_{HP}(z)$ is a maximum-phase filter, and vice versa. The key problem in designing power-complementary filters satisfying the criteria given by (8.24a – d) and (8.25) is to determine the minimum-phase lowpass filter $H_{LP}(z) = H_{min}(z)$. This problem can be solved by designing first a separable linear-phase FIR filter $H_{SEP}(z)$, which is a product of the minimum-phase and maximum-phase spectral factors, i.e., $H_{SEP}(z) = H_{min}(z) \times H_{max}(z)$. The linear-phase filter $H_{SEP}(z)$ has to be specified for the same passband/stopband edge frequencies as the desired minimum-phase filter, but the stopband attenuation should be twice the attenuation requested for the minimum-phase filter. The transfer function zeros of the linear-phase separable filter $H_{SEP}(z)$, which are located on the unit circle are of a double multiplicity, and the out unit-circle zeros exhibit the reciprocal symmetry. The minimum-phase filter $H_{LP}(z) = H_{min}(z)$ is considered as a minimum-phase spectral factor of the separable linear-phase filter $H_{SEP}(z)$. Hence, the extraction of the minimum-phase spectral factor consists of picking-up the inside-unit circle zeros and one of each double zeros from the unit circle.

Thereby, the first step in the design of a power-complementary filter pair is the design of a separable linear-phase lowpass filter $H_{SEP}(z)$. This can be done by using *Steps 1 – 3* of the procedure given for the magnitude complementary filter pair as explained in the previous subsection. Notice that $H_{SEP}(z)$ has to be designed for the twice the stopband attenuation requested for the minimum-phase filter $H_{LP}(z) = H_{min}(z)$. In the sequel, the lowpass minimum-phase filter $H_{LP}(z)$ is obtained by extracting the minimum-phase spectral factor $H_{min}(z)$ from $H_{SEP}(z)$.

The design procedure is the following:

Step 1: Design the separable linear-phase lowpass filter $H_{SEP}(z)$.

Substep 1.1: Compute the modified pasband/stopband ripple δ_{ELP},

$$\delta_{ELP} = \frac{\delta^2}{2(1-\delta^2)}. \tag{8.26}$$

Substep 1.2: Use the McClellan, Parks and Rabiner algorithm to design an even-order intermediate optimal FIR filter $E_{LP}(z)$ for the passsband/stopband ripple δ_{ELP} determined in *Substep 1.1*, and for the desired crossover frequency ω_c.

Substep 1.3: Modify the transfer function of the intermediate filter $E_{LP}(z)$ from *Substep 1.2* to determine the transfer function of the separable lowpass filter $H_{SEP}(z)$,

$$H_{SEP}(z) = (1-\delta^2)E_{LP}(z) + (\delta^2/2)z^{-(N-1)/2}. \tag{8.27}$$

Step 2: Compose the minimum-phase low-pass filter $H_{LP}(z)$ by extracting the minimum-phase spectral factor from $H_{SEP}(z)$. This is the most critical operation of the algorithm, since the very high computation accuracy is requested for determining the transfer function zeros of $H_{SEP}(z)$.

Step 3: Use equation (8.25) to generate the highpass filter $H_{HP}(z)$, which is conjugate symmetric and power-complementary to the filter $H_{LP}(z)$. Since the magnitude response of $H_{LP}(z)$ satisfies the criteria given in (8.24a – d) for the lowpass filter, the filter $H_{HP}(z)$ satisfies those criteria for the highpass filter. Therefore, the resulting filter pair $[H_{LP}(z) \ H_{HP}(z)]$ is a power-complementary filter pair.

The following example demonstrates MATLAB application of the procedure explained above. We show here the design and properties of the power-complementary filter pair where the lowpass and higpass filters in the pair are conjugate quadratic filters, i.e. their transfer functions are related by equation (8.25). In the design procedure, we use the existing MATLAB function firceqrip for the linear-phase FIR filter design, and for extracting the minimum-phase spectral factor $H_{LP}(z) = H_{min}(z)$, we use the m-file minphase, published by Orchard and Wilson (2003) and included in Appendix A of this book.

Example 8.2

Design the power-complementary halfband FIR filter pair to meet the specifications given below.

Specifications:
Filter order N_{ord} = 15, the passband/stopband ripple δ = 0.1, the crossover frequency ω_c = 0.5π.

Solution:
```
Nord = 15; Nord = 2*Nord;  delta = 0.1;        % Setting the input papameters
dfir =(delta^2)/(2*(1 - (delta^2)));           % Adjusting the ripple value, Substep 1.1
elp = firceqrip(Nord,0.5,[dfir,dfir]);         % Design of the intermediate filter Elp(z), Substep 1.2
eh = (1-(delta^2))* elp;                        % Modifying the intermediate filter Elp(z), Substep 1.3
eh(Nord/2+1) = eh(Nord/2 + 1) + (delta^2)/2;   % Modifying the intermediate filter Elp(z), Substep 1.3
[hlp,ssp,iter] = minphase(eh(Nord/2 + 1:Nord+1));% Generating the minimum-phase lowpass filter Hlp(z), Step 2
n = 0:length(hlp) - 1;                          % Indexing the filter coefficients
hhp = fliplr(hlp).*((-1).^n);                   % Complementary highpass filter, Step 3
```

Figures 8.10 and 8.11 display the characteristics of the power-complementary filters $H_{LP}(z)$ and $H_{HP}(z)$. Figure 8.10 illustrates the impulse response and the pole-zero locations of the filter pair. Figure 8.11 plots the magnitude responses of $H_{LP}(z)$ and $H_{HP}(z)$ with the passband details. Since both filters are halfband filters, the crossover frequency is located at the middle of the baseband at ω_c = 0.5π, and the magnitudes at this frequency amount to $\sqrt{2}/2$ (approximately 0.7). Reader can verify that the power-complementary property defined by equation (8.3) is satisfied for all frequencies.

The algorithm explained above may sometimes face serious numerical problems in practice. When it is requested to design a power-complementary filter pair exhibiting a high stopband attenuation (60 dB, for example), the initial linear-phase filter $H_{SEP}(z)$ has to achieve the minimal stopband attenuation of 120 dB. This may produce inaccuracies in the optimization algorithm because the computations are to be carried out with very small numbers. Additionally, the extraction of the minimum-phase spectral

Figure 8.10. Impulse responses and pole-zero plot for power-complementary FIR halfband filter pair $[H_{LP}(z), H_{HP}(z)]$. Left side – lowpass filter. Right side – highpass filter

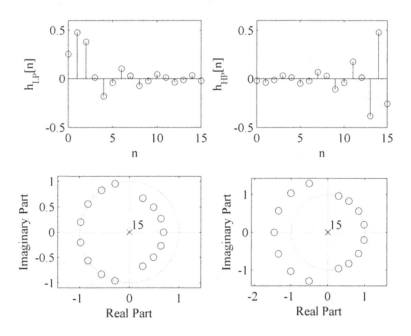

Figure 8.11. Magnitude responses for the power-complementary FIR filter pair $[H_{LP}(z), H_{HP}(z)]$

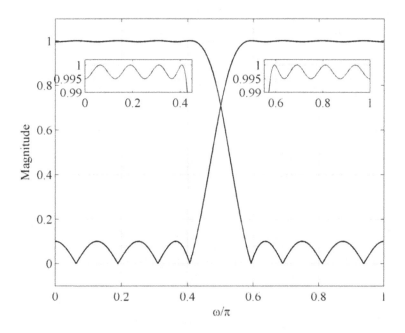

factor from $H_{SEP}(z)$ becomes insecure. For solving this type of problems, special computer programs should be used (Saramäki and Milić, Submitted for publication).

Power-complementary conjugate quadrature filter pairs (a halfband filter pair whose transfer functions are related by (8.25)) become particularly attractive when implemented in the lattice form. This structure implements simultaneously both the lowpass and highpass filters of the pair. The lattice structure consists of the cascade of lattice stages, each of them determined by the lattice coefficients k_i. The values of the lattice coefficients are computed recursively from stage to stage starting from the output stage. The lattice structure of the conjugate quadrature filter pair has two favourable characteristics (Fliege, 1994):

1. In each stage of the lattice, one coefficient is positive, and the other is negative, but both with the same magnitude.
2. All coefficients with the even-valued indices are zero.

Those two properties lead to a very efficient implementation of the complementary halfband filter pair shown in Figure 8.12 (a). The procedures for computing the lattice coefficients values from the coefficients of the minimum-phase filter $H_{LP}(z)$ are given in (Fliege, 1994; Mitra, 2006; Vaidyanathan, 1993). In Appendix A of this book, we give the MATLAB program qmflattice, which computes the lattice coefficients using the following code

k = qmflattice(hlp); % Computing the lattice coefficients

For the given lowpass halfband filter whose coefficients are stored in vector hlp, the program returns lattice coefficients in vector k.

The lattice structure is particularly efficient in multirate systems when the power-complementary filter pair is associated with the sampling rate conversion by a factor-of-2. As Figure 8.12 (b) illustrates,

Figure 8.12. Lattice structure for the efficient implementation of the conjugate-symmetric halfband filter pair

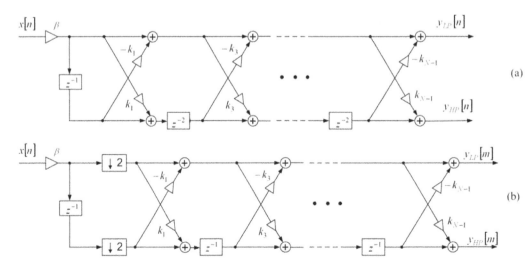

the down-sampling-by-2 is moved to the input, and the double-delays z^{-2} are replaced with the simple delays z^{-1}. In this way, all the computations are evaluated at the lower sampling rate.

The power-complementary conjugate quadrature filter pairs are used for the signal decomposition and reconstruction as will be demonstrated in Chapter XII of this book.

IIR COMPLEMENTARY FILTER PAIRS

IIR filter pairs can satisfy either all-pass complementary, power-complementary, or magnitude-complementary conditions as defined in (8.2), (8.3) – (8.4), and (8.5) – (8.6), respectively. IIR filter pairs achieve the requested magnitude characteristics with a lower computational complexity than their FIR counterparts. Their advantage is in applications where the computational efficiency and (or) low propagation delay are the main requirements.

Regarding the complementary properties, IIR filter pairs are classified into three classes (Milić and Saramäki, 2003; Saramäki, and Milić, Submitted for publication). The term *Class I* denotes the IIR filter pair satisfying all-pass complementary and power-complementary properties, which is implemented as a parallel connection of two all-pass subfilters. The terms *Class II* and *Class III* are associated with power-complementary and magnitude/all-pass complementary filter pairs both implemented as a tapped cascaded interconnection of two identical all-pass subfilters. This section considers the IIR complementary filter pairs mentioned above.

Class I: Power-Complementary and All-Pass Complementary Filter Pairs Implemented as a Parallel Connection of Two All-Pass Subfilters

Class I complementary filter pair [$G_{LP}(z)$, $G_{HP}(z)$] is characterized by two main properties:

1. The lowpass (highpass) filter transfer function $G_{LP}(z)$ [$G_{HP}(z)$] is an odd-order minimum-phase IIR transfer function whose magnitude response meets the following criteria,

$$\sqrt{1-\delta^2} \leq \left|G_{LP}\left(e^{j\omega}\right)\right| \leq 1 \quad \text{and} \quad \left|G_{HP}\left(e^{j\omega}\right)\right| \leq \delta \quad \text{for} \quad \omega \in \left[0,\omega_p\right] \tag{8.28a}$$

$$\sqrt{1-\delta^2} \leq \left|G_{HP}\left(e^{j\omega}\right)\right| \leq 1 \quad \text{and} \quad \left|G_{LP}\left(e^{j\omega}\right)\right| \leq \delta \quad \text{for} \quad \omega \in \left[\omega_s,\pi\right] \tag{8.28b}$$

$$\left|G_{LP}\left(e^{j\omega}\right)\right| = \left|G_{HP}\left(e^{j\omega}\right)\right| = \frac{\sqrt{2}}{2} \quad \text{for} \quad \omega = \omega_c \tag{8.28c}$$

$$\delta = 10^{-a_s/20}. \tag{8.28d}$$

Here, ω_p and ω_s are passband/stopband edge frequencies, ω_c is the crossover frequency, and a_s denotes minimal attenuation in the stopband.

2. The implementation structure of the filter pair [$G_{LP}(z)$, $G_{HP}(z)$] is based on the parallel connection of two all-pass subfilters.

The filter pairs satisfying the properties stated above are both power complementary and all-pass complementary. The first property expressed in equations (8.28a – d) is satisfied with Butterworth filters and with elliptic filters with minimal Q factors (EMQF filters) introduced by Lutovac and Milić (1997 and 1999), see Chapter 5 of this book, section IIR Structures with two All-Pass Subfilters: Applications of EMQF filters.

The second property means that the lowpass and highpass transfer functions $G_{LP}(z)$ and $G_{HP}(z)$ can be represented as a sum and difference of two all-pass subfilters $A_0(z)$ and $A_1(z)$,

$$G_{LP}(z) = \frac{1}{2}\left(A_0(z) + A_1(z)\right),$$ (8.29a)

$$G_{HP}(z) = \frac{1}{2}\left(A_0(z) - A_1(z)\right).$$ (8.29b)

Here, $A_0(z)$ and $A_1(z)$ are stable all-pass transfer functions. Figure 8.13 depicts an efficient implementation of equations (8.29a – b).

Class I IIR halfband filter pairs are of particular interest since they achieve a high computational efficiency in multirate systems. Let us consider the minimum-phase halfband filter pairs. When $G_{LP}(z)$ and $G_{HP}(z)$ are minimum-phase halfband filters, their transfer function poles are placed on the imaginary axis. Accordingly, equations (8.29a – b) can be expressed in the form

$$G_{LP}(z) = \frac{1}{2}\left(A_0^{HB}(z^2) + z^{-1} A_1^{HB}(z^2)\right),$$ (8.30a)

$$G_{HP}(z) = \frac{1}{2}\left(A_0^{HB}(z^2) - z^{-1} A_1^{HB}(z^2)\right),$$ (8.30b)

where the all-pass functions $A_0^{HB}(z)$ and $A_1^{HB}(z)$ are expressible as the products of the first-order all-pass sections,

$$A_0^{HB}(z) = \prod_{l=2,4,\ldots}^{(N+1)/2} \frac{\beta_l^{HB} + z^{-1}}{1 + \beta_l^{HB} z^{-1}} \quad \text{and} \quad A_1(z) = \prod_{l=3,5,\ldots}^{(N+1)/2} \frac{\beta_l^{HB} + z^{-1}}{1 + \beta_l^{HB} z^{-1}}.$$ (8.31)

Figure 8.13. Implementation structure for Class I complementary IIR filter pair

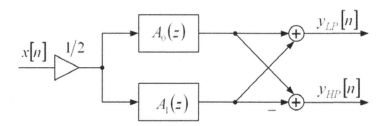

Recall that the constants β_l^{HB} in (8.31) are the squared magnitudes of the filter poles,

$$\beta_l = \left(r_l\right)^2, \ \ \text{with } \beta_l < \beta_{l+1}, \ l = 2, 3, \ldots, (N+1)/2. \tag{8.32}$$

Computation of the constants β_l^{HB} and their distribution among the all-pass subfilters $A_0^{HB}(z)$ and $A_1^{HB}(z)$ are explained in Chapter VII, subsection IIR Halfband Filter Design in MATLAB.

Figure 8.14 illustrates the efficient implementation of *Class I* complementary halfband filter pair with decimation-by-2 in both branches. Notice that the all-pass subfilters $A_0^{HB}(z)$ and $A_1^{HB}(z)$ are implemented with the first-order all-pass sections, and computations in both all-pass branches are evaluated at a half of the input sampling rate.

In *Example 8.3*, we illustrate the design procedure and efficiency of *Class I* halfband filter pair.

Example 8.3

Design the double-complementary elliptic halfband IIR filter pair of *Class I* to meet the specifications given below.

Specifications:
Filter order $N = 5$. The stopband edge frequency of the lowpass filter $\omega_s = 0.6\pi$.

Solution:
The halfband filter passband edge frequency ω_p is to be determined to satisfy the frequency symmetry condition, and therefore is given by $\omega_p = \pi - 0.6\pi = 0.4\pi$. For the halfband filter design, we use the program halfbandiir from Appendix A of this book. Then, we compute the constants of $A_0^{HB}(z)$ and $A_1^{HB}(z)$, and the frequency responses of the lowpass and highpass filters $G_{LP}(z)$ and $G_{HP}(z)$.

```
N = 5;  fp = 0.4;                                  % Setting input parameters for lowpass filter
[b,a,z,p,k] = halfbandiir(N,fp);                   % Lowpass halfband filter design
pabs = sort(abs(p));                               % Order of the filter poles
beta0 = (abs(pabs(2)))^2;   beta1 = (abs(pabs(4)))^2;   % Coefficients of allpass branches
[A0,f] = freqz([beta0,0,1],[1,0,beta0],512,2);     % Frequency response of A0(z)
[A1,f] = freqz([0,beta1,0,1],[1,0,beta1],512,2);   % Frequency response of A1(z)
GLp = (A0 + A1)/2;                                 % Lowpass filter frequency response
GHp = (A0 - A1)/2;                                 % Highpass filter frequency response
```

Figure 8.14. Implementation structure for Class I halfband complementary filter pair with decimation-by-2

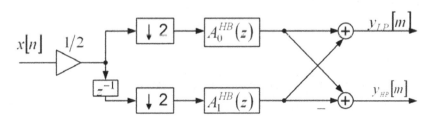

Figure 8.15 shows the characteristics of the double-complementary filters $G_{LP}(z)$ and $G_{HP}(z)$. Since both filters are halfband filters, the crossover frequency is located at the middle of the baseband at $\omega_c = 0.5\pi$, and at this frequency magnitudes of both filters amount to $\sqrt{2}/2$ (≈ 0.7). Reader can verify that the power-complementary and all-pass complementary properties defined by equations (8.2) and (8.3) are satisfied for all frequencies.

The lowpass/highpass *Class I* halfband filter pair of *Example 8.3* is implemented with only two multiplication constants. Moreover, the implementation structure of Figure 8.14 enables the arithmetic operation in all-pass branches $A_0^{HB}(z)$ and $A_1^{HB}(z)$ to be evaluated at the half of the input sampling rate.

In this subsection, we have presented the design and implementation of the *Class I* complementary filter pairs when the lowpass and highpass filters in the pair are halfband filters with the crossover frequency at the middle of the baseband, at $\omega_c = 0.5\pi$. Designs of Butterworth and EMQF *Class I* filter pairs with the crossover frequency located at an arbitrary position have been developed by Milić and Saramäki and presented in (Milić and Saramäki, 2003) and in (Saramäki and Milić, Submitted for publication). For the application of transfer functions having a desired number of zeros at $z = -1$, see (Damjanović, Milić & Saramäki, 2005), and (Milić, Damjanović & Nikolić, 2006).

Class I filter pairs being power complementary and all-pass complementary are used as building blocks in constructing complex filtering structures and also in constructing multilevel multi-channel filter banks.

Solutions based on *Class I* filter pairs enable the extremely efficient implementation of a double-complementary IIR filter pairs. The drawback of this class is a high stopband sensitivity, which can cause derogations of the stopband performances especially when the fixed-point arithmetic is used (Ćertić & Milić, 2005, 2006, 2007). The sensitivity is considerably reduced when using the implementation structures based on a tapped cascaded interconnection of identical all-pass subfilters, which are denoted as *Class II* and *Class III* complementary filter pairs.

Figure 8.15. Magnitude responses for Class I double-complementary filter pair $[G_{LP}(z), G_{HP}(z)]$

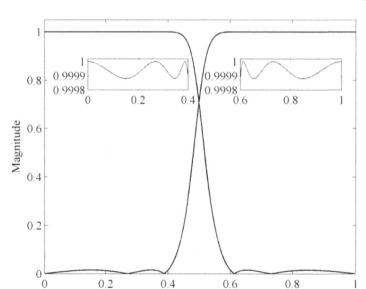

Class II: Power-Complementary Filter Pairs Implemented as a Tapped Cascaded Interconnection of Two Identical All-Pass Subfilters

A lowpass/highpass *Class II* complementary filter pair $[H_{LP}(z), H_{HP}(z)]$ is characterized by two important properties:

1. The filter pair exhibits power-complementary property defined by equations (8.5) and (8.6) with K = 1. Therefore, the design specifications for two filters $H_{LP}(z)$ and $H_{HP}(z)$ are those stated by (8.21a – d) for power-complementary filter pairs, i.e.,

$$\sqrt{1-\delta^2} \leq \left| H_{LP}\left(e^{j\omega}\right) \right| \leq 1 \quad \text{and} \quad \left| H_{HP}\left(e^{j\omega}\right) \right| \leq \delta \quad \text{for} \quad \omega \in \left[0, \omega_p\right] \qquad (8.33a)$$

$$\sqrt{1-\delta^2} \leq \left| H_{HP}\left(e^{j\omega}\right) \right| \leq 1 \quad \text{and} \quad \left| H_{LP}\left(e^{j\omega}\right) \right| \leq \delta \quad \text{for} \quad \omega \in \left[\omega_s, \pi\right] \qquad (8.33b)$$

$$\left| H_{LP}\left(e^{j\omega}\right) \right| = \left| H_{HP}\left(e^{j\omega}\right) \right| = \frac{\sqrt{2}}{2} \quad \text{for} \quad \omega = \omega_c \qquad (8.33c)$$

$$\delta = 10^{-a_s/20}. \qquad (8.33d)$$

Here, ω_p and ω_s are passband/stopband edge frequencies, ω_c is the crossover frequency, and a_s denotes minimal attenuation in the stopband.

2. The filter pair $[H_{LP}(z), H_{HP}(z)]$ is constructed as a tapped cascaded interconnection of two identical all-pass subfilters. The overall filtering task of $H_{LP}(z)$ $[H_{HP}(z)]$ is shared between two filters: a very low-order minimum-phase FIR prototype $F_{LP}(w)$ $[F_{HP}(w)]$, and a very low-order IIR prototype $G_{LP}(z)$ $[G_{HP}(z)]$.

The solution for the *Class II* filter pair $[H_{LP}(z), H_{HP}(z)]$ is achieved with the aid of the following transfer functions:

a) FIR prototype filter pair $[F_{LP}(w), F_{HP}(w)]$
b) IIR prototype filter pair $[G_{LP}(z), G_{HP}(z)]$

The role of FIR prototype is to provide the requested passband/stopband ripple and also the power-complementary property of the overall filter pair $[H_{LP}(z), H_{HP}(z)]$, whereas the IIR prototype provides the passband/stopband edge frequencies of the pair $[H_{LP}(z), H_{HP}(z)]$. This approach, usually called the *tapped cascaded interconnection of two identical all-pass filters,* was first introduced in filter design by Saramäki and Renfors (1987), and later extended to the complementary filter pairs by Saramäki and Milić, see (Milić & Saramäki, 2003), and also the recent article by Saramäki and Milić (Submitted for publication).

When the filters $H_{LP}(z)$ and $H_{HP}(z)$ are constructed as a tapped cascaded interconnection of two all-pass functions $A_0(z)$ and $A_1(z)$, their transfer functions are expressible in terms of tap values $a_{LP}[n]$ and $a_{HP}[n]$, and those two all-pass functions. The transfer functions for *Class II* filter pair $[H_{LP}(z), H_{HP}(z)]$ are given by

$$H_{LP}(z) = \sum_{n=0}^{N-1} a_{LP}[n] \left[A_0(z) \right]^n \left[A_1(z) \right]^{N-1-n},$$ (8.34a)

$$H_{HP}(z) = \sum_{n=0}^{N-1} a_{HP}[n] \left[A_0(z) \right]^n \left[A_1(z) \right]^{N-1-n}.$$ (8.34b)

The tap values, denoted by $a_{LP}[n]$ and $a_{HP}[n]$ in equations (8.34a – b), are the constants of the minimum-phase FIR power-complementary filter pair $[F_{LP}(w) \, F_{HP}(w)]$. Here, in the FIR prototype, the complex variable is denoted by w to make a distinction from the complex variable z, which is used for the overall filter pair. The FIR prototype $[F_{LP}(w), F_{HP}(w)]$ is an FIR filter pair of a very low order with the same minimal stop-band attenuation as the overall filter pair $[H_{LP}(z), H_{HP}(z)]$. All-pass filters $A_0(z)$ and $A_1(z)$ in equations (8.34a – b) are the solutions for the all-pass branches in *Class I* lowpass/highpass IIR filter pair $[G_{LP}(z), G_{HP}(z)]$ presented in the preceding subsection. Filters $G_{LP}(z)$ and $G_{HP}(z)$ compose the IIR prototype filter pair, which has the same transition band as the overall filter pair $[H_{LP}(z), H_{HP}(z)]$. Most importantly, the stopband ripples of $[G_{LP}(z), G_{HP}(z)]$ are related with the edge frequencies of the FIR prototype pair $[F_{LP}(w), F_{HP}(w)]$.

The design problem for *Class II* filter pair is to find the constants $a_{LP}[n]$ and $a_{HP}[n]$, and the all-pass functions $A_0(z)$ and $A_1(z)$ such that the overall filter pair $[H_{LP}(z), H_{HP}(z)]$ from (8.34a – b) satisfies the criteria given in (8.33a – d). The theoretical background for solving this design problem can be found in (Saramäki & Renfors, 1987) and in (Saramäki & Milić, Submitted for publication). In this subsection, we concentrate on the explanation of the resulting design procedure.

From the practical point of view, the most attractive *Class II* filter pairs are those where the lowpass and highpass filters $H_{LP}(z)$ and $H_{HP}(z)$ are both halfband filters. The reasons are threefold: (i) halfband filter pairs are widely used building blocks in multirate signal processing, (ii) *Class II* halfband filter pairs achieve extremely efficient implementation, (iii) starting from the halfband filter pair and with the aid of very simple transformation formulae, a complementary filter pair with an arbitrary crossover frequency can be obtained in a single step, see (Milić & Saramäki, 2003) and (Saramäki & Milić, Submitted for publication).

For the overall filter pair $[H_{LP}(z), H_{HP}(z)]$ being a halfband filter pair, the prototypes: FIR prototype $[F_{LP}(w), F_{HP}(w)]$, and IIR prototype $[G_{LP}(z), G_{HP}(z)]$ should also be the halfband filter pairs. Below, we describe the four-step procedure for designing the halfband filter pair $[H_{LP}(z), H_{HP}(z)]$ specified with the minimal stopband attenuation a_s, and the passband/stopband edge frequencies ω_p and ω_s satisfying the halfband filter symmetry condition, $\omega_s = \pi - \omega_p$.

Step 1: Design of the halfband FIR filter pair $[F_{LP}(w), F_{HP}(w)]$.

1)   For the requested minimal attenuation in the stopband of the overall filter pair denoted by a_s, compute the stopband ripple δ from the expression

$$\delta = 10^{-a_s/20}$$

Choose the filter order N_{FIR}, an odd number.

2) Design a minimum-phase lowpass halfband filter $F_{LP}(w)$ for the given values of N_{FIR} and δ. The magnitude response $|F_{LP}(e^{j\theta})|$ should satisfy

$$\sqrt{1-\delta^2} \le \left| F_{LP}\left(e^{j\theta}\right) \right| \le 1 \text{ for } \theta \in \left[0, \theta_p\right], \text{ and } \left| F_{LP}\left(e^{j\theta}\right) \right| \le \delta \text{ for } \theta \in \left[\theta_s, \pi\right]$$

with $\theta_p = \pi - \theta_s$.

Design can be evaluated according to the procedure exposed in subsection Power-Complementary FIR Filter Pairs presented earlier in this Chapter.

3) Use expression (8.25) to generate the power-complementary conjugate quadratic highpass filter $F_{HP}(w)$

$$F_{HP}\left(w\right) = z^{-N_{FIR}} F_{LP}\left(-w^{-1}\right).$$

Step 2: Determine the design parameters for the halfband IIR prototype filter pair $[G_{LP}(z), G_{HP}(z)]$. In this step, we compute the stopband ripple of the IIR filter pair δ_{IIR}, and choose the filter order with the goal to achieve the prescribed transition bandwidth of the overall filter pair $\omega_s - \omega_p$.

1) Compute the stopband ripple δ_{IIR} from the expression

$$\delta_{IIR} = \cos\left(\theta_s/2\right).$$

2) Determine the order of an elliptic halfband filter $G_{LP}(z)$ which guarantees the passband/stopband edge frequencies ω_p and ω_s, with the stopband ripple of exactly δ_{IIR} as determined in the previous substep. The filter order N_{IIR} should be an odd number.

Step 3: Design IIR prototype $[G_{LP}(z), G_{HP}(z)]$.

1) Design an elliptic halfband filter of order N_{IIR} with the exact stopband ripple δ_{IIR} as determined in *Step 2*.
2) Compose the all-pass functions $A_0(z)$ and $A_1(z)$.

Step 4: Compose the overall filter pair $[H_{LP}(z), H_{HP}(z)]$ according to (8.34a − b) by using the coefficients of the FIR prototype $[F_{LP}(w), F_{HP}(w)]$ determined in *Step 1*, and the all-pass functions $A_0(z)$ and $A_1(z)$ determined in *Step 3*.

On the basis of the design procedure explained above, selective filters with a high stop-band attenuation can be achieved by using very low-order FIR and IIR prototypes.

Figure 8.16 depicts the efficient lattice structure that implements simultaneously both halfband filter transfer functions $H_{LP}(z)$ and $H_{HP}(z)$. The lattice coefficients k_i can be computed recursively from the constants of the FIR prototype $[F_{LP}(w) F_{HP}(w)]$ by using the procedure given in (Fliege, 1994; Mitra, 2006; Vaidyanathan, 1993). For computing the lattice coefficients we give the MATLAB program qmflattice in Appendix A of this book.

Figure 8.16. Efficient implementation of Class II complementary filter pair

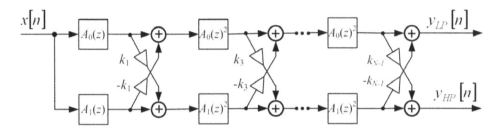

Using *Class II* complementary filter pairs we achieve efficient and robust solutions for the specifications requesting high stopband attenuation and narrow transition bands. This is illustrated by means of an example halfband filter pair with highpassband/stopband performances.

Example 8.4

Design the *Class II* power-complementary halfband IIR filter pair $[H_{LP}(z), H_{HP}(z)]$ to meet the specifications given below.

Specifications: Minimal attenuation in the stopband a_s = 60 dB. The edge frequencies of the filter pair are $\omega_s = 0.6\pi$ and $\omega_p = \pi - \omega_s = 0.4\pi$.

Solution: We use the four-step procedure for *Class II* halfband filter pairs, explained above in this subsection.

Step 1: Design of the halfband prototype FIR filter pair $[F_{LP}(w), F_{HP}(w)]$.

1) For a_s = 60 dB, we compute the stopband ripple $\delta = 0.001$, and we chose the filter order N_{FIR} = 5.
2) We utilize the procedure presented in subsection Power-Complementary FIR Filter Pairs of this Chapter to design the 5th-order lowpass filter $F_{LP}(w)$ with the stopband ripple $\delta = 0.001$, and obtain
   $F_{LP}(w) = 0.23921626118 + 0.57004438287 \ w^{-1} + 0.32265465634 \ w^{-2} - 0.09601119349 \ w^{-3} - 0.06186597755 \ w^{-4} + 0.02596187065 \ w^{-4}$
3) We determine $F_{HP}(w)$ as $F_{HP}(w) = z^{-N_{FIR}} F_{LP}(-w^{-1})$,
   $F_{HP}(w) = 0.02596187065 + 0.06186597755 \ w^{-1} - 0.09601119349 \ w^{-2} - 0.32265465634 \ w^{-3} \ 0.57004438287 \ w^{-4} \ -0.23921626118 \ w^{-5}$
   The impulse responses and the pole-zero plots of $F_{LP}(w)$ and $F_{HP}(w)$ are displayed in Figure 8.17, and their gain responses are shown in Figure 8.18(a).

Step 2:

1) We compute the stopband ripple of the IIR prototype as

   $\delta_{IIR} = \cos(0.93141767\pi \, / \, 2)$

The corresponding stopband attenuation amounts to 19.37 dB

2)   We find that the 3rd-order elliptic halfband filter can achieve the passband/stopband edge frequencies $\omega_p = 0.4\pi$, $\omega_s = 0.6\pi$ with the minimal stopband attenuation of exactly 19.37 dB. Hence, we choose $N_{IIR} = 3$.

Step 3: We design the 3rd-order prototype IIR halfband filter pair $[G_{LP}(z), G_{HP}(z)]$, and determine the all-pass functions $A_0(z)$ and $A_1(z)$,

$$A_0(z) = A_0^{HB}(z^2) = \frac{0.5455789 + z^{-2}}{1 + 0.5455789 z^{-2}} \quad \text{and} \quad A_1(z) = z^{-1}$$

The gain response of the IIR prototype $[G_{LP}(z), G_{HP}(z)]$ is shown in Figure 8.18(b).

Step 4: Using the coefficients of the FIR prototype $[F_{LP}(w), F_{HP}(w)]$ determined in *Step 1*, and the all-pass functions $A_0(z)$ and $A_1(z)$ from *Step 3*, we compose the overall *Class II* filter pair $[H_{LP}(z), H_{HP}(z)]$ according to equations (8.34a – b). Figure 8.18 (c) displays the gain response of the resulting filter pair.

Figures 18(a) – (c) illustrate the benefits that are achieved with *Class II* complementary filter pairs. Two very simple prototype filter pairs are combined to construct the high performance overall filter pair. In this way, the sensitivity to the filter constants is significantly decreased if compared with the solution obtainable with *Class I* filter pairs.

The efficiency and simplicity of implementation becomes apparent when the filter pair includes the sampling rate reduction by the factor-of-two. Figure 8.19 shows the lattice implementation structure for the filter pair $[H_{LP}(z), H_{HP}(z)]$ of example 8.4 which includes the decimation-by-2 in both channels.

Figure 8.17. Impulse responses and pole-zero plot of $F_{LP}(w)$ – left side, and $F_{HP}(w)$ – right side

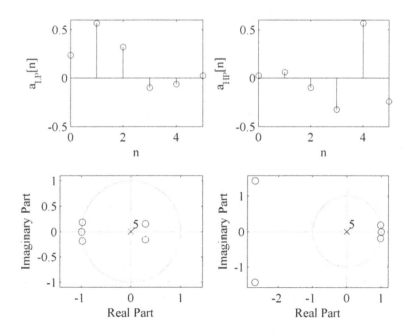

Figure 8.18. Gain responses of (a) FIR prototype [$F_{LP}(w)$, $F_{HP}(w)$], (b) IIR prototype [$G_{LP}(z)$, $G_{HP}(z)$], (c) overall Class II filter pair [$H_{LP}(z)$, $H_{HP}(z)$]

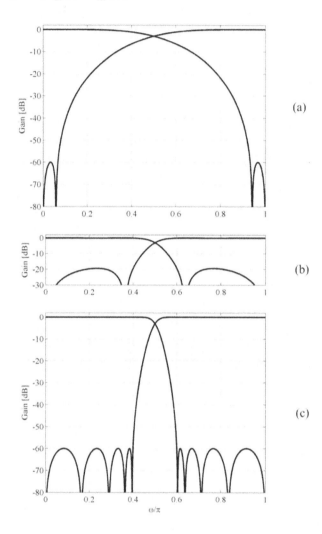

The five identical sections $A_0^{HB}(z)$ are the first-order allpass sections

$$A_0^{HB}(z) = \frac{0.5455789 + z^{-1}}{1 + 0.5455789z^{-1}}$$

from *Example 8.4* determined in *Step 3* of the procedure. The interconnecting coefficients k_i are the lattice coefficients of the FIR prototype. One can utilize program qmflattice from Appendix A to compute their values.

The implementation structure of *Class II* filter pairs is highly modular and thus suitable for VLSI applications. It is also suitable for multiplierless realizations with increased computation speed, see (Johansson & Wanhammar, 1999) and (Milić & Lutovac, 2001).

Figure 8.19. Class II complementary filter pair with factor-of-two sampling rate reduction. Lattice implementation for the 5th-order FIR prototype and the 3rd-order IIR prototype filters.

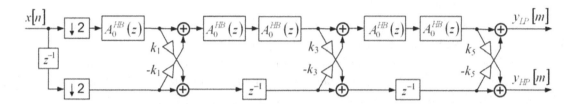

We have presented here the design and implementation of the *Class II* complementary filter pairs when the lowpass and highpass filters in the pair are halfband filters with the crossover frequency at the middle of the baseband, at $\omega_c = 0.5\pi$. The design procedure for *Class II* filter pairs with the crossover frequency located at an arbitrary position have been developed by Milić and Saramäki and presented in (Milić and Saramäki, 2003a,b) and in (Saramäki and Milić, Submitted for publication).

Class III: Magnitude-Complementary and All-Pass Complementary Filter Pair Implemented as a Tapped Cascaded Interconnection of Two Identical All-Pass Subfilters

Class III complementary filter pairs satisfy the all-pass complementary property as defined by equation (8.2), and also the magnitude-complementary property as defined by equations (8.5) – (8.6) with $K = 1$. Their implementation is based on the tapped cascaded interconnection of two identical all-pass filters.

The concept of magnitude-complementary IIR filters implemented as tapped cascaded interconnection of two identical all-pass filters have been first introduced by Johansson and Saramäki (1999) and extended later by Saramäki and Milić, see (Milić and Saramäki, 2003a), and also the recent article by Saramäki and Milić (Submitted for publication).

The magnitude complementary IIR filter pair $[H_{LP}(z), H_{HP}(z)]$ can be constructed on the basis of two prototype filters:

a)   FIR prototype filter pair $[F_{LP}(w), F_{HP}(w)]$
b)   IIR prototype filter pair $[G_{LP}(z), G_{HP}(z)]$

The role of FIR prototype is to provide the requested passband/stopband ripple and also the magnitude-complementary property of the overall filter pair $[H_{LP}(z)\,H_{HP}(z)]$, whereas the IIR prototype provides the passband/stopband edge frequencies of the pair $[H_{LP}(z)\,H_{HP}(z)]$.

The transfer functions for *Class II* filter pair $[H_{LP}(z)\,H_{HP}(z)]$ are given by

$$H_{LP}(z) = \sum_{n=0}^{N-1} a_{LP}[n]\left[A_0\left(z\right)\right]^n\left[A_1\left(z\right)\right]^{N-1-n}, \qquad (8.35a)$$

$$H_{HP}(z) = \sum_{n=0}^{N-1} a_{HP}[n] \left[A_0(z) \right]^n \left[A_1(z) \right]^{N-1-n}. \tag{8.35b}$$

The tap values, denoted by $a_{LP}[n]$ and $a_{HP}[n]$ in equations (8.34a – b), are the constants of the linear-phase FIR magnitude-complementary filter pair $[F_{LP}(w), F_{HP}(w)]$. The linear-phase FIR prototype $[F_{LP}(w), F_{HP}(w)]$ is an FIR filter pair of a very low order with the same minimal stop-band attenuation as the overall filter pair $[H_{LP}(z), H_{HP}(z)]$. All-pass filters $A_0(z)$ and $A_1(z)$ in equations (8.35a – b) are the solutions for the all-pass branches in *Class I* low-pass/high-pass IIR filter pair $[G_{LP}(z), G_{HP}(z)]$. Filters $G_{LP}(z)$ and $G_{HP}(z)$ compose the IIR prototype filter pair, which has the same transition band as the overall filter pair $[H_{LP}(z), H_{HP}(z)]$. The frequency responses of $[F_{LP}(w), F_{HP}(w)]$ and $[G_{LP}(z), G_{HP}(z)]$ are related in a manner that the stopband ripples of $[G_{LP}(z), G_{HP}(z)]$ are related with the edge frequencies of the FIR prototype pair $[F_{LP}(w), F_{HP}(w)]$.

The magnitude-complementary property of the *Class III* overall filter pair $[H_{LP}(z), H_{HP}(z)]$ is provided by the magnitude-complementary property of the FIR prototype filter pair $[F_{LP}(w), F_{HP}(w)]$, whereas the all-pass complementary property is provided by the IIR prototype pair $[G_{LP}(z), G_{HP}(z)]$.

As in the case of *Class II* filter pairs, the most attractive *Class III* filter pairs are those where the lowpass and highpass filters $H_{LP}(z)$ and $H_{HP}(z)$ are both halfband filters. A particularly attractive property of *Class III* filter pairs is the possibility of changing the crossover frequency in a very simple manner. Namely, starting from the magnitude-complementary halfband filter pair and with the aid of very simple transformation formulae, a magnitude-complementary filter pair with an arbitrary crossover frequency can be obtained in a single step, see (Milić and Saramäki, 2003a) and (Saramäki and Milić, Submitted for publication).

The design procedure for *Class III* filter pairs is similar to that presented in the previous section for *Class II* filter pairs. The only difference in the procedure is in the design of the FIR prototype filter pair $[F_{LP}(w), F_{HP}(w)]$. Therefore, in the four-step procedure for *Class II* filter pairs, *Step 1* has to be replaced with the procedure for designing magnitude-complementary FIR filter pairs presented earlier in this Chapter, subsection Delay-Complementary and Magnitude-Complemntary FIR Filter Pairs. The remaining *Steps 2 – 4* remain unchanged.

Figure 8.20 indicates an efficient implementation structure of *Class III* complementary filter pair. The FIR prototype is an even-order halfband filter pair, $N_{ord} = 2K$, where K is an odd number. All-pass filters $A_0(z)$ and $A_1(z)$ are the solutions for the all-pass branches in *Class I* low-pass/high-pass prototype IIR filter pair $[G_{LP}(z), G_{HP}(z)]$.

It was shown recently that *Class III* complementary filter pairs may be used in digital audio to construct loudspeaker crossover networks. For design and application see (Salom, Todorović, and Milić, 2007).

MATLAB EXERCISES

8.1 Design the 7th-order elliptic lowpass filter with the passband ripple $a_p = 0.5$ dB, the stoppband attenuation $a_s = 50$ dB. Use the function tf2ca to decompose the filter transfer functions into two allpass functions $A_0(z)$ and $A_1(z)$. Combine $A_0(z)$ and $A_1(z)$ to generate the transfer function of the complementary highpass filter. Plot the magnitude responses of the two complementary filters. Verify numerically the all-pass complementary and power-complementary properties.

Figure 8.20. Efficient implementation structure of Class III complementary filter pair

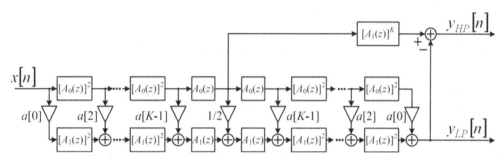

8.2 Design a double-complementary (delay-complementary and magnitude complementary) lowpass/highpass FIR filter pair $[H_{LP}(z), H_{HP}(z)]$ to meet the following specifications: filter order $N_{ord} = 18$, the passband/stopband ripple $\delta = 0.04$, the crossover frequency $\omega_c = 0.4\pi$. Plot the following characteristics of $H_{LP}(z)$, $H_{HP}(z)$: (i) impulse responses, (ii) locations of poles and zeros, (iii) magnitude responses. Compare the results with those of *Example 8.1*.

8.3 Modify *Example 8.2* to design a power-complementary FIR halfband filter pair for the following specifications: Filter order $N_{ord} = 31$, the passband/stopband ripple $\delta = 0.01$, the crossover frequency $\omega_c = 0.5\pi$. Plot the following characteristics of $H_{LP}(z)$, $H_{HP}(z)$: (i) impulse responses, (ii) locations of poles and zeros, (iii) gain responses. Plot the impulse response, pole-zero locations and the gain response of the linear-phase separable filter $H_{SEP}(z)$.

8.4 Modify *Example 8.3* to design a double-complementary IIR halfnand filter pair for the following specifications: Filter order $N = 7$, the passband edge frequency $\omega_p = 0.5\pi$. Plot the following characteristics of lowpass and highpass filters $H_{LP}(z)$, $H_{HP}(z)$: (i) the first 50 samples of impulse responses, (ii) locations of poles and zeros, (iii) gain responses. For the designed filter pair verify magnitude complementary and all-pass complementary properties.

REFERENCES

Ćertić, J., & Milić, L., (2005). Signal processor implementation of a low-pass/high-pass IIR digital filter with variable cutoff frequency. *Proc. IEEE EUROCON 2005 – The International Conference on "Computer as a Tool,"* 1618 – 1621.

Ćertić, J., & Milić, L. (2006). Fixed-point implementation of two-channel IIR filter banks with variable crossover frequency. *Proceedings of The 2006 International Workshop on Spectral Methods and Multirate Signal Processing – SMMSP2006,* 23-37.

Ćertić, J., & Milić, L. (2007). On the sensitivity of two-channel IIR filter banks with variable crossover frequency. *Proc. 5th International Symposium on Image and Signal Processing and Analysis, ISPA 2007,* 86-91.

Damjanović, S., Milić, L., & Saramäki, T. (2005). Frequency transformations in two-band wavelet IIR filter banks. *Proc. IEEE EUROCON 2005 – The International Conference on "Computer as a Tool,"* 87-90.

Filter design toolbox for use with MATLAB. User's guide. Version 6. (2006). Natick: MathWorks.

Fliege, N. J. (1994). *Multirate digital signal processing*. New York, NY: John Wiley & Sons, Inc.

Johansson H., & Saramäki, T. (1999). A class of complementary IIR filters. *Proc. IEEE Int. Symp. Circuits Syst, ISCAS 1999*, 3, 299–302.

Johansson, H., & Wanhammar, L. (1999). High-speed recursive filter structures composed of identical all-pass subfilters for interpolation, decimation, and QMF banks with perfect magnitude reconstruction. *IEEE Trans. Circuits and Systems-II: Analog and Digital Signal Processing*, 46(1), 16–28.

Lutovac, M. D., & Milić, L. D. (1997). Design of computationally efficient elliptic IIR filters with a reduced number of shift-and add operations in multipliers, *IEEE Transactions on Signal Processing*, 45(10), 2422-2430.

Milić, L., Damjanović, S., & Nikolić, M. (2006). Frequency transformations of IIR filters with filter bank applications. *Proc. APCAS 2006 IEEE Asia Pacific Conference on Circuits and Systems*, 1053-156.

Milić, L. D., & Lutovac, M. D. (1999). Design of multiplierless elliptic IIR filters with a small quantization error. *IEEE Transactions on Signal Processing*, 47(2), 469-479.

Milić, L. D., & Lutovac, M. (2001). High Speed IIR Filters for QMF Banks. *Proc. International Conference on Telecommunications in Modern Satellite, Cable and Broadcasting Services, TELSIKS 2001*, Niš, 171-174.

Milić, L. D., & Lutovac, M.D. (2003). Efficient algorithm for the design of high-speed elliptic IIR filters. *AEÜ Int. J. Electronics and Communications*, 57(4), 255-262.

Milić, L. D., & Saramäki, T. (2003, a). Three classes of IIR complementary filter pairs with an adjustable crossover frequency. *Proc. IEEE Int. Symp. Circuits Syst. ISCAS 2003*, 4, 145–148.

Milić, L. D., & Saramäki, T. (2003, b). Power-complementary IIR filter pairs with an adjustable crossover frequency. *Facta Universitatis, Ser.: Elec. Energ.* 16(3), 295–304.

Mitra, S. K. (2006). *Digital signal processing: A computer based approach*. 3rd edition. New York, NY: The McGraw-Hill Companies, Inc.

McClellan, J. H., Parks, T. W., & Rabiner, L. R. (1973). A computer program for designing optimum FIR linear-phase digital filters. *IEEE Transactions on Audio Electroacoustics*, 21(6), 506-526.

Orchard, H. J., & Willson, A. N. (2003). On the computation of a minimum- phase spectral factor. *IEEE Trans. Circuits and Systems-I: Fundamental Theory and Application*. 50(3), 365–375.

Regalia, P. A., & Mitra, S. K. (1987). A Class of Magnitude Complementary Loudspeaker Crossovers. IEEE Trans. on Acoustics, Speech and Signal Processing. 35(11), 1509-1516.

Salom, I. M., Todorović, D. Z., & Milić, L. D. (2007). The influence of impulse response length and transition bandwidth of magnitude complementary crossovers on perceived sound quality. *Journal of Audio Engineering Society*, 56(11), 941-954.

Saramäki, T. (1993). Finite impulse response filter design., Chapter 4 in *Handbook for Digital Signal Processing*. Edited by S. K. Mitra and J. F. Kaiser, New York: John Wiley Interscience, 155 – 277.

Saramäki, T. *Multirate Signal Processing*. (2001). Lecture notes for a graduate course, the Institute of Signal Processing, Tampere University of Technology, Finland.

Saramaki, T., & Bregovic, R. (2002). Multirate systems and filter banks. In Gordana Jovanović-Doleček, (ed.), *Multirate Systems: Design & Applications*. Hershey, PA: Idea Group Publishing, 27-85.

Saramäki, T., & Milić, L. (Submitted for publication). Tree Classes of Complementary Recursive Filter Pairs with Variable Crossover Frequency. *Submitted to Circuits Systems and Signal Processing*, Birkhäuser.

Saramäki, T., & Renfors, M. (1987). A novel approach for the design IIR filters as a tapped cascaded interconnection of identical allpass subfilters. *Proc. 1987 IEEE Int. Symp. Circuits Syst., ISCAS 1987.* 2, 629–632.

Signal processing toolbox for use with MATLAB. User's guide. Version 6. (2006). Natick: Math-Works.

Schüssler, H. W., & Stefen, P. (1998). Halfband filters and Hilbert transformers. *Circuits Systems Signal Processing.* 17(2), 137–164.

Schüssler, H.W., & Steffen, P. (2001). Recursive halfband-filters. *AEÜ Int. J. Electron. Commun.* 55(6), 377-388.

Vaidyanathan, P.P., (1993). *Multirate systems and filter banks*. Englewood Cliffs, NJ: Prentice Hall.

Chapter IX
Multirate Techniques in Filter Design and Implementation

INTRODUCTION

Digital filters with sharp transition bands are difficult, sometimes impossible, to be implemented using single-stage structures. A serious problem with a single-stage sharp FIR filter is its complexity. The FIR filter length is inversely proportional to the transition–width and complexity becomes prohibitively high for sharp filters, (Lim, 1986). IIR filters with sharp transition bands suffer from extremely high sensitivities of transfer function poles. In many practical cases, the multirate approach is the promising solution that could be applied for implementation of a sharp FIR or IIR filter.

In this chapter, we present two methods for designing filters having narrow transition bandwidths: multistage filtering suitable for narrowband filters, and the method based on multirate and complementary filtering, which may be used for filters of arbitrary bandwidths.

SOLVING COMPLEX FILTERING PROBLEMS USING MULTIRATE TECHNIQUES

Multirate techniques provide efficient tools for constructing digital filters with equal input and output sampling rates. Multirate techniques are shown to be an appropriate solution for the specifications that require a single-rate filter of a very high order. Instead of designing the single-rate filter, one can achieve a more effective solution with a *multirate filter* constructed as a combination of several low-order subfilters and sampling rate alteration devices. The benefits that can be achieved in comparison with the single-rate design are threefold:

1) The overall filtering problem can be shared between several lower-order subfilters.
2) The final wordlength effects on the overall filter performances are considerably reduced.
3) The arithmetic operations in subfilters can be evaluated at the reduced sampling rate thus reducing the overall computational complexity.

On the other hand, the multirate solutions introduce the unwanted effects that should be taken into account and minimized. As shown earlier in Chapter II, the signal distortions are produced when applying two basic operations of sampling rate conversion, down-sampling and up-sampling:

1) Each down-sampler produces aliasing.
2) Each up-sampler produces imaging.

Therefore, the subfilters in a multirate filter should sufficiently suppress the aliasing and imaging caused by all down-samplers and up-samplers that are incorporated in the multirate filtering structure.

The expressions that relate the *z*-transforms and the Fourier transforms of down-sampled and up-sampled signals with the *z*-transforms and the Fourier transforms of original signals have been shown earlier in Chapter II. Those expressions will also be used through this Chapter for the analysis of aliasing and imaging effects.

Let us denote with $\{x[n]\}$ and $X(z)$ the original signal and its *z*-transform, respectively. For the down-sampled-by-*M* signal $\{x_D[n]\}$, the *z*-transform $X_D(z)$ is given by (see equation (2.17)),

$$X_D(z) = \frac{1}{M} \sum_{k=0}^{M-1} X\left(z^{1/M} W_M^{-k}\right), \tag{9.1}$$

and, accordingly, the corresponding Fourier transform follows, equation (2.18),

$$X_D\left(e^{j\omega}\right) = \frac{1}{M} \sum_{k=0}^{M-1} X\left(e^{j(\omega - 2\pi k)/M}\right). \tag{9.2}$$

The *z*-transform relation for up-sampling as given in equation (2.22), is expressed by,

$$X_I(z) = X\left(z^L\right) \tag{9.3}$$

where with $X_I(z)$ we denote the *z*-transform of the up-sampled-by-*L* signal, $\{x_I[n]\}$. The Fourier transform relations for up-sampler according to equation (2.23) is given by,

$$X_I\left(e^{j\omega}\right) = X\left(e^{j\omega L}\right). \tag{9.4}$$

Relations (9.1) – (9.4) will be used later in this chapter for the computations of aliasing and imaging spectra that appear in multirate filters.

When considering the passband/stopband specifications for individual subfilters, one should take into account that their role in a multirate filter is twofold:

1) Subfilters have to provide the overall passband/stopband requirements of the multirate filter.
2) Subfilters have to attenuate sufficiently aliasing and imaging spectra that appear in the multirate filter structure.

In this chapter, we concentrate on two methods for constructing complex multirate filters. In the next section, we consider the application of multistage filtering to constructing narrow band filters. In the sequel, we show the application of multistage and complementary filters to multirate filters with arbitrary bandwidths and narrow transition bands.

MULTISTAGE NARROWBAND FILTERS

This method is convenient for the lowpass, highpass and bandpass filters having the bandwidths lower than one fourth of the sampling rate. Those filters are considered as narrowband filters.

Let us examine first the multistage filtering approach in the case of a lowpass filter with a low cutoff frequency. A general structure of the multistage multirate filter, consisting of a decimator, kernel filter $H_K(z)$, and interpolator, is shown in Figure 9.1. The sampling rate F_0 of the input signal $\{x[n]\}$ is first reduced to a lower rate $F_{0,1} = F_0/M$, the actual filtering with the kernel filter $H_K(z)$ is performed at this lower rate, and the original rate F_0 is restored again by interpolation.

The transfer function of the overall multirate filter of Figure 9.1,

$$H(z) = \frac{Y(z)}{X(z)}, \tag{9.5}$$

is composed of three transfer functions $H_D(z)$, $H_K(z)$ and $H_I(z)$. Under the assumption that $H_D(z)$ and $H_I(z)$ eliminate the aliasing effects, the overall transfer function $H(z)$ is expressible by

$$H(z) = H_D(z) H_K(z^M) H_I(z), \tag{9.6}$$

and, accordingly, with the replacement $z = e^{j\omega}$, we obtain the frequency response of the multirate filter,

$$H(e^{j\omega}) = H_D(e^{j\omega}) H_K(e^{jM\omega}) H_I(e^{j\omega}). \tag{9.7}$$

Figure 9.2 indicates the magnitude responses of filters from Figure 9.1. The passband and stopband edge frequencies of the kernel filter, F_p and F_s, respectively, are identical to those of the requested overall filter. The frequency F_p is also the passband edge of the decimation filter $H_D(z)$ and of the interpolation filter $H_I(z)$. For a fixed conversion factor M, which determines the lower sampling rate $F_{0,1} = F_0/M$, the stopband edge frequency for $H_D(z)$ and $H_I(z)$ is located at $F_{0,1} - F_s$, as indicated in Figure 9.2.

Figure 9.2 illustrates that the chain consisting of a decimation filter, kernel filter and interpolation filter achieves a narrowband characteristic with the use of three filters of significantly relaxed specifications. The overall magnitude response is the product of the magnitude responses $|H_K(e^{jM\omega})|$, $|H_D(e^{j\omega})|$

Figure 9.1 Multistage filter: decimator, kernel filter, and interpolator

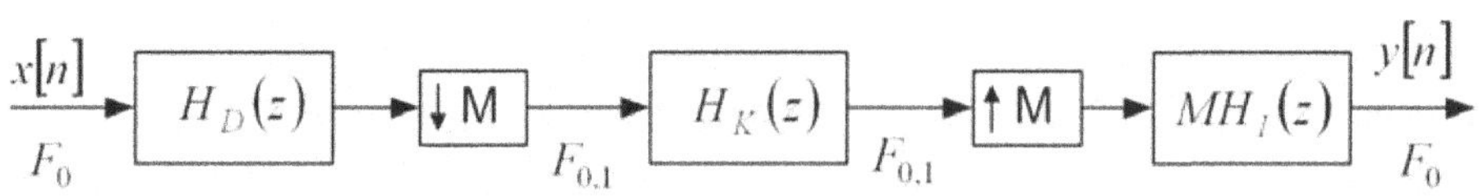

and $|H_I(e^{j\omega})|$. The minimum stopband attenuation of the overall filter is determined from the stopband attenuations of $H_K(z^M)$, $H_D(z)$ and $H_I(z)$. In the range $F_s \leq F \leq F_{0,1} - F_s$, the minimum stopband attenuation of the overall filter is determined from the attenuation of the kernel filter $H_K(z^M)$. For frequencies above $F_{0,1} - F_s$, the minimum stopband attenuation is determined by the cascade of decimation and interpolation filters $H_D(z)$ and $H_I(z)$. It is important to underline that the stopband attenuation of $H_D(z)$ and $H_I(z)$ is not only to ensure the stopband of the overall filter. Additionally, $H_D(z)$ has to suppress aliasing in decimation, and $H_I(z)$ to remove imaging in interpolation.

Multistage Filtering with the Halfband Decimation and Interpolation Filters

In has been shown so far that FIR and IIR halfband filters are desirable components for multistage systems because of the reduced computational complexity which is achieved when the halfband filters are used in decimation and interpolation. In this subsection, we investigate the transfer functions and the frequency response of the three-stage multirate filter of Figure 9.1 when $H_D(z)$ and $H_I(z)$ are the halfband filters. We concentrate particularly on the aliasing characteristics that are generated in the multistage structure of Figure 9.1 for the conversion factor $M = 2$. Additionally, we demonstrate by means of examples the computational efficiency for the solutions based on FIR and IIR filters.

Equations (9.6) and (9.7) describe the transfer function and the frequency response of the overall multirate filter neglecting aliasing and imaging effects, which in some cases can cause serious derogation of the signal. In order to study the interdependence of the filter frequency responses and the levels of aliasing and imaging components, we will develop the corresponding transfer functions and frequency responses for the multistage filter with the conversion factor $M = 2$. Figure 9.3 depicts the structure of this multirate filter with the notification of the intermediate signals of interest. In the structure of Figure 9.3 the decimation and interpolation filters are the halfband filters.

Let us express the z-transform relations for the signals $\{v[n]\}$, $\{r[m]\}$, $\{s[m]\}$, $\{u[n]\}$, and $\{y[n]\}$. Starting from the z-transform of the input signal $X(z)$, we express first the z-transform for $\{v[n]\}$,

Figure 9.2. (a) Magnitude responses of decimation and interpolation filters $H_D(z)$ and $H_I(z)$, and of the kernel filter $H_K(z)$. (b) Magnitude response of the overall filter $H(z)$.

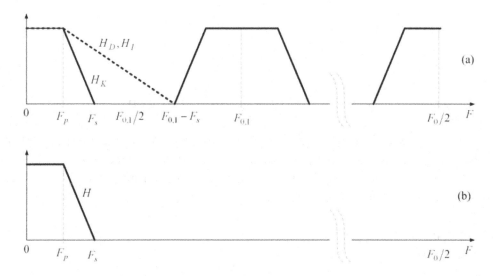

Figure 9.3. Multistage filter for M = 2: decimator, kernel filter, and interpolator

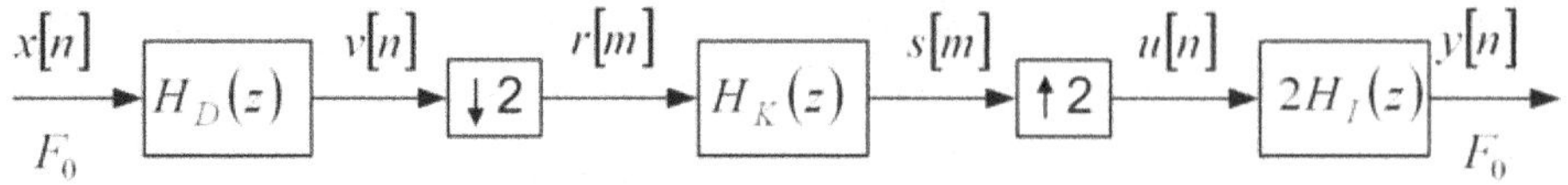

$$V(z) = H_D(z)X(z). \tag{9.8}$$

According to equation (9.1), the z-transform of the decimated-by-2 signal $\{r[m]\}$ is given by

$$R(z) = \frac{1}{2}\left[V\left(z^{1/2}\right) + V\left(-z^{1/2}\right) \right]. \tag{9.9}$$

At the output of the kernel filter $H_k(z)$ we obtain,

$$S(z) = H_K(z)R(z) \tag{9.10}$$

From (9.3) follows the z-transform of the up-sampled-by-2 signal $\{u[n]\}$,

$$U(z) = S\left(z^2\right) \tag{9.11}$$

And finally, after filtering with $H_I(z)$, we obtain the z-transform of the output signal $\{y[n]\}$,

$$Y(z) = 2H_I(z)U(z) \tag{9.12}$$

Using equations (9.8) – (9.12), we express the output $Y(z)$ in terms of $X(z)$ and $X(-z)$,

$$Y(z) = H_K\left(z^2\right)H_D(z)H_I(z)X(z) + H_K\left(z^2\right)H_D(-z)H_I(z)X(-z), \tag{9.13}$$

or concisely,

$$Y(z) = Q_1(z)X(z) + Q_2(z)X(-z), \tag{9.14}$$

where

$$Q_1(z) = H_K\left(z^2\right)H_D(z)H_I(z) \quad \text{and} \quad Q_2(z) = H_K\left(z^2\right)H_D(-z)H_I(z). \tag{9.15}$$

The first term in (9.13) and (9.14) represents the input/output relations of the multistage filter without aliasing, whereas the second term represents the aliasing characteristic of the system. Comparing equation (9.6) with equations (9.14) and (9.15), we observe that the first term in (9.14), $Q_1(z)$, represents the transfer function of the system without aliasing, i.e.,

$$Q_1(z) = H(z). \tag{9.16}$$

The second term, $Q_2(z)$, represents the aliased transfer function,

$$Q_2(z) = H_{al}(z),$$ (9.17)

which is desired to be minimized. The frequency responses of $Q_1(z)$ and $Q_2(z)$ when expressed as functions of the "real" frequency variable F are given by

$$Q_1\left(e^{j2\pi F/F_0}\right) = H\left(e^{j2\pi F/F_0}\right) = H_K\left(e^{j4\pi F/F_0}\right)H_D\left(e^{j2\pi F/F_0}\right)H_I\left(e^{j2\pi F/F_0}\right),$$ (9.18a)

$$Q_2\left(e^{j2\pi F/F_0}\right) = H_{al}\left(e^{j2\pi F/F_0}\right) = H_K\left(e^{j4\pi F/F_0}\right)H_D\left(e^{j2\pi (F+F_0/2)/F_0}\right)H_I\left(e^{j2\pi F/F_0}\right).$$ (9.18b)

In two examples that follow, we demonstrate the contribution of subfilters of the three-stage structure of Figure 9.3 to the overall filter magnitude response and to the aliasing characteristic, as well.

Example 9.1

We design subfilters $H_K(z)$, $H_D(z)$ and $H_I(z)$ using the following design parameters:

```
F0 = 2000;                                          % Sampling frequency [Hz]
Fp = 230;                                           %  Passband edge frequency [Hz]
Fs = 300;                                           % Stopband edge frequency [Hz]
delta = 0.001;                                      % Peak stopband ripple
Nord_k = 50;                                        % Setting the kernel filter order
hk = firgr(50,[0,2*Fp/(F0/2),2*Fs/(F0/2),1],[1,1,0,0]);   % Designing the kernel filter
hd = firhalfband('minorder',(2*Fs + Fp)/(3*F0/2),delta);  % Designing the decimation filter
hi = hd;                                            % Interpolation filter
```

For the given design parameters, program firhalfband with the option 'minorder' returns the decimation filter coefficients in vector hd. Here, the kernel filter is an FIR filter of the length $N = 51$, and decimation and interpolation filters are halfband filters of the length $N = 15$.

Figures 9.4 (a) and (b) illustrate the behaviours of the magnitude responses of the subfilters, overall filter $H(z)$ and aliasing characteristics (dashed lines). The plots of Figures 9.4 (a) and (b) are generated with the use of two FIR halfband filters as decimation and interpolation filters $H_D(z)$ and $H_I(z)$, and the FIR kernel filter $H_K(z)$.

The magnitude responses of $H_K(z^2)$, and that of $H_D(z)$ and $H_I(z)$ are shown in the upper subfigure of Figure 9.4. Dashed line displays the magnitude response of the aliased characteristic $H_D(-z)$. The bottom subfigure of Figure 9.4 plots the magnitude response of the overall filter $H(z)$, and the aliasing characteristic of the overall filter $H_{al}(z)$ indicated with the dashed line.

The implementation structure of the cascade: decimator with $H_D(z)$, kernel filter $H_K(z)$, and interpolator with $H_I(z)$ is indicated in Figure 9.5. The implementation of linear-phase FIR halfband filters $H_D(z)$ and $H_I(z)$ is given in the polyphase form, which achieves significant savings in computational complexity, see Chapter VII, subsection Efficient Implementation of Linear-Phase Halfband Filters. In $H_D(z)$ and

$H_I(z)$, each odd-indexed coefficient is zero valued, except for the central coefficient, which amounts to 0.5. Moreover, the nonzero coefficients exhibit coefficient symmetry property. According to this, for the halfband filter length of $N = 15$, the polyphase component $E_0(z)$ has 8 nonzero constants, which exhibit coefficient symmetry. Therefore, exploiting the coefficient symmetry only 4 multiplication constants are needed to implement $E_0(z)$. The kernel filter $H_K(z)$ is a linear-phase filter with 51 coefficients, and can be implemented with 26 multiplication constants.

Since all the arithmetic operations in the structure of Figure 9.5 are to be evaluated at the reduced sampling rate ($F_0/2 = 1000$ Hz), the total multiplication rate for $2 \times E_0(z)$ and $H_K(z)$ amounts to,

$$R_{M\ -\mathrm{MS}} = (2 \times 4 + 26) \times 1000 = 34000 \text{ multiplications per second.}$$

The benefits of multirate multistage filtering structure become evident when comparing the solution of Figure 9.5 with the single-rate FIR filter. The design specifications are met with the linear-phase single-rate filter of the length $N = 94$. The arithmetic operations are to be evaluated at the input/output sampling rate (2000 Hz), and the number of multiplication constants can be halved due to the coefficient symmetry. Thereby the multiplication rate for the single-rate design is given by

$$R_{M\ -\mathrm{SS}} = 47 \times 2000 = 94000 \text{ multiplications per second.}$$

The multiplication rate achieved with the multirate solution is nearly one third of that of the single-rate design.

Figure 9.4. Characteristics of multirate three stage FIR filter. Upper subfigure, solid lines: gain responses of $H_K(z^2)$, $H_D(z)$ and $H_I(z)$; dashed line: gain response of $H_D(-z)$. Lower subfigure, solid line: gain response of $H(z)$, dashed line gain response of $H_{al}(z)$.

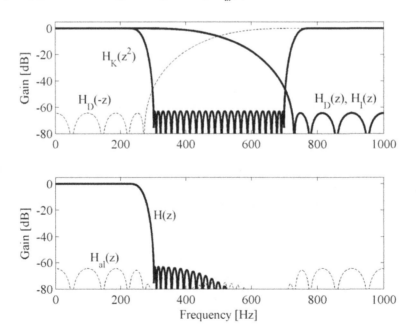

Figure 9.5. Configuration of multirate three-stage filter with halfband FIR filters in decimator and interpolator stages

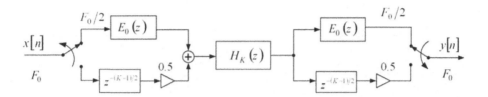

Figures 9.4 and 9.5 present the multirate three-stage filter based on FIR subfilters. The multirate multi-stage filtering can also be realized with the use of IIR filters. In the following, we demonstrate how the magnitude responses of Figure 9.4 can be achieved with IIR filters.

Example 9.2

We use two identical IIR halfband filters for decimation and interpolation filters $H_D(z)$ and $H_I(z)$, and an elliptic minimal-Q factors (EMQF) filter for the kernel filter $H_K(z)$. The design specifications are the same as those given for FIR filters in *Example 9.1*. The filter order for IIR halfband filters $H_D(z)$ and $H_I(z)$ is only 5, whereas $H_K(z)$ is the 9th-order EMQF filter.

Figure 9.6 presents the magnitude characteristics achieved with IIR filters. The magnitude responses of $H_K(z^2)$, and that of $H_D(z)$ and $H_I(z)$ are shown in the upper subfigure of Figure 9.6 . Dashed line displays the magnitude response of the aliased characteristic $H_D(-z)$. The bottom subfigure of Figure 9.6 shows the magnitude response of the overall filter $H(z)$, and the aliasing characteristic of the overall filter $H_{al}(z)$ indicated with the dashed line.

Figure 9.7 shows the efficient configuration for the three-stage multirate filter based on the parallel connection of two all-pass branches.

The realization structure depicted in Figure 9.7 presents the efficient implementation of the three-stage filter consisting of halfband decimation and interpolation filters, and an EMQF kernel filter.

Solution for $H_D(z)$ and $H_I(z)$.

Decimation and interpolation filters $H_D(z)$ and $H_I(z)$ are the 5th-order halfband filters, and accordingly, the allpass branches $A_0^{HB}(z)$ and $A_1^{HB}(z)$ are the first-order all-pass sections

$$A_0^{HB}(z) = \frac{\beta_2^{HB} + z^{-1}}{1 + \beta_2^{HB} z^{-1}} \quad \text{and} \quad A_1^{HB}(z) = \frac{\beta_3^{HB} + z^{-1}}{1 + \beta_3^{HB} z^{-1}}$$

where $\beta_2^{HB} = 0.14331399719644$, and $\beta_3^{HB} = 0.59309694413159$

Solution for $H_K(z)$.

The kernel filter $H_K(z)$ is the 9th-order EMQF filter. According to the design method presented in Chapter V, section IIR Structures with Two All-Pass Subfilters: Applications of EMQF Filters, the all-pass

Figure 9.6. Characteristics of multirate three-stage IIR filter. Upper subfigure, solid lines: gain responses of $H_K(z^2)$, $H_D(z)$ and $H_I(z)$; dashed line: gain response of $H_D(-z)$. Lower subfigure, solid line: gain response of $H(z)$, dashed line gain response of $H_{al}(z)$.

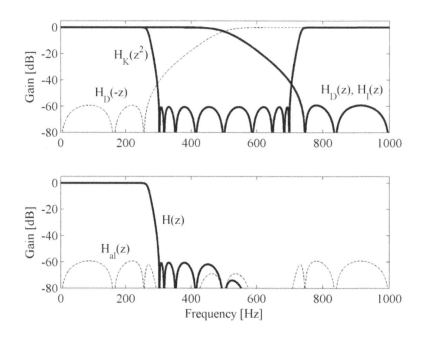

branch $A_0^K(z)$ is the cascade of two second-order sections, and the all-pass branch $A_1^K(z)$ is the cascade of a single first-order section and two second-order all-pass sections,

$$A_0^K(z) = \prod_{l=2,4,...}^{5} \frac{\beta_l + \alpha(1+\beta_l)z^{-1} + z^{-2}}{1 + \alpha(1+\beta_l)z^{-1} + \beta_l z^{-2}}, \quad A_1^K(z) = \frac{\alpha_1 + z^{-1}}{1 + \alpha_1 z^{-1}} \prod_{l=3,5,...}^{5} \frac{\beta_l + \alpha(1+\beta_l)z^{-1} + z^{-2}}{1 + \alpha(1+\beta_l)z^{-1} + \beta_l z^{-2}}$$

The values of the constants are the following:

The constant α, which has the same value in all second-order sections is adjusted to the simple combination of powers-of-two $\alpha = 1/16 + 1/32 + 1/256$.

The remaining constants are given by

$\alpha_1 = 0.048945099$
$\beta_2 = 0.10216831532199$, $\beta_3 = 0.33987190412921$
$\beta_4 = 0.60915479904033$ $\beta_5 = 0.86598448079941$

The overall multirate structure of Figure 9.7 contains 9 multiplication constants: two constants in decimation filter $H_D(z)$ and two constants in interpolation filter $H_I(z)$, and five constants in the kernel filter $H_K(z)$. Notice that the second-order sections of $A_0^K(z)$ and $A_1^K(z)$ have a common constant α which is a simple combination of shift-and-add operations and thereby implementable without a general multiplier. This property halves the number of multiplication constants in the EMQF filter $H_K(z)$.

The arithmetic operations in the structure of Figure 9.7 are evaluated at the half of the input sampling frequency F_0. Hence, the multiplication rate for the structure of Figure 9.7 for $F_0 = 2000$ Hz is determined by

$$R_{M-MS} = (2 \times 2 + 5) \times 1000 = 9000 \text{ multiplications per second.}$$

This result shows that the solution based on IIR subfilters achieves the highest computational efficiency. The drawback of this design is a nonlinear phase characteristic. The problem of phase nonlinearity could be solved by using the approximately linear-phase IIR filter transfer functions. But in that case, the order of subfilters $H_K(z)$, $H_D(z)$ and $H_I(z)$ should be increased. Therefore the improvement of the phase characteristic can be achieved at the cost of increased computational complexity.

In *Examples* 9.1 and 9.2, we have shown the construction of lowpass frequency response. To obtain the overall filter $H(z)$ as a highpass filter, the decimation and interpolation filters $H_D(z)$ and $H_I(z)$ should be highpass, whereas the kernel filter $H_K(z)$ is a lowpass.

In this subsection, we have considered the three-stage multirate filter with halfband filters in decimators and interpolators. When a very narrow passband is requested for the overall multirate filter, decimation and interpolation can be realized as a cascade of halfband decimators and interpolators as shown in Figure 9.8. This structure is sometimes called the *dyadic cascading*. In this structure, the kernel filter $H_K(z)$ operates at the sampling frequency of $F_0/(2^p)$, where p is the number of decimation (interpolation) stages.

Estimation of the Conversion Factor

The reduction of computational complexity is the main goal in multirate filter design. For the given specifications, one can make a choice between various combinations for decimation/interpolation factor, and kernel filter bandwidth. In this subsection, we consider selection of the adequate conversion factor for the specified overall filter, which has to be implemented as a cascade of a decimator, kernel filter and interpolator, see Figure 9.1.

An approach based on the estimation of the FIR filter computational complexity has been derived by Fliege (1994). It was shown that for the cascade factor-of-M decimator, kernel filter, and factor-of-M interpolator, the optimal value for M is a real solution of the equation,

$$(F_s^2 - F_p^2)M^3 - (F_s + F_p)^2 M^2 + 2F_0(F_s + F_p)M - F_0^2 = 0 \qquad (9.19)$$

Figure 9.7. Configuration of multirate three-stage filter with halfband IIR filters in decimator and interpolator stages

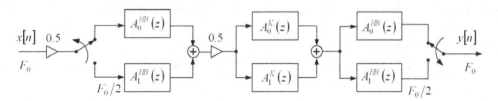

Figure 9.8. Dyadic cascade structure of a multirate filter

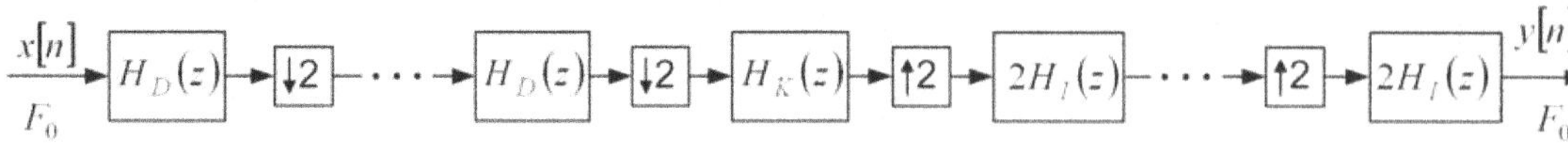

Notice that the above equation is derived under the assumption that the decimation, kernel and interpolation filters are all FIR filters. Generally, the real root of equation (9.19) is not an integer. Therefore, in practice we choose for *M* a near-by integer.

We demonstrate by means of example the choice of the conversion factor and corresponding design of the multirate filter.

Example 9.3

The specifications for an FIR lowpass filter are: bandpass edge frequency $F_p = 50$ Hz, stopband edge frequency $F_s = 100$ Hz, and the sampling frequency $F_0 = 2000$ Hz. The ripple tolerance in the passband is $\delta_1 = 0.01$, and the minimal stopband attenuation is 60 dB ($\delta_2 = 0.001$).

The specifications can be met with a single stage optimal FIR filter of the length $N = 103$.

Multistage solution:
Instead of the single stage FIR filter with $N = 103$, we can apply the multirate implementation from Figure 9.1.

The first step is to determine the optimal conversion factor *M*. We use MATLAB to compute the optimal value of *M* by solving equation (9.19);

```
F0 = 2000;                              % Sampling frequency
Fp = 50;                                % Passband edge frequency
Fs = 100;                               % Stopband edge frequency
D = [Fs^2-Fp^2,-(Fs+Fp)^2,2*F0*(Fs+Fp),-F0^2];   % Coefficient vector of equation (9.19)
Mopt = roots(D)                         % Solution of equation (9.19)
```

The only real solution for Mopt is the noninteger number 5.6270. Hence, we choose to implement the multirate filter with the sampling rate conversion factor $M = 5$.

For the input/output sampling frequency of $F_0 = 2000$ Hz and the conversion factor $M = 5$, the kernel filter $H_K(z)$ is designed for sampling frequency of $F_{0,1} = 400$ H, and edge frequencies are those of the overall filter, i.e. $F_p = 50$ Hz, $F_s = 100$ Hz. The stopband attenuation of $H_K(z)$ is equal to the requested minimal attenuation of the overall filter and amounts to 60 dB.

The decimation and interpolation filters $H_D(z)$ and $H_I(z)$ are designed for the sampling frequency $F_0 = 2000$ Hz, and the edge frequencies $F_p = 50$ Hz, $F_s = 350$ Hz. The minimal stopband attenuation amounts to 60 dB with don't care bands in the range of frequencies where the periodic characteristic of the kernel filter $H_K(z)$ has the stopbands.

We choose one third of the requested passband ripple tolerance for each of three filters, i.e., $\delta_p = 0.01/3$.

The requirements are met with the kernel filter of the length $N_k = 25$, while for interpolation and decimation filters we obtain $N_D = N_I = 19$.

The MATLAB code for designing $H_K(z)$, $H_D(z)$ and $H_I(z)$ is the following

```
% Kernel filter design
fk = [0,50/200,100/200,1]; mk=[1,1,0,0];      % Setting the design parameters
w = [1,3.333];                                 % Passband/stopband weighting coefficients
hk = firpm(24,fk,mk,w);                        % Computation of filter coefficients


% Decimation/ interpolation filter design
f0 = [0,50/1000,350/1000,450/1000,750/1000,850/1000];   % Setting the design parameters
m0 = [1,1,0,0,0,0];                                      % Setting the design parameters
w0 = [w,w(2)];                                           % Passband/stopband weighting coefficients
hd = firpm(18,f0,m0,w0);                                 % Computation of filter coefficients
```

Figure 9.9 plots the magnitude responses

The specifications of *Example 9.3* are fulfilled with a total of $N_k + N_D + N_I = 63$ coefficients in multistage design instead of 103 coefficients needed for the single stage design. The number of multiplications per second for single stage implementation amounts to $R_{M\text{-}SS} = 103 \times 2000/2 = 103000$, whereas the presented solution for multistage filter requires $R_{M\text{-}MS} = 63 \times 2000/(2 \times 5) = 12600$ multiplications per second.

Highpass Multirate Filter

Highpass filters with edge frequencies close to half the sampling rate can be realized in a similar way as narrowband lowpass filters. We only have to realize decimation and interpolation filters as highpass

Figure 9.9. Multistage lowpass filter. Magnitude characteristic of the kernel filter (upper subfigure). Magnitude characteristic of the decimation and interpolation filters (lower subfigure).

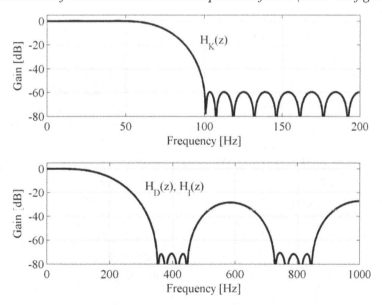

Figure 9.10. Multistage lowpass filter. Magnitude characteristic of the overall filter (upper subfigure). Aliasing characteristics (lower subfigure).

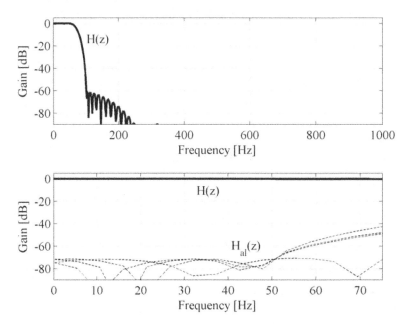

filters, but the kernel filter is a lowpass. Therefore, if $H_D(z)$ and $H_I(z)$ are the decimation and interpolation filters of the lowpass multirate filter, the highpass multirate filter makes use of $H_D(-z)$ and $H_I(-z)$ in interpolation and decimation stages, respectively. We demonstarate this simple approach by means of *Example 9.4*.

Example 9.4

In this example we show how the lowpass multirate filter of *Example 9.1* is modified into a highpass multirate filter.

For the highpass filter design parameters:

F0 = 2000;	% Sampling frequency [Hz]
Fp = 1000 - 230;	% Passband edge frequency
Fs = 1000 - 300;	% Stopband edge frequency
delta = 0.001;	% Peak stopband ripple

we use the lowpass filters $H_K(z)$, $H_D(z)$ and $H_I(z)$ from *Example 9.1*, and modify the lowpasses $H_D(z)$ and $H_I(z)$ into the highpasses. In *Example 9.1*, the coefficients of $H_D(z)$ are stored in vector hd. We generate the desired highpass filters, using the code:

n = 0:length(hd)-1;	% Time index n
hd = hd.*(-1).^n;	% Decimation filter
hi = hd;	% Interpolation filter

The resulting solution for the desired multirate highpass filter is shown in Figure 9.11. Notice that the characteristics shown in Figure 9.11 exhibit the mirror symmetry to the characteristics of Figure 9.4.

STRUCTURES BASED ON COMPLEMENTARY FILTERS AND MULTIRATE TECHNIQUES

The multirate approach may be used to design broadband filters by subtracting output of the narrowband filter from the delayed version of the input signal (Ramstad, and Saramäki, 1988 and 1990; Fliege, 1994). This method can be used in designing filters with any passband bandwidth. In this approach, we combine the complementary property and the narrow passband filter design method to develop highpass and lowpass filters with wide passbands. The multirate techniques are included to reduce the computational complexity.

Let us examine the multirate filter stage shown in Figure 9.12. As a part of this structure, we recognize the narrowband three-stage filter of Figure 9. 3 composed of a cascade of a factor-of-2 decimator, kernel filter, and factor-of-2 interpolator. The novelty is the delay element z^{-D} introduced to provide the delayed replica of the input signal. Filters $H_K(z)$, $H_D(z)$ and $H_I(z)$ are all linear-phase FIR filters. The output of the narrowband filter $\hat{y}[n]$ is subtracted from the delayed-by-D version of the input signal $x[n]$. In this way, the complementary wideband filter is obtained. If the narrowband filter is a highpass filter, the resulting filter is a lowpass filter, and vice versa. The delay D has to be selected to exactly equal the group delay of the multirate filter chain: decimator, kernel filter and interpolator.

For determining the z-transform of the signal $\hat{y}[n]$ of Figure 9.12, we can utilize the procedure

Figure 9.11. Highpass FIR three-stage multirate filter. Upper subfigure, solid lines: gain responses of $H_K(z^2)$, $H_D(z)$ and $H_I(z)$; dashed line: gain response of $H_D(-z)$. (b) Lower subfigure, solid line: gain response of $H(z)$, dashed line: gain response of $H_{al}(z)$.

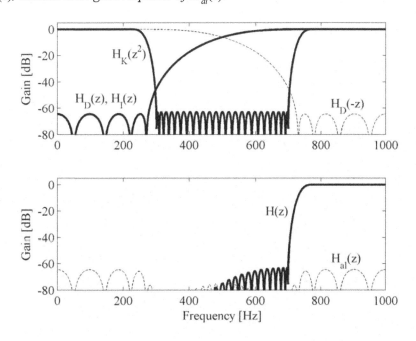

developed for the structure of Figure 9.3. Hence, utilizing equation (9.13), we can write the expression for $\hat{Y}(z)$

$$\hat{Y}(z) = H_K\left(z^2\right)H_D(z)H_I(z)X(z) + H_K\left(z^2\right)H_D(-z)H_I(z)X(-z),$$ (9.20)

where

$$Q_1(z) = H_K\left(z^2\right)H_D(z)H_I(z) \quad \text{and} \quad Q_2(z) = H_K\left(z^2\right)H_D(-z)H_I(z).$$ (9.21)

Thereby, equation (9.20) can be represented in the concise form

$$\hat{Y}(z) = Q_1(z)X(z) + Q_2 X(-z).$$ (9.22)

Finally, the *z*-transform of the output *Y(z)* is obtained by the complementing operation,

$$Y(z) = z^{-D}X(z) - \hat{Y}(z).$$ (9.23)

In order to express *Y(z)* in terms of *X(z)*, we replace $\hat{Y}(z)$ with expression given in (9.22),

$$Y(z) = z^{-D}X(z) - Q_1(z)X(z) - Q_2 X(-z),$$ (9.24)

and introducing expressions (9.21) for $Q_1(z)$ and $Q_2(z)$, we express *Y(z)* as a function of the input *X(z)*, and the aliased term *X(−z)*,

$$Y(z) = \left[z^{-D} - H_K\left(z^2\right)H_D(z)H_I(z)\right]X(z) - \left[H_K\left(z^2\right)H_D(-z)H_I(z)\right]X(-z).$$ (9.25)

From equation (9.25), we define the transfer function of the system without aliasing

$$H(z) = z^{-D} - H_K\left(z^2\right)H_D(z)H_I(z),$$ (9.26)

and also the aliased transfer function,

Figure 9.12. Multirate filter stage with complementing operation

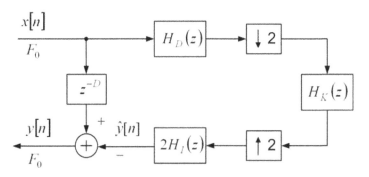

$$H_{al}(z) = H_K\left(z^2\right)H_D(-z)H_I(z).$$ (9.27)

The construction of the wideband filter is demonstrated by means of the following example.

Example 9.5

In this example, we demonstrate the application of the multirate complementary structure of Figure 9.12 to construct a wideband lowpass filter with the following design parameters:

Sampling frequency: $F_0 = 2000$ Hz, passband edge frequency $F_p = 700$ Hz, stopband edge frequency $F_s = 770$ Hz, minimal stopband attenuation $a_s = 60$ dB.

Solution:

We first design the narrowband multirate highpass filter with the passband edge at $F_p = 700$ Hz, and the stopband edge at $F_s = 770$ Hz using the procedure of *Example 9.4*. Here, we specify subfilters $H_K(z)$, $H_D(z)$ and $H_I(z)$ for the passband/stopband ripple of $\delta = 0.0003$.

The above specifications are met with the following filter lengths:

- For Halfband filters, $H_D(z)$ and $H_I(z)$, $N_D = N_I = 19$.
- For the kernel filter $H_K(z)$, $N_K = 63$.

The magnitude responses of $H_K(z)$, $H_D(z)$ and $H_I(z)$ are given in the upper subfigure of Figure 9.13, whereas the lower subfigure plots the magnitude response of the narrowband highpass filter with the corresponding aliasing characteristic.

In the second step, we determine the desired wideband lowpass filter as a delay-complementary filter to the highpass narrowband filter of Figure 9.13. The complementing operation is evaluated according to Figure 9.13 and equation (9.23). We find the equivalent impulse response of the narrowband filter,

```
hy = conv(downsample(hd,2),hk);
hy = conv(upsample(hy,2),2*hi);        % Equivalent impulse response of the narrowband multirate filter

hdel = zeros(size(1:length(hy)));
D = (length(hdel))/2;
hdel(D) = 1;                           % Delay sequence
h = hdel-hy;                           % Equivalent impulse response of the resulting lowpass filter
```

Figure 9.14 displays the magnitude response of the desired lowpass filter, and also the aliasing component that is generated in the multirate filter. The plots of Figure 9.14 demonstrate that combining the complementary operation with the multirate technique, the wideband filter with a sharp transition band is achieved with the subfilters of a moderate complexity.

Complementary multirate filtering can be used for constructing filters with very sharp transition bands. This may be achieved by cascading complementary multirate filter stages of Figure 9.12. We illustrate the cascade connection of two complementary stages in Figure 9.15. The cascade configuration enables one to replace the kernel termination filter $H_K(z)$ from Figure 9.12 with the new complementary multirate filter stage. This process can be continued leading to the very narrow transition bandwidth.

Figure 9.13. Highpass FIR three-stage narrowband filter. Upper subfigure, solid lines: gain responses of $H_K(z^2)$, $H_D(z)$ and $H_I(z)$; dashed line: gain response of $H_D(-z)$. (b) Lower subfigure, solid line: gain response of $H(z)$, dashed line: gain response of $H_{al}(z)$.

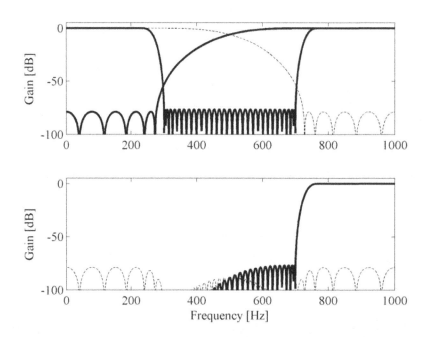

Let us consider the transfer function of the cascade multirate filter of Figure 9.15. The second stage is actually the complementary multirate filter of Figure 9.12, and therefore, we can use equation (9.26) to express its transfer function, which we denote by $H_K^{eq}(z)$,

$$H_K^{eq}(z) = z^{-D_2} - H_K(z^2)H_{D_2}(z)H_{I_2}(z). \tag{9.28}$$

For the first stage $H_K^{eq}(z)$ represents the equivalent kernel filter. With this assumption, the overall transfer function of the two-stage filter of Figure 9.15 is determined by the following expression

$$H_K^{eq}(z) = z^{-D_2} - H_K(z^2)H_{D_2}(z)H_{I_2}(z). \tag{9.29}$$

We illustrate the performances of the two stage complementary multirate filter structure of Figure 9.15, in the following example where we exploit the results of *Example 9.5*.

Example 9.6

In this example, we demonstrate the application of the two-stage multirate complementary structure of Figure 9.15 to construct the lowpass filter with the following design parameters:

Sampling frequency: $F_0 = 4000$ Hz, passband edge frequency $F_p = 700$ Hz, stopband edge frequency $F_s = 770$ Hz, minimal stopband attenuation $a_s = 60$ dB.

Figure 9.14. Wideband lowpass filter. Complementary multirate structure of Figure 9.12. Solid line: resulting lowpass filter. Dashed line: Aliasing component.

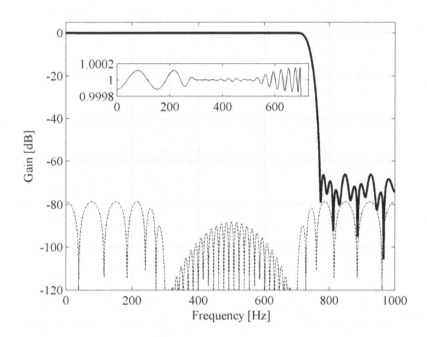

Figure 9.15. Cascade of two complementary multirate filter stages

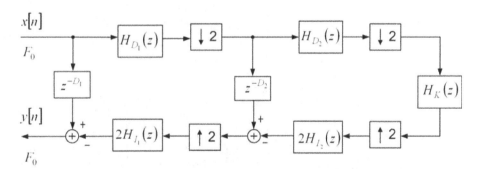

Solution:

In the first step, the filters from the second stage of the cascade need to be determined. In this example, we exploit the resulting filter from *Example 9.5* whose performances are shown in Figure 9.14. Here, this filter has a role of the equivalent kernel filter $H_K^{eq}(z)$, see equations (9.28) and (9.29).

In the second step, we design decimation and interpolation filters $H_{D1}(z)$ and $H_{I1}(z)$ with the goal to attenuate the upper passband of the periodic equivalent kernel filter $H_K^{eq}(z)$ as shown in Figure 9.16.

Figure 9.16. Gain responses of $H_K^{eq}(z^2)$, $H_{D1}(z)$ and $H_{J1}(z)$; dashed line: gain response of $H_{D1}(-z)$

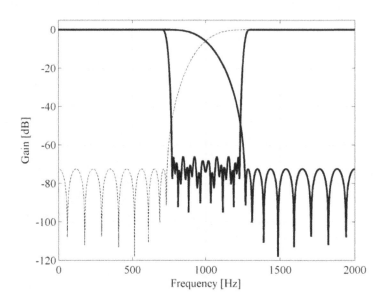

Figure 9.17. Multirate lowpass filter, gain responses: Solid line: resulting lowpass filter. Dashed line, and dashed-doted line: aliasing components.

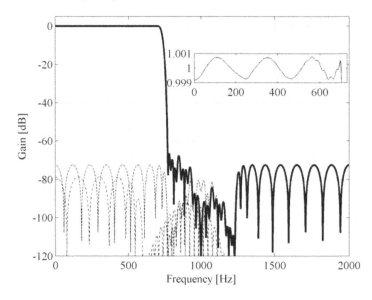

In the third step, we determine the group delay D_1 of the two stage multirate filter, and compute the overall response according to equation (9.29). Figure 9.17 displays the gain response of the two-stage multirate filter with aliasing components generated in the multirate structure of Figure 9.15.

It is of interest to examine the complexity of the solutions based on complementary multirate filtering. For the two-stage filter of *Example 9.6*, the number of multiplication constants is calculated as:

$$N_{\text{total}} = (N_k + N_{D2} + N_{I2} + N_{D1} + N_{I1})/2, \tag{9.30}$$

where N_k is the number of multiplication constants of the kernel filter $H_k(z)$, and N_{D2}, N_{I2}, N_{D1} and N_{I1} denote the numbers of multiplication constants in the halfband filters $H_{D1}(z)$, $H_{I1}(z)$, $H_{D2}(z)$ and $H_{I2}(z)$, respectively. Since all subfilters in the multirate structure are the linear-phase FIR filters, we exploit the coefficient symmetry property when expressing the number of multiplication constants in equation (9.30).

For the filters of *Example 9.6*, we have:

$N_k = 63$, $N_{D2} = N_{I2} = 9$, and $N_{D1} = N_{I1} = 15$, resulting in $N_{\text{total}} = (87+1)/2 = 44$ multiplication constants.

One should notice here that the multistage solution exploits the halfband filter property that half of the coefficients are zero valued.

The equivalent linear-phase FIR single-stage filter meets the specifications with $N = 211$ of the filter length. Because of the coefficient symmetry, this filter requires 106 multiplication constants for implementation. Recall that the two-stage complementary multirate filter needs only 44 multiplication constants for implementation.

In this section, we have considered the complementary multirate filtering structures based on FIR filters. Recently, the complementary filtering approach has been extended to IIR filters (Johansson, 2003). In this method, the overall filter makes use of an IIR filter as a kernel filter, the periodic all-pass filters for constructing complementary pair, and linear phase FIR filters for the sampling rate alterations.

MATLAB EXERCISES

9.1 Design a lowpass FIR filter composed of a cascade of a factor-of-2 decimator, kernel filter and factor-of-2 interpolator. The specifications for the overall filter are the following:

Sampling frequency: $F_0 = 10000$ Hz, passband edge frequency $F_p = 800$ Hz, stopband edge frequency $F_s = 1000$ Hz, minimal attenuation in the stopband $a_s = 50$ dB, peak passband ripple $\delta_p = 0.01$. Decimation and interpolation filters are identical halfband filters.

Compute and plot:

- Magnitude responses of decimation and interpolation filters
- Magnitude response of the kernel filter
- Magnitude response of the overall multirate filter
- Aliasing characteristic of the overall filter.

9.2 Compose the multirate highpass filter consisting of a cascade of a factor-of-2 decimator, kernel filter and factor-of-2 interpolator. The specifications for the overall filter are the following:

Sampling frequency: $F_0 = 10000$ Hz, passband edge frequency $F_p = 4200$ Hz, stopband edge frequency $F_s = 4000$ Hz, minimal attenuation in the stopband $a_s = 50$ dB, peak passband ripple $\delta_p = 0.01$. Decimation and interpolation filters are identical halfband filters.

Compute and plot:

- Magnitude responses of decimation and interpolation filters
- Magnitude response of the kernel filter
- Magnitude response of the overall multirate filter
- Aliasing characteristic of the overall filter.

REFERENCES

Crochiere, R.E., & Rabiner, L.R. (1983). *Multirate digital signal processing.* Englewood Cliffs, NJ: Prentice-Hall.

Filter design toolbox for use with MATLAB. User's guide. Version 6. (2006). Natick: MathWorks.

Fliege, N. J. (1994). *Multirate digital signal processing.* New York, NY: John Wiley & Sons.

Johansson, H. (2003). Multirate IIR filter structures for arbitrary bandwidth. *IEEE Transactions on Circuits and Systems-I: Fundamental Theory and Applications, 50*(12), 1515-1529.

Lim, Y.C. (1986). Frequency-response masking approach for the synthesis of sharp linear phase digital filters. *IEEE Transactions on Circuits and Systems, 33*(4), 357-364.

Milić, Lj., & Lutovac, M.D. (2002). Efficient multirate filtering. In Gordana Jovanović-Doleček, (ed.), *Multirate Systems: Design & Applications.* Hershey, PA: Idea Group Publishing, 105-142.

Mitra, S. K. (2006). *Digital signal processing: A computer based approach.* 3rd edition. New York, NY: The McGraw-Hill Companies, Inc.

Ramstad, T.A., & Saramäki, T. (1988, June). Multistage, multirate FIR Filter structures for narrow transition-band filters. *Proc. 1988 IEEE Int. Symp. Circuits and Systems – ISCAS*, 2019 – 2022.

Ramstad, T.A., & Saramäki, T. (1990, May). Multistage, multirate FIR filter structures for narrow transition-band filters. *Proc. 1990 IEEE Int. Symp. Circuits and Systems – ISCAS*, New Orleans, Louisiana, 2017 – 2021.

Saramäki, T. (2001). *Multirate Signal Processing.* Lecture notes for a graduate course, the Institute of Signal Processing, Tampere University of Technology, Finland.

Signal processing toolbox for use with MATLAB. User's guide. Version 6. (2006). Natick: MathWorks.

Chapter X
Frequency–Response Masking Techniques

INTRODUCTION

The initial concept of the frequency-response masking technique was introduced by Neuvo, Cheng-Yu and Mitra (1984). It was shown that the complexity of a linear phase FIR filter can be considerably reduced by using the cascade connection of an interpolated FIR (IFIR) filter and a properly designed FIR filter. The IFIR filter transfer function is obtained by replacing the unit delay z^{-1} with the delay block z^{-M}, where M is an integer. In this way, the frequency response of the IFIR filter is made periodic. The FIR filter in the cascade is used to eliminate (mask) the images from the IFIR filter frequency response. Two years later, Lim (1986) proposed a complete approach for the application of frequency-response masking technique in designing narrow-band and arbitrary-band linear phase FIR filters. It was shown that the approach given in (Lim, 1986) results in a linear phase FIR filter with a small fraction of nonzero coefficients, and thus is suitable for implementing sharp filters with arbitrary bandwidths. The arithmetic complexity is considerably smaller in comparison with the arithmetic complexity of an optimal FIR filter having the equivalent frequency response.

This approach is applied later to IIR filters by Johansson and Wanhammar (1997, 2000). The overall filter is composed of an IIR periodic model filter and its complementary periodic filter, and FIR linear-phase masking filters. In this way, the arbitrary-band filter can be designed. For a narrowband filter, the cascade of a periodic filter and masking filter can be used.

The frequency-response masking approach is suitable for digital filters with sharp transition bands. Compared to the classical single-filter design, this technique offers the advantage of lower coefficients' sensitivity, higher computation speed and lower power consumption.

Recently, the application of frequency-response masking approach has been extended to filter banks to achieve a sharp band-separation with reduced computational complexity (Furtado, Diniz, Netto, and Saramäki, T. 2005; Rosenbaum, Lövenborg, and Johansson, 2007).

In this chapter, we review the frequency-response masking techniques for narrow-band and arbitrary bandwidth IIR filters. We demonstrate through examples that very selective characteristics can be obtained using relatively low-order sub-filters. In this way, stable, low-sensitive filters are obtained.

NARROWBAND FILTER DESIGN

The frequency-response masking technique can be used for a narrowband filter design. The principle is very simple: the narrow-band filter is obtained as a cascade of a periodic model filter and a masking filter. Figure 10.1 illustrates the cascade connection of the periodic model filter $G(z^M)$ and the masking filter $F(z)$, and Figure 10.2 indicates the concept of the narrowband filter design.

Design starts from the model filter $G(z)$ and its frequency response $G(e^{j\omega})$, illustrated in Figure 10.2 (a). We call this filter a model filter. Replacing each delay in the model filter by M delays, the periodic model filter $G(z^M)$ is obtained. The frequency response of the periodic model filter $G(e^{jM\omega})$ is sketched in Figure 10.2 (b). The periodic spectra (images) produced by $G(e^{jM\omega})$ can be eliminated by the masking filter. For a low-pass filter design, the masking filter is a lowpass, indicated by $F_L(e^{j\omega})$ in Figure 10.2 (c). The cascade of $G(e^{jM\omega})$ and $F_L(e^{j\omega})$ produces a desired narrowband lowpass filter $H_L(e^{j\omega})$, Figure 10.2 (d). For a bandpass filter design, the masking filter has to be the bandpass as shown in Figures 10.2 (e) and 10.2 (f). If a highpass characteristic is required, only $F(z)$ has to be a highpass. Therefore, for narrow-band filters, the transfer function $H(z)$ is expressed in the form

$$H(z) = G(z^M) F(z)$$

(10.1)

We can arbitrarily choose FIR or IIR transfer functions for $G(z)$ and $F(z)$.

The important outcome of the proposed approach is that the transition band of the overall filter is M times smaller than that of the model filter. This effect is produced by the replacement of every delay in $G(z)$ by M delays. Consequently, the passband bandwidth is also reduced by the same factor. Hence, this method is only suitable for narrowband design.

This masking technique provides the implementation of a sharp narrowband characteristic using filters with much wider transition bands. The specifications for masking filtering may be additionally relaxed by using the multistage implementation of $F(z)$, as shown in Figure 10.3.

The overall transfer function for the structure of Figure 10.3 is given by

$$F(z) = \prod_{k=1}^{R} F_k\left(z^{M_k}\right)$$

(10.2)

Periodic filters F_1, F_2, ..., F_R from Figure 10.3 are designed to subsequently remove the images from the frequency response. It is shown in (Johansson & Wanhammar, 1996) that for masking filters $F_1(z)$, $F_2(z)$, ..., $F_R(z)$ the halfband filters can be used. Configurations based on the use of halfband masking filters achieve an improvement in the computational efficiency.

Figure 10.1. Cascade connection of periodic model filter and masking filter

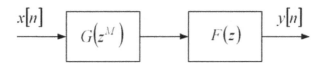

Figure 10.2. Frequency-response masking approach. Narrowband filter design.

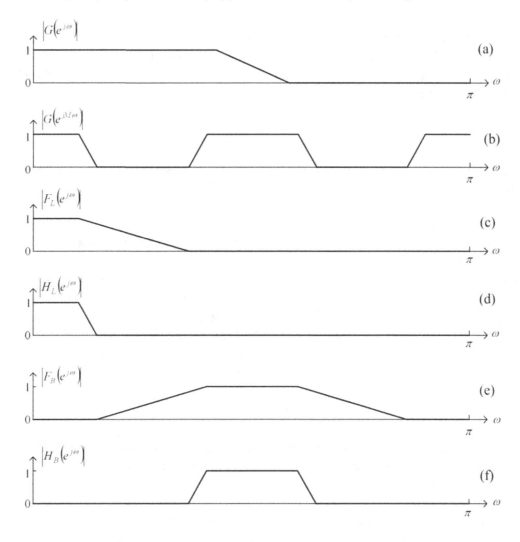

Figure 10.3. Multistage implementation of masking filter

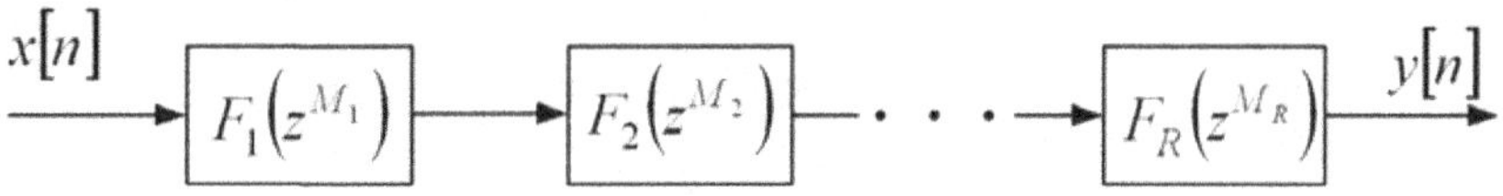

By means of example that follows, we illustrate the application of the frequency-response masking technique in the narrowband FIR filter design.

Example 10.1

Specifications:
A lowpass FIR filter is designed for $F_p = 100$ Hz, $F_s = 150$ Hz, and the sampling frequency of $F_0 = 2000$ Hz. The passband and stopband ripples are $\delta_p = 0.01$, and $\delta_s = 0.01$ ($a_s = 40$ dB).

Solution:
For the frequency-response masking approach, a suitable choice is $M = 4$. Therefore, we design the model filter $G(z)$ with $F_p = 100$ Hz, $F_s = 150$ Hz, and the sampling frequency $F_0/M = 500$ Hz. The specifications can be met with a linear-phase halfband FIR filter of the length $N_M = 23$. We use the MATLAB function firhalfband to design the model filter $G(z)$:

```
g = firhalfband(22,100/250);              % Generating model filter
```

The impulse response of the periodic model filter $G(z^M)$ is obtained by inserting $M–1$ zeros between two consecutive samples,

```
gm = zeros(size(1:4*length(g)));
gm([1:4:length(gm)]) = g;                 % Generating periodic model filter, M = 4
```

The magnitude response of the periodic model filter is shown in Figure 10.4 (a).

The role of the masking filter $F(z)$ is to provide the attenuation in the frequency bands where $G(z^M)$ has the unwanted passbands. We use the MATLAB function firpm from the *Signal Processing Toolbox* to design the masking filter. According to Figure 10.4 (a), we specify the passband/stopband edges $F(z)$: $F_p = 100$ Hz, $F_s = 350$ Hz, and the don't care band in the range $650 – 850$ Hz. Hence, the vector of frequencies f0, and the vector of amplitudes m0 are specified by

```
f0 = [0,100/1000,350/1000,650/1000,850/1000,1]; m0 = [1,1,0,0,0,0];
```

It was found experimentally that a better solution can be obtained when slightly shifting the boundary frequencies in the vector f0 defined above. So, in the masking filter design we use the following vector of frequencies f0,

```
f0 = [0,100/1000,375/1000,625/1000,875/1000,1]; m0 = [1,1,0,0,0,0];
```

We choose $N_F = 19$ for the filter length. Finally, we use the MATLAB function firpm to compute the vector of coefficients ff of the masking filter $F(z)$,

```
ff = firpm(18,f0,m0,[1,0.5,0.5]);          % Generating the masking filter
```

The magnitude response of the masking filter is shown in Figure 10.4 (b), whereas Figure 10.4 (c) presents the magnitude response of the resulting narrowband filter.

In this design, the model filter of the length $N_M = 23$, being a halfband filter, has only 13 nonzero coefficients. Since the masking filter has 19 coefficients, the total number of nonzero coefficients is 32. The model filter and the masking filter are both linear-phase FIR filters exhibiting the coefficient symmetry property. When exploiting the coefficient symmetry property, the number of nonzero coefficients in the model filter reduces to 7, and in the masking filter the number of coefficients reduces to 10. That means that the overall solution based on the frequency-response masking approach whose magnitude response is shown in Figure 10.4 (c) can be implemented with only 17 nonzero coefficients.

The specifications of *Example 10.1* can be met with an optimal single-stage linear-phase FIR filter of the length $N = 79$, which requires 79 nonzero coefficients for implementation. This number is reduced to 40 when exploiting the coefficient symmetry property. Evidently, the frequency-response masking approach considerably reduces the computational complexity of the overall narrowband filter.

However, the concept of frequency response masking technique leads to an increase of the overall filter delay. Let us consider the solution of *Example 10.1*. The length of the periodic model filter is $N_P = M \times N_M - (M-1) = 4 \times 23 - 3 = 89$, whereas the length of the masking filter is $N_F = 19$. Thus, the cascade of periodic model filter and masking filter has a total delay of $N_T = 48$ samples, whereas the alternative realization consisting of only one optimal FIR filter ($N = 79$) has a delay of 39 samples.

In *Example 10.1*, we have demonstrated in detail the design of model and masking filters. In MATLAB, one can also use the function ifir, which, for the given specifications, returns the coefficients of the model and masking filters.

Efficient narrowband filters can be achieved when IIR transfer functions are used for model and masking filters (Johansson & Wanhammar 1997, 2000; Lutovac & Milić, 2001). It will be shown here

Figure 10.4. Narrowband FIR filter: Gain responses of periodic model filter, masking filter and the narrow-band filter

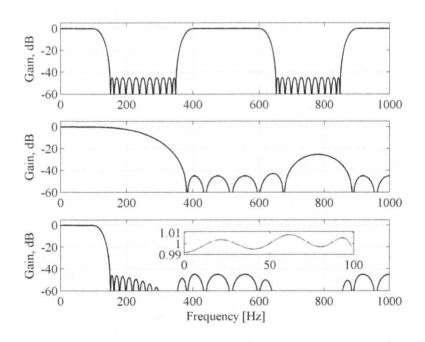

that the frequency-response masking approach provides the implementation of a sharp narrowband filter using simple IIR subfilters.

The low-sensitivity realisation structure for model and masking filters is essential. For this technique, the attractive solutions can be achieved with elliptic minimal Q-factors (EMQF) transfer functions and the realization structures based on the parallel connection of two all-pass subfilters such as lattice wave digital filters (WDFs). Hence, the transfer function of the model filter and that of masking filter are represented as a combination of all-pass functions,

$$G(z)=\frac{1}{2}\left(A_0^G(z)\pm A_1^G(z)\right), \qquad F(z)=\frac{1}{2}\left(A_0^F(z)\pm A_1^F(z)\right) \tag{10.3}$$

where $A_0(z)$ and $A_1(z)$ with upper-scripts G and F are the all-pass subfilters that compose the model filter $G(z)$ and the masking filter $F(z)$, respectively. When using "+" in equation (10.3) filter $G(z)$ [$F(z)$] is a lowpass filter, and sign "$-$" is used for the highpass filters.

Design and implementation of EMQF filters have been exposed in Chapter V, section IIR Structures with Two All-Pass Subfilters: Applications of EMQF Filters. In the following example, we illustrate the application of a simple EMQF filter in constructing the lowpass narrowband IIR filter.

Example 10.2

Specifications:
Stopband edge frequency $\omega_s = 0.1\pi$, passband edge frequency $\omega_p = 0.065\pi$, and the minimal stopband attenuation $a_s = 30$ dB.

Solution:
In this example, we show a solution for the narrowband filter based on the identical EMQF subfilters, i.e. we choose $G(z) = F(z)$.

For both the model and the masking filter ($G(z)$ and $F(z)$) we use the 5th order EMQF filter with the 3dB cutoff $\omega_{3dB} = (1/3)\pi$, and the stopband edge frequency $\omega_s = 0.4\pi$. The minimal stopband attenuation amounts to $a_s = 31.4372$ dB, and the passband ripple is only $a_p = 0.0031$dB with the passband edge frequency at $\omega_p = 0.2738\pi$,

Using the 6-step procedure given in Chapter V, section IIR Structures with Two All-Pass Subfilters: Applications of EMQF Filters, we compute the constants for the all-pass branches.

The resulting all-pass functions are the following

$$A_0(z)=\frac{0.3445-0.5(1+0.3445)z^{-1}+z^{-2}}{1-0.5(1+0.3445)z^{-1}+0.3445z^{-2}}$$

$$A_1(z)=\frac{-0.2679+z^{-1}}{1-0.2679z^{-1}}\times\frac{0.7871-0.5(1+0.7871)z^{-1}+z^{-2}}{1-0.5(1+0.7871)z^{-1}+0.7871z^{-2}}$$

We can notice that the second-order sections have the common constant of the value $\alpha = -0.5$, which can be implemented as a binary shift. This is due to the choice of the 3dB cutoff $\omega_{3dB} = (1/3)\pi$, see equation (5.43). Hence, this 5th order filter requires only three multipliers in implementation. The gain characteristic of the 5th order filter is plotted in Figure 10.5, upper subfigure.

Figure 10.5. Narrowband IIR filter: Gain responses of EMQF filter used as model and masking filter, periodic model filter and resulting narrow-band filter.

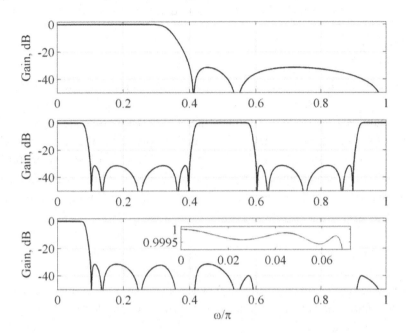

In this example, we use the same 5th order EMQF filter for masking filter, and the upper subfigure of Figure 10.5 represents also the gain characteristic of the masking filter.

The periodic model filter is obtained with $M = 4$, see the second subfigure in Figure 10.5.

The narrowband lowpass filter is formed as a cascade of the periodic model filter and the masking filter. The bottom subfigure of Figure 10.5 plots the overall gain response. Notice that the overall narrowband filter requires only 6 multiplication constants: 3 for the model filter, and 3 for the masking filter.

ARBITRARY BANDWIDTH DESIGN

In this section, we present the principle of the frequency-response masking technique suitable for arbitrary bandwidth IIR filters.

A combination of a complementary filter pair and a properly designed pair of masking filters may result in a digital filter with an arbitrary bandwidth. The lowpass/highpass complementary filter pair is used to generate a periodic model filter pair. The role of masking filters is to remove the unwanted periodic spectra.

FIR and IIR Complementary Filter Pairs that are Used as Model Filters

The complementary filter pair $[G(z), G_c(z)]$ that is used as a pair of model filters in the frequency-response masking structure should satisfy the following complementary property

$$\left| G\left(e^{j\omega}\right) + G_c\left(e^{j\omega}\right) \right| = 1. \tag{10.4}$$

Recall that the properties of the complementary filter pairs are exposed in Chapter VIII.

For the solution based on linear-phase FIR filters, we use the delay-complementary filter pair $[G(z), G_C(z)]$ whose transfer functions $G(z)$ and $G_C(z)$ are related according to equation (8.7) by,

$$G_C(z) = z^{-(N-1)/2} - G(z), \tag{10.5}$$

where N is the filter length. Usually, $G(z)$ is the lowpass and $G_C(z)$ the highpass filter. The implementation structure of the filter pair is sketched in Figure 10.6.

For the IIR filter pair, we use a double-complementary filter pair composed of the parallel connection of two all-pass subfilters, see subsection *Class I:* Power-Complementary and All-Pass Complementary Filter Pairs Implemented as a Parallel Connection of Two All-Pass Subfilters in Chapter VIII. This filter pair is both an all-pass complementary and a power complementary filter pair. The transfer functions of $G(z)$ and its complementary $G_C(z)$ are expressible as a sum and difference of two all-pass functions $A_0^G(z)$ and $A_1^G(z)$, respectively,

$$G(z) [G_C] = \frac{1}{2}\left(A_0^G(z) \pm A_1^G(z)\right). \tag{10.6}$$

The sign "+" corresponds to the lowpass filter $G(z)$, whereas the sign "−" is used for the complementary highpass filter $G_C(z)$. The efficient implementation structure for the lowpass/highpass IIR filter pair $[G(z), G_C(z)]$ is shown in Figure 10.7.

Filter Synthesis Based on a Complementary Filter Pair and Two Masking Filters

In the frequency-response masking technique, the complementary pair of periodic filters is used with two masking filters. The transfer function of the overall filter $H(z)$ is represented in the form

$$H(z) = G\left(z^M\right)F_0(z) + G_C\left(z^M\right)F_1(z) \tag{10.7}$$

where M is a positive integer. We refer to $G(z)$ and $G_C(z)$ as a model filter and complementary model filter, respectively, and to $G(z^M)$ and $G_C(z^M)$ as a periodic model filter and periodic complementary model

Figure 10.6. Implementation of the lowpass/highpass delay-complementary FIR filter pair $[G(z), G_C(z)]$

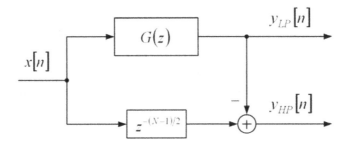

Figure 10.7. Implementation of the lowpass/highpass double-complementary IIR filter pair [G(z), $G_C(z)$]

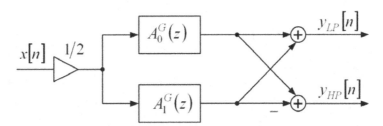

filter. Masking filters $F_0(z)$ and $F_1(z)$ extract one or several passbands of the periodic model filter $G(z^M)$ and the complementary periodic model filter $G_C(z^M)$.

The block diagram in Figure 10.8 shows the structure of the frequency-response masking approach when the linear-phase FIR filters are used to implement equation (10.7). The complementary pair of periodic model filters is implemented according to Figure 10.6. The masking filters $F_0(z)$ and $F_1(z)$ are two distinct FIR filters, which should be adjusted to the same length, i.e. $N_0 = N_1$. Since $F_0(z)$ and $F_1(z)$ are two distinct filters, the original design may result in different filter lengths, $N_0 \neq N_1$. In that case, the shorter impulse response has to be extended (zero-padded) up to that of the longer one.

The frequency-response masking filter can also be synthesized with the use of IIR complementary filter pair as model filters. The model filter pair is a double-complementary filter pair, which is realizable as a parallel connection of two all-pass subfilters, see equation (10.6) and Figure 10.7. The masking filters are linear-phase FIR filters. It is to be emphasized that the impulse responses of the masking filters should be adjusted to the same length.

Introducing expression (10.6) for model filters into equation (10.7), we express the overall transfer function $H(z)$ in terms of the all-pass functions $A_0{}^G(z)$ and $A_1{}^G(z)$, and masking filters $F_0(z)$ and $F_1(z)$,

$$H(z) = \frac{A_0^G\left(z^M\right) + A_1^G\left(z^M\right)}{2} F_0(z) + \frac{A_0^G\left(z^M\right) - A_1^G\left(z^M\right)}{2} F_1(z). \tag{10.8}$$

The block diagram in Figure 10.9 shows the realization structure that implements the frequency-response masking expressed by equation (10.8).

Figure 10.8. Block diagram of the FIR filter synthesized using the frequency-response masking approach

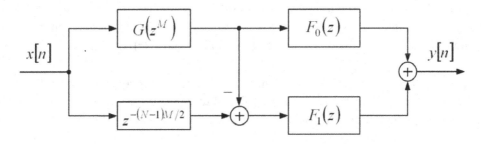

Figure 10.9. Block diagram of the IIR filter synthesized using the frequency-response masking approach

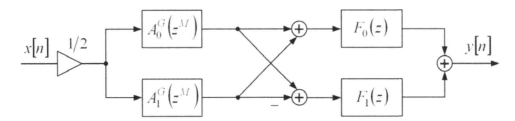

Typical magnitude responses for model, periodic model, masking, and overall filters are shown in Figure 10.10 for the case of a wide-band lowpass filter.

Model filters $G(z)$ and $G_C(z)$ are complementary half-band filters with band edge frequencies ω_p^G and ω_s^G, Figure 10.10(a). Replacing each delay in $G(z)$ and $G_C(z)$ by M delays, the periodic model filter $G(z^M)$ and the periodic complementary model filter $G_C(z^M)$ are obtained, Figure 10.10(b). Notice that either FIR or IIR complementary filter pair can be used for $[G(z), G_C(z)]$.

Two masking filters $F_0(z)$ and $F_1(z)$, whose frequency responses are shown in Figure 10.10(c), may be used to mask $G(e^{jM\omega})$ and $G_C(e^{jM\omega})$, respectively. The passband and the stopband edge frequencies for $F_0(e^{j\omega})$ and $F_1(e^{j\omega})$ are indicated in Figure 10.10(c). If the outputs of filters $F_0(z)$ and $F_1(z)$ are added as in Figures 10.8 and 10.9, the overall frequency response $H(e^{j\omega})$ of the resulting filter is obtained, Figure 10.10(d). The resulting filter $H(z)$ is a wideband filter with a sharp transition band.

If ω_p and ω_s are the passband and the stopband edge frequencies of $H(e^{j\omega})$, respectively, it can be shown that

$$\omega_p = \frac{2k\pi + \omega_p^G}{M}, \quad \omega_s = \frac{2k\pi + \omega_s^G}{M}, \tag{10.9}$$

where k is an integer, $k < M$. It apparently follows from equation (10.9) that $\omega_p - \omega_s = (\omega_s^G - \omega_p^G)/M$, i.e., the transition band of $H(e^{j\omega})$ is equal to that of $G(e^{jM\omega})$.

In a filter design problem, ω_p and ω_s are specified and k, M, ω_p^G and ω_s^G are the parameters to be determined. Usually M is chosen first to minimize the overall filter complexity, (Lim, 1986). To ensure that equation (10.9) yields a solution with $0 < \omega_p^G < \omega_s^G$, parameters k, ω_p^G and ω_s^G are determined according to (Lim, 1986),

$$k = \lfloor \omega_p M / (2\pi) \rfloor, \quad \omega_p^G = 2k\pi - \omega_p M, \quad \omega_s^G = \omega_s M - 2k\pi \tag{10.10}$$

where $\lfloor \omega_p M / (2\pi) \rfloor$ denotes the largest integer less than $\omega_p/(2\pi)$.

For a proper filter design, the following points should be noted. First, the group delay of $F_0(z)$ and that of $F_1(z)$ must be equal. If this is not the case, leading delays must be added to either $F_0(z)$ or $F_1(z)$ to equalize their group delays. Obviously, the lengths of $F_0(z)$ and $F_1(z)$ must either be both even or both odd.

The transition band of the overall filter $H(z)$ can be selected to equal one of the transition bands of either $G(z^M)$ or $G_C(z^M)$. These two cases are referred to as *Case 1* and *Case 2*, respectively (Johansson, 2000). The approach described above and indicated in Figure 10.10 is the *Case 1* design since the

Figure 10.10. Frequency-response masking of complementary filters

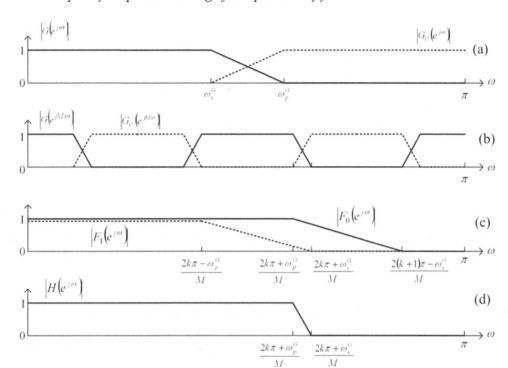

Table 10.1. Boundary Frequencies: Case 1

Filter	ω_p	ω_s
$H(z)$	$(2k\pi + \omega_p{}^G)/M$	$(2k\pi + \omega_s{}^G)/M$
$F_0(z)$	$(2k\pi + \omega_p{}^G)/M$	$[2(k+1)\pi - \omega_s{}^G]/M$
$F_1(z)$	$(2k\pi - \omega_p{}^G)/M$	$(2k\pi + \omega_s{}^G/M$

Table 10.2. Boundary Frequencies: Case 2

Filter	ω_p	ω_s
$H(z)$	$(2k\pi - \omega_s{}^G)/M$	$(2k\pi - \omega_p{}^G)/M$
$F_0(z)$	$[2(k-1)\pi + \omega_s{}^G]/M$	$(2k\pi - \omega_p{}^G)/M$
$F_1(z)$	$(2k\pi - \omega_s{}^G)/M$	$(2k\pi + \omega_p{}^G)/M$

transition band of $G(z^M)$ is utilized for the transition band of $H(z)$. Table 10.1 summarizes the relations from Figure 10.10.

In the *Case II* design, the transition band of the complementary filter $G_C(z^M)$ is utilized as the transition band of the overall filter $H(z)$. The boundary frequencies for overall filter $H(z)$ and masking filters $F_0(z)$ and $F_1(z)$ for *Case II* design are given in Table 10.2.

For *Case II* design, the constant k is determined by

$$k = \lceil M\omega_s / 2\pi \rceil \tag{10.11}$$

where $\lceil \omega_S M / (2\pi) \rceil$ denotes the smallest integer larger than $\omega_s/(2\pi)$.

The two examples that follow illustrate the frequency-response masking approach for the design of wide band filters. In *Example 10.3*, we illustrate the *Case 1* design for a linear-phase FIR filter. The *Case 1* design for the IIR filter is illustrated by means of *Example 10.4*.

Example 10.3

Specifications:
A wideband linear-phase lowpass filter is specified with the passband edge frequency $F_p = 600$ Hz, the stopband edge frequency $F_s = 650$ Hz and the sampling frequency $F_0 = 2000$ Hz. The maximal ripple in the passband is $a_p \pm 0.1$ dB, and minimal stopband attenuation amounts to $a_s = 40$ dB.

Solution:
We find $\omega_p = 2\pi F_p/F_0 = 0.6\pi$, and $\omega_s = 2\pi F_s/F_0 = 0.650\pi$. We choose $M = 4$, and using Equation 10.10, we determine k, $\omega_p{}^G$ and $\omega_s{}^G$: $k = 1$, $\omega_p{}^G = 0.4\pi$, $\omega_s{}^G = 0.6\pi$.

The model filter $G(z)$ is designed as a half-band linear-phase FIR filter of the length $N = 23$ using the MATLAB function firhalfband,

```
N = 23; fp = 0.4;          % Setting the design parameters for firhalfband
g = firhalfband(N-1,fp);   % Model filter design
```

and to obtain the periodic model filter $G(z^M)$, we insert $M-1$ zeros between two consecutive samples

```
M = 4;
gm = zeros(size(1:M*length(g)));   gm([1:M:length(gm)]) = g;   % Generating the periodic model filter
```

Following the block diagram of Figure 10.8, we compute the complementary periodic model filter $G_c(z^M)$,

```
d = zeros(size(gm)); d(45) = 1;   % Delay branch
gmc = d - gm;                      % Complementary periodic model filter
```

The masking filters $F_0(z)$ and $F_1(z)$ are lowpass optimal FIR filters: $F_0(z)$ is designed with the boundary frequencies 0.6π for the passband and 0.850π for the stopband, and the length of $N = 19$; $F_1(z)$ has the boundary frequencies 0.4π for the passband and 0.650π for the stopband, and the filter length is $N = 21$. We use the MATLAB function firpm to design linear-phase masking filters $F_0(z)$ and $F_1(z)$

```
ap = 0.1;  % Passband ripple, dB
as = 40;   % Stopband ripple, dB
delta_p = (10^(ap/20)-1)/(10^(ap/20)+1);   % Passband ripple
```

```
delta_s = 10^(-as/20);    % Stopband ripple
% Setting design parameters for masking filter f0
N0 = 19; freq = [0,0.6,0.85,1]; m0 = [1,1,0,0]; w = [delta_p/delta_s,1];
f0 = firpm(N0-1,freq,m0,w);  % Designing the masking filter f0
% Setting design parameters for masking filter f1
N1 = 21; freq = [0,0.4,0.625,1]; m1=[1,1,0,0];  w = [delta_p/delta_s,1];
f1 = firpm(N1-1,freq,m1,[0.5756,1]);    % Designing the masking filter f1
f0 = [0,f0,0];      % Equalization of the masking filters' delays
```

Figure 10.11 plots the gain responses of the model filter, complementary pair of model filters and those of two masking filters.

According to the implementation structure of Figure 10.8, and equation (10.7), we compute the impulse response of the overall filter $H(z)$

```
h = conv(gm,f0) + conv(gmc,f1);          % Impulse response of the overall filter
```

The gain response of the resulting overall filter $H(z)$ is displayed in Figure 10.12.
Let us consider the computational complexity of the overall filter $H(z)$:

- The complementary pair of model filters is a halfband filter pair and as such is implemented at the cost of a single filter, see Figures 10.6 and 10.8. The length of the halfband model filter is 23, and thereby the model filter pair has 13 nonzero coefficients.

Figure 10.11. Gain responses of model filter, complementary pair of periodic model filters, and masking filters

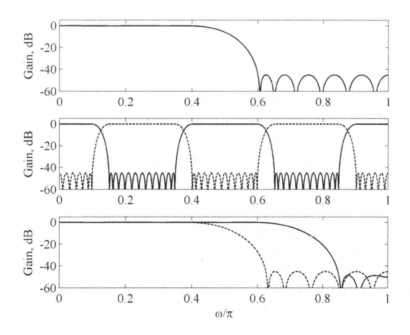

Figure 10.12. Gain response of wideband FIR filter synthesized using frequency-response masking approach. Case 1 design.

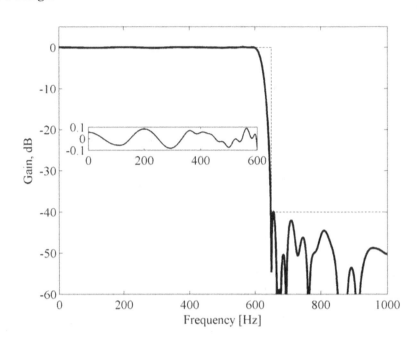

- The masking filters are of the lengths $N_0 = 19$, and $N_1 = 21$
- The total number of the nonzero coefficients amounts to $N_T = 13+19+21 = 53$.

In implementation, one can exploit the coefficient symmetry property in the model filter pair and in the masking filters, as well. In this way, the number of multiplication constants can be halved. Note that the group delay of the overall filter $H(z)$ is 54 samples.

Example 10.4

In this example, we demonstrate the filter synthesis based on the complementary IIR model filter pair, and two FIR masking filters according to the structure of Figure 10.9.

The specifications are the same as in *Example 10.3*.

For the model filter, we use the 5th-order IIR half-band IIR filter with passband edge at $\omega_p^G = 0.4\pi$, and the stopband edge at $\omega_s^G = 0.6\pi$. We use the program halfbandiir from the Appendix A of this book to compute the poles of the model filter $G(z)$. Then, we compute the frequency responses of the all-pass branches $A_0^G(z)$ and $A_1^G(z)$, and in the next step we determine the frequency responses of the complementary filter pair $[G(z), G_C(z)]$.

```
[b,a,z,p,k] = halfbandiir(5,0.4);    % Halfband filter design
beta2 = (abs(p(2)))^2;   beta3 = (abs(p(4)))^2;  % Coefficients of all-pass branches
[A0,w1] = freqz([beta2,0,1],[1,0,beta2],512);    % Model filter: all-pass branch A0(z)
[A1,w1] = freqz([0,beta3,0,1],[1,0,beta3],512);   % Model filter: all-pass branch A1(z)
G = (A0+A1)/2; Gc = (A0-A1)/2;   % Model filter G(z), and complementary model filter G_C(z)
```

We use the following MATLAB code to compute the frequency responses for the periodic model filters $[G(z^M)\ G_C(z^M)]$,

```
[A0m,w1] = freqz(upsample([beta2,0,1],4),upsample([1,0,beta2],4),512);  % All-pass branch A0(z^M)
[A1m,w1] = freqz(upsample([0,beta3,0,1],4),upsample([1,0,beta3],4),512);  % All-pass branch A1(z^M)
% Periodic model filter and complementary periodic model filter
Gm = (A0m+A1m)/2;   Gcm =( A0m-A1m)/2;
```

We use the masking filters $F_0(z)$ and $F_1(z)$ from *Example 10.3* as the masking filters for this example too. Since the impulse responses $f_0[n]$ and $f_1[n]$ are already determined in *Example 10.3*, we compute only the frequency responses,

```
[F0,w1] = freqz(f0,1,512);            % Frequency response of masking filter F0(z)
[F1,w1] = freqz(f1,1,512);            % Frequency response of masking filter F1(z)
```

Finally, we compute the frequency response of the resulting filter $H(z)$,

```
H = Gm.*F0 + Gcm.*F1;              % Frequency response of the overall filter H(z)
```

Figure 10.13 plots the gain of the resulting overall filter $H(z)$. Since the model filter is a nonlinear-phase IIR filter, the phase characteristic of $H(z)$ is nonlinear too. Note that the 5th order halfband filter is implemented with only two nonzero coefficients. Therefore, the total number of nonzero coefficients for this implementation is 42. This number can be reduced when exploiting the coefficient symmetry property of the linear-phase masking filters $F_0(z)$ and $F_1(z)$.

Figure 10.13. Gain response of wideband IIR filter synthesized using frequency-response masking approach. Case 1 design.

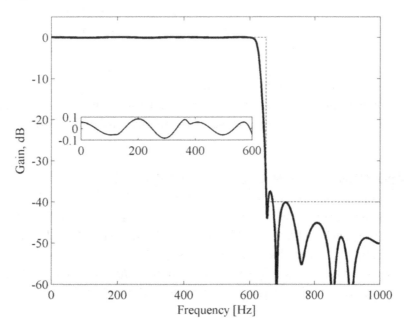

PHASE CHARACTERISTICS

Narrowband Filters

The phase characteristic of the narrowband filter $H(z)$, which is synthesized as a cascade of the periodic model filter $G(z^M)$ and the masking filter $F(z)$ according to equation (10.1) is the sum of the phase characteristics of $G(z^M)$ and $F(z)$, i.e.,

$$\phi(\omega) = \arg\left[H\left(e^{j\omega}\right) \right] = \arg\left[G\left(e^{jM\omega}\right)F\left(e^{j\omega}\right) \right] = \arg\left[G\left(e^{jM\omega}\right) \right] + \arg\left[F\left(e^{j\omega}\right) \right] \tag{10.12}$$

The phase characteristic $\phi(\omega)$ is a linear function of ω if the model filter and masking filter are both linear-phase FIR filters. Obviously, when either the model filter or masking filter (or both) are nonlinear phase filters, the overall phase characteristic $\phi(\omega)$ is nonlinear.

Wideband Filters

For the wideband frequency-response-masking filter synthesized according to equation (10.7), the phase characteristic of the overall filter $H(z)$ is given by

$$\phi(\omega) = \arg\left[H\left(e^{j\omega}\right) \right] = \arg\left[G\left(e^{j\omega M}\right)F_0\left(e^{j\omega}\right) + G_C\left(e^{j\omega M}\right)F_1\left(e^{j\omega}\right) \right] \tag{10.13}$$

Here, the masking filters $F_0(z)$ and $F_1(z)$ are linear-phase FIR filters, and therefore, the linearity/nonlinearity property of the phase characteristic $\phi(\omega)$ depends on the choice of the model filter pair $[G(z), G_C(z)]$. If the linear phase FIR filters are used for the model filter pair, the phase characteristic of the overall filter $H(z)$ is a linear function of frequency. When an IIR filter pair is used for $[G(z), G_C(z)]$, the overall phase characteristic $\phi(\omega)$ becomes a nonlinear function of frequency. The application of elliptic transfer function for the model filter pair may cause the significant phase nonlinearity in the overall filter.

In a wideband filter design approach, several bands of the periodic model filter and the complementary periodic model filter are used to produce the passband of the overall filter, see Figure 10.10. It is to be noticed that the transition bands of periodic filters are in the passband of the overall filter. It is well known that the group delay reaches its maximum at the crossover frequency of the complementary filter pair. This fact may produce the higher passband phase nonlinearity in the wideband frequency-response-masking filter when compared with the passband phase nonlinearity of the model filter.

CONSTRAINED DESIGN FOR WIDEBAND FILTERS

The frequency-response masking approach for wideband filters presented above in this Chapter is based on the usage of two FIR masking filters, see Figures 10.8 and 10.9. Therefore, the overall filter complexity may become rather high, especially for large values of M.

When the passband and stopband edges of the overall filter $H(z)$ are in the neighborhood of $\pi/2$, the total complexity can be reduced by imposing constraints on the masking filters.

One method for IIR filters, which is suitable for filters with edge frequencies in the vicinity of $\pi/2$ is proposed by Lutovac and Milić (2001). This method is developed for the realization structure of Figure 10.9. The complementary model filter pair $[G(z), G_C(z)]$ is an elliptic minimal Q-factors (EMQF) filter pair. The EMQF filter pair is both all-pass complementary and power-complementary, and as such satisfies the complementary criterion of equation (10.4). The point is that by adjusting the crossover frequency of the model filter pair $[G(z), G_C(z)]$ one of the FIR masking filters $F_0(z)$ or $F_1(z)$ can become a halfband FIR filter. Since every second coefficient in an odd-length halfband filter is zero valued, the overall complexity of the resulting frequency-response masking filter $H(z)$ is reduced.

According to Figure 10.10, we can choose $F_0(z)$ or $F_1(z)$ to be a halfband filter if the passband and stopband edges of the overall filter $H(z)$ are in the neighborhood of $\pi/2$. Therefore, letting one of the masking filters to be an odd-length halfband filter reduces the overall filter complexity. Regarding $F_0(z)$ and $F_1(z)$, it is desirable to provide that the filter with a smaller transition band is designed as a halfband filter. The optimal overall filter design is a trade-off between the complexity of masking filters, the value of M, and the transition band of the EMQF model filter.

We can use the above method for designing *Case 1* and *Case 2* frequency-response masking filters, see Tables 10.1 and 10.2. In the following example, we show the solution for a very sharp transition lowpass filter using *Case 2* frequency-response masking design, Table 10.2.

Example 10.5

Lowpass wideband filter design.

Specifications:
We synthesize the lowpass filter $H(z)$ with the following specifications: $\omega_p = 0.4\pi$, $\omega_s = 0.43\pi$, the minimal stopband attenuation $a_s = 40$ dB, and the maximal passband ripple $\delta_p = \pm 0.1$.

Solution:
In this example, we illustrate *Case 2* design with $M = 4$. For the model filter $G(z)$, the 7th order EMQF filter with boundary frequencies $\omega_p^G = 0.2738\pi$ and $\omega_s^G = 0.4\pi$ is used.

Using the 6-step procedure given in Chapter V, section IIR Structures with Two All-Pass Subfilters: Applications of EMQF Filters, we compute the constants for the all-pass branches of the 7th-order EMQF filter, $A_0^G(z)$ and $A_1^G(z)$. The resulting all-pass functions are the following

$$A_0^G(z) = \frac{0.2247 - 0.5(1 + 0.2247)z^{-1} + z^{-2}}{1 - 0.5(1 + 0.2247)z^{-1} + 0.2247z^{-2}} \times \frac{0.8465 - 0.5(1 + 0.8465)z^{-1} + z^{-2}}{1 - 0.5(1 + 0.8465)z^{-1} + 0.8465z^{-2}}$$

$$A_1^G(z) = \frac{-0.2679 + z^{-1}}{1 - 0.2679z^{-1}} \times \frac{0.5394 - 0.5(1 + 0.5394)z^{-1} + z^{-2}}{1 - 0.5(1 + 0.5394)z^{-1} + 0.5394z^{-2}}$$

The gain response for the lowpass/highpass complementary filter pair $[G(z)\ G_C(z)]$ is shown in the upper-subfigure of Figure 10.14. Note that only 4 multiplication constants are needed to implement the complementary filter pair.

The gain responses of the complementary periodic filter pair for the factor $M = 4$ are plotted in the second subfigure of Figure 10.14.

The masking filter $F_1(z)$ is the 35-length half-band filter, whereas $F_0(z)$ is the 13-length optimal FIR filter. Since $F_0(z)$ and $F_1(z)$ have to be of the same length, the length of $F_0(z)$ is extended with the corresponding number of zeros.

```
fp = 0.2738;      M = 4;                              % Design parameters for masking filters
f1 = firhalfband(34,0.5-fp/M);                        % Designing the halfband masking filter f1
f0 = firgr(12,[0,fp/M,0.5-fp/M,1],[1,1,0,0]);         % Designing the masking filter f0
f0 = [0,0,0,0,0,0,0,0,0,0,0,0,f0,0,0,0,0,0,0,0,0,0,0,0,0,0];  % Adjusting the length of masking filter f0
```

The bottom subfigure of Figure 10.14 illustrates the gain characteristics of FIR masking filters $F_0(z)$ and $F_1(z)$.

Gain response of the overall filter $H(z)$ implemented according to Figure 10.9 is plotted in Figure 10.15. The resulting boundary frequencies are placed at $\omega_p = 0.4\pi$, $\omega_s = 0.4315\pi$, and the minimal stopband attenuation exceeds the requested value $a_s = 40$ dB.

The total number of nonzero constants in the overall filter $H(z)$ is 36: (i) 4 constants for complementary IIR filter pair, (ii) 19 for the halfband masking filter $F_1(z)$, (iii) 13 for the masking filter $F_0(z)$. Since the masking filters $F_0(z)$ and $F_1(z)$ are linear-phase FIR filters, the number of constants can be halved in the masking filters when exploiting the coefficient symmetry property in implementation.

Figure 10.14. Gain responses. Upper subfigure: solid line –EMQF model filter, dashed line – EMQF complementary model filter. Second subfigure: Periodic model filters for M = 4, solid line- periodic model filter, dashed line – complementary periodic model filter. Bottom subfigure: masking filters, solid line – $F_0(z)$, dashed line – $F_1(z)$.

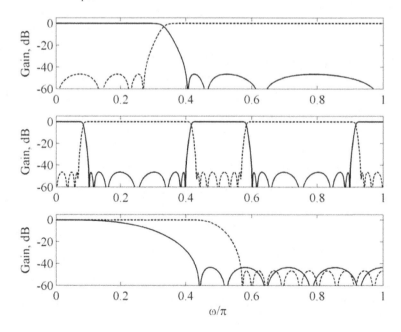

Figure 10.15. Gain response of IIR wideband filter synthesized using frequency-response masking approach. Case 2 design.

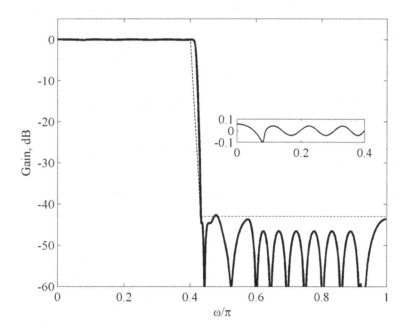

MATLAB EXERCISES

10.1 Use the MATLAB function ifir to design a lowpass FIR filter to satisfy the specifications of *Example10.1*. Compute and plot the magnitude response of the overall lowpass filter and compare the results with that of *Example 10.1*.

10.2 Design a narrowband filter for the specifications of *Example 10.1*. The model filter $G(z)$ is an IIR halfband filter, and masking filter $F(z)$ is linear-phase FIR filters. Compute and plot the magnitude responses for $G(z)$, $G(z^M)$, $F(z)$, and for the overall filter $H(z)$. Compute the total number of multiplication constants and compare with the solution of *Example 10.1*.

10.3 Using the frequency-response masking approach design a highpass FIR filter to satisfy the following specifications: F_p = 900 Hz, F_s = 850 Hz, and the sampling frequency of F_0 = 2000 Hz. The passband and stopband ripples are δ_p = 0.01, and δ_s = 0.01 (a_s = 40 dB). Compute and plot the magnitude responses for the following filters: model filter, periodic model filter, masking filter, and for the resulting highpass filter.

10.4 Design a wideband linear-phase lowpass filter specified with the passband edge frequency F_p = 1100 Hz, the stopband edge frequency F_s = 1350 Hz and the sampling frequency F_0 = 4000 Hz. The maximal ripple in the passband is a_p ± 0.1 dB, and minimal stopband attenuation amounts to a_s = 40 dB. Compute and plot the magnitude responses for the following filters: model filter, periodic model filter, masking filters, and for the resulting lowpass filter.

REFERENCES

Filter design toolbox for use with MATLAB. User's guide. Version 6. (2006). Natick: MathWorks.

Furtado, M. B. Jr., Diniz, P. S. R., Netto, S. L., & Saramäki, T. (2005). On the design of high-complexity cosine-modulated transmultiplexers based on the frequency-response masking approach. *IEEE Transactions on Circuits and Systems I: Regular papers, 52*(11), 2413-2426.

Johansson, H., & Wanhammar, L. (1996). High speed narrow-band lattice wave digital filters. *ICECS'96 Conference Proceedings,* 390-393.

Johansson, H., & Wanhammar, L. (1997, June). High-speed recursive filtering using the frequency-response masking approach, *IEEE International Symposium on Circuits and Systems – ISCAS,* 2208-2211.

Johansson, H., and Wanhammar, L. (1997). High-speed recursive filtering using the frequency-response masking approach. *Proceedings of the IEEE Int. Symposium on Circuits and Systems – ISCAS,* 2208-2211.

Johansson, H., & Wanhammar, L. (1997). A digital filter structure composed of allpass filters and an FIR filter for wide-band filtering. *ICECS'97 Conference Proceedings,* 249-253.

Johansson, H., & Wanhammar, L. (2000). High-speed recursive digital filters based on the frequency-response masking approach. *IEEE Transactions on Circuits and Systems-II: Analog and Digital Signal Processing, 47*(1), 48-61.

Lim, Y.C. (1986). Frequency-response masking approach for the synthesis of sharp linear phase digital filters. *IEEE Transactions on Circuits and Systems, 33*(4), 357-364.

Lim, Y. C., & Lian, Y. (1994). Frequency-response masking approach for digital filter design: complexity reduction via masking filter factorization. *IEEE Transactions on Circuits and Systems-II: Analog and Digital Signal Processing, 41*(8), 518-525.

Lim, Y. C., & Yang, R. (2005). On the synthesis of very sharp decimators and interpolators using the frequency-response masking technique. *IEEE Transactions on Signal Processing, 53*(4), 1387-1397.

Lutovac, M. D., & Milić, L. D. (1997). Design of computationally efficient elliptic IIR filters with a reduced number of shift-and-aadd operations in multipliers. *IEEE Transactions on Signal Processing, 45*(10), 2422-2430.

Lutovac, M. D. Tošić, D. V., & Evans, B. L. (2001). *Filter Design for Signal Processing Using MATLAB and Mathematica,* Upper Saddle River, NJ, Prentice Hall.

Lutovac, M. D., & Milić, L. D. (2001). IIR filters based on frequency-response masking approach. *International Conference on Telecommunications in Modern Satelite, Cable and Broadcasting Services, TELSIKS 2001,* 163-170.

Milić, Lj., & Lutovac, M.D. (2002). Efficient multirate filtering. In Gordana Jovanović-Doleček, (ed.), *Multirate Systems: Design & Applications.* Hershey, PA: Idea Group Publishing, 105-142.

Mitra, S. K. (2006). *Digital signal processing: A computer based approach*. 3rd edition. New York, NY: The McGraw-Hill Companies, Inc.

Neuvo, Y., Cheng Yu, D., & Mitra, S.K. (1984). Interpolated finite fmpulse response filters. *IEEE Transactions on Acoustics, Speech, and Signal Processing, 32*(3), 563-570.

Rosenbaum, L., Lövenborg, P., & Johansson, H. (2007). An approach for synthesis of modulated M-channel FIR filter banks utilizing the frequency-response masking technique. *EURASIP Journal on Advances in Signal Processing*. 2007, Article ID 68285, 13 pages.

Saramäki, T., Lim, Y. C., & Yang, R. (1995). The synthesis of halfband filter using frequency-response masking technique. *IEEE Transactions on Circuits and Systems-II: Analog and Digital Signal Processing, 42*(1), 58-60.

Signal processing toolbox for use with MATLAB. User's guide. Version 6. (2006). Natick: MathWorks.

Chapter XI
Comb–Based Filters for Sampling Rate Conversion

INTRODUCTION

Comb filters are developed from the structures based on the moving average (boxcar) filter. The comb-based filter has unity-valued coefficients and, therefore, can be implemented without multipliers. This filter class can operate at high frequencies and is suitable for a single-chip VLSI implementation. The main applications are in communication systems such as software radio and satellite communications.

In this chapter, we introduce first the concept of the basic comb filter and discuss its properties. Then, we present the structures of the comb-based decimators and interpolators, discuss the corresponding frequency responses, and demonstrate the overall two-stage decimator constructed as the cascade of a comb decimator and an FIR decimator. In the next section, we expose the application of the polyphase implementation structure, which is aimed to reduce the power dissipation. We consider techniques for sharpening the original comb filter magnitude response and emphasize an approach that modifies the filter transfer function in a manner to provide a sharpened filter operating at the lowest possible sampling rate. Finally, we give a brief presentation of the modified comb filter based on the zero-rotation approach. Chapter concludes with several MATLAB Exercises for the individual study. The reference list at the end of the chapter includes the topics of interest for further research.

COMB-BASED FILTER SECTIONS

The simplest lowpass FIR filter is a *boxcar* or a *moving-average* (MA) filter whose impulse response is of a rectangle shape,

$$g_C[n] = \frac{1}{N} \begin{cases} 1, & \text{for } 0 \le n \le N-1 \\ 0, & \text{otherwise} \end{cases}, \tag{11.1}$$

where N is an integer. The z-domain representation results in the following transfer function $G_C(z)$,

$$G_C(z) = \frac{1}{N} \sum_{k=0}^{N-1} z^{-k}. \tag{11.2}$$

Equation (11.2) is recognized as the first N terms of the geometric series, whose closed form expression,

$$G(z) = \frac{1}{N} \frac{1 - z^{-N}}{1 - z^{-1}} \tag{11.3}$$

is fond suitable for an efficient implementation. According to equation (11.3), the filter $G(z)$ can be implemented by cascading the *comb section* $(1 - z^{-N})$ and the *integrator section* $1/(1 - z^{-1})$ thus leading to an extremely efficient device which performs the filtering task employing only two additions regardless of the filter length N. The term *CIC filter* (cascade-integrator-comb) is frequently used for this filter class. As will be shown later in this chapter, the various realization structures have been developed for the implementation of the transfer function (11.3). In this chapter, we generally use terms comb filter or comb-based filter for the class of digital filters based on the transfer function (11.3). The term CIC filter we use only for the cascade-integrator-comb implementation scheme.

The frequency response $G_C(e^{j\omega})$ following from equation (11.3) is given by,

$$G_C(e^{j\omega}) = \frac{1}{N} \frac{\sin\left(\frac{\omega N}{2}\right)}{\sin\left(\frac{\omega}{2}\right)} e^{-j\omega[(N-1)/2]}. \tag{11.4}$$

Hence, the comb filter section is a linear-phase lowpass filter, which exhibits $\sin(Nx)/\sin(x)$ amplitude characteristic. Due to its particular amplitude response, this filter is also called the *sinc filter*.

Figure 11.1 shows the gain response of the comb-based filter for $N = 10$. As Figure 11.1 illustrates, the $G_C(e^{j\omega})$ exhibits the comb-like magnitude response. It has $N/2$ natural nulls distributed along the normalized frequency axis at integer multiples of $(2/N)$. Notice that for N being odd, the number of natural nulls is $(N-1)/2$. The filter has a very wide transition band and the stop band attenuation at the first side-lobe amounts only to 13 dB.

A very poor magnitude characteristic of the comb filter is improved by cascading several identical comb filters. The transfer function $G_C^K(z)$ of the multistage comb filter composed of K identical single-stage comb filters is given by

$$G_C^K(z) = \left[\frac{1}{N} \frac{1 - z^{-N}}{1 - z^{-1}} \right]^K. \tag{11.5}$$

Figures 11.2 plots of the magnitude responses for the comb filter of the order $N = 10$, and $K = 1, 2, 3$, and 4, and the passband zoom displays the passband zoom.

Figure 11.1. Gain response of the single comb filter for N = 10

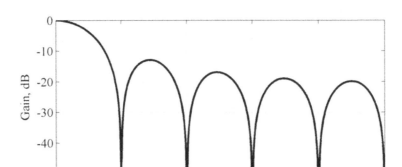

Figure 11.2. Comb filter gain responses: single-stage K = 1, two- stage K = 2, three- stage K = 3, and four-stage K = 4

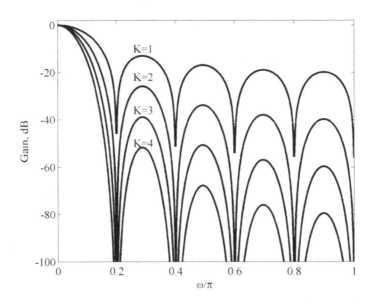

Figure 11.2 shows how the multistage realization improves the selectivity and the stop-band attenuation of the overall filter: the selectivity and the stopband attenuation are augmented with the increase of the number of comb filter sections. The filter has multiple nulls with multiplicity equal to the number of the sections (*K*). Consequently, the stopband attenuation in the null intervals is very high. Figure 11.3 illustrates a monotonic decrease of the magnitude response in the passband, called the *passband droop*.

Figure 11.3. Comb filter gain responses: Passband details for Figure 11.2

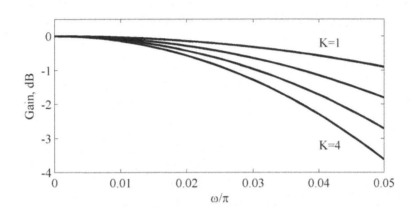

CASCADE INTEGRATOR-COMB (CIC) FILTERS IN DECIMATORS AND INTERPOLATORS

The comb-based filters from the previous section are utilized in multirate systems for constructing efficient decimators and interpolators. The comb filter ability to perform filtering without multiplications is very attractive to be applied to high rate signals. Moreover, comb-based filters are convenient for large conversion factors since the lowpass bandwidth is very small. In multistage decimators with a large conversion factor, the comb filter is the best solution for the first decimation stage, whereas in interpolators, the comb filter is convenient for the last interpolation stage.

The multirate application of comb filters has been proposed first by Hogenauer (1981), and since that time, the so-called *Hogenauer filters* have attracted many researchers and practicing engineers.

The basic concept of a comb-based decimator is explained in Figure 11.4. Figure 11.4(a) shows the factor-of-N decimator consisting of the K-stage CIC filter and the factor-of-N down-sampler. Applying the Third Identity, the factor-of-N down-sampler is moved and placed behind the integrator section and before the comb section, see Figure 11.4(b). Finally, the CIC decimator is implemented as a cascade of K integrators, factor-of-N down-sampler, and the cascade of K differentiator (comb) sections. The integrator portion operates at the input data rate, whereas the differentiator (comb) portion operates at the N times lower sampling rate.

The basic concept of a comb-based interpolator is shown in Figure 11.5. Figure 11.5 (a) depicts the factor-of-N interpolator consisting of the factor-of-N up-sampler and the K-stage CIC filter. Applying the Sixth Identity, the factor-of-N up-sampler is moved and placed behind the differentiator (comb) section and before the integrator section, see Figure 11.5 (b). Finally, the CIC interpolator is implemented as a cascade of K differentiator (comb) sections, factor-of-N down-sampler, and the cascade of K integrators. The comb portion operates at the input data rate, whereas the integrator portion operates at the N times higher sampling rate.

Figure 11.4. Block diagram representation of CIC decimator: (a) Cascade of CIC filter and down-sampler. (b) Cascade of integrator section, down-sampler, and comb section. (c) Implementation structure consisting of the cascade of K integrators, down sampler, and the cascade of K differentiators.

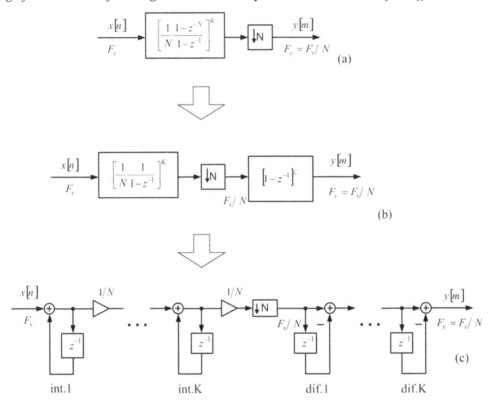

The configuration composed as a cascade of interpolators and differentiators (differentiators and interpolators) separated by a down-sampler (up-sampler) is called recursive realization structure, or a CIC realization structure.

The advantage of CIC decimators and interpolators is the ability of sampling rate conversion without multiplying operations. This is of particular interest when operating at high frequencies.

Considering the implementation aspects of CIC filters, one should expect the register overflow, since the integrator has a unity feedback. Actually, the register overflow is a reality in all integrator stages. It is shown that the register overflow is of no consequence if the two's complement arithmetic is used and the range of the number system is equal to or exceeds the maximum magnitude expected at the output of the composite filter. In order to avoid register overflow in the integrator section, the word-length has to be equal to or greater than $(W_0 + K \cdot \log_2 N)$ bits, where W_0 is the word-length in bits of the input signal. For studying the overflow and the register word-length in CIC filters, see for example Hogenauer (1981), Harris (2004), *Filter Design Toolbox for Use with* MATLAB (2006).

MATLAB *Filter DesignToolbox* includes two design objects that construct CIC decimators and interpolators: mfilt.cicdecim for decimator and mfilt.cicinterp for interpolator. The decimation filter object mfilt. cicdecim implements a fixed-point arithmetic CIC decimator according to Figure 11.4(c), whereas mfilt. cicinterp constructs a fixed-point arithmetic CIC interpolator according to Figure 11.5(c).

Figure 11.5. Block diagram representation of CIC interpolators: (a) Cascade of up-sampler and CIC filter. (b) Cascade of comb section, down-sampler, and integrator section. (c) Implementation structure consisting of the cascade of K differentiators, up-sampler, and the cascade of K integrators.

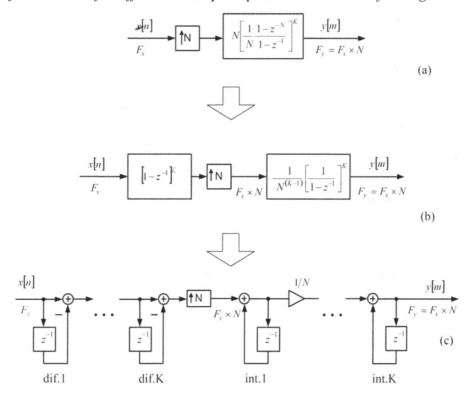

MAIN PERFORMANCES OF A COMB-BASED DECIMATOR

The comb-based filters are frequently used as the first decimation stage of the multistage decimators. Because of the simplicity of implementation, the comb filters are suitable for operating at high frequencies. Most importantly, the comb-based filters provide the best solution for the implementation of the first decimation stage in the multistage down-sampling conversion for the oversampled $\Sigma\Delta$ A/D converters.

For modifying the comb filter magnitude response one can use only two parameters: the comb filter order N, and the number of the comb filter sections K, see equation (11.5). Figures 11.2 and 11.3 on example filters ($N = 10$, and $K = 1, 2, 3,$ and 4) illustrate the typical advantages and disadvantages associated with the Nth order comb filter when applied as the antialising filter in a factor-of-N decimator. Firstly, the natural nulls of the comb filter occur exactly at the integer multiples of F_x/N (F_x – input sampling frequency) thus providing the maximum alias suppression at those frequencies. Secondly, the aliasing bandwidths around the nulls are narrow, and usually too small to provide sufficient suppression of aliasing in the entire baseband of the signal. Thirdly, the monotonic passband characteristic produces an inevitable passband droop, which for many applications should be compensated.

Figure 11.6 indicates the main performances of the comb-based decimation filter as presented in (Laddomada, 2007b). Only the magnitude response is considered since the phase response is linear.

The signal band at the decimator input occupies the frequency range $[0, F_m]$. For the input signal sampling frequency F_x, and for the decimation factor N, the aliasing bands of the bandwidths $2F_m$ are located around the natural null frequencies F_x/N, $2F_x/N$, ..., JF_x/N, where $J = N/2$ for N even, and $J = (N−1)/2$ for N odd. The main parameters that characterize the comb filter performances are the passband droop and the selectivity factor:

- The passband droop denoted by d_C^K in Figure 11.6 indicates the maximum attenuation at the edge of the useful signal bandwidth compared to the ideal lowpass filter.
- The selectivity factor denoted by Φ_C^K in Figure 11.6 is defined as a ratio between the exact values of the filter magnitude response achieved at the passband edge frequency (F_m in Figure 11.6) and at the lower edge frequency of the first aliasing band ($F_x/N−F_m$) in Figure 11.6 indicated as the worst case attenuation).

It is straightforward to develop the closed form expressions for the passband droop d_C^K and the selectivity factor Φ_C^K in the case of the K-stage comb filter. Using equations (11.4) and (11.5) and notations from Figure 11.6 one obtains the following expression for d_C^K

$$d_C^K = \left|\frac{G_C^K(F_m)}{G_C^K(0)}\right| = \left|\frac{\sin(\pi N F_m/F_x)}{N\sin(\pi F_m/F_x)}\right|^K. \qquad (11.6)$$

Figure 11.6. Main performances of comb-based decimation filter

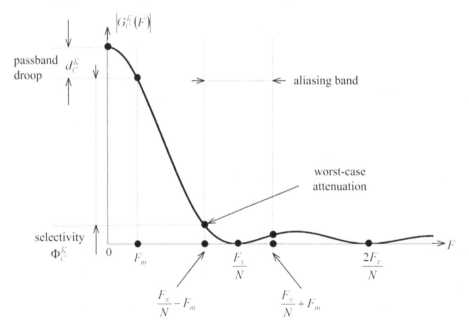

Introducing the normalization to the half of the sampling frequency $f = F/(F_x/2)$, we express (11.6) in terms of digital frequencies,

$$d_C^K = \left| \frac{\sin(\pi N f_m/2)}{N \sin(\pi f_m/2)} \right|^K . \tag{11.7}$$

In the same manner, the selectivity factor for the K-stage comb filter is determined by

$$\Phi_C^K = \frac{G_C^K(F_m)}{G_C^K(F_x/N - F_m)} = \left| \frac{\sin(\pi(1/N - F_m/F_x))}{\sin(\pi F_m/Fx)} \right|^K , \tag{11.8}$$

and in terms of digital frequencies

$$\Phi_C^K = \left| \frac{\sin(\pi(1/N - f_x/2))}{\sin(\pi f_x/2)} \right|^K . \tag{11.9}$$

CASCADING CIC FILTER AND FIR FILTER

A CIC filter can be used as a first stage in decimation when the overall conversion ratio M is factorizable as

$$M = N \times R. \tag{11.10}$$

The overall factor-of-M sampling rate conversion system can be implemented by cascading a factor-of-N CIC decimator and a factor-of-R FIR decimator as shown in Figure 11.7(a). The corresponding single-stage equivalent is given in Figure 11.7(b).

When constructing an interpolator with a conversion factor L factorizable as

$$L = R \times N, \tag{11.11}$$

Figure 11.7. Two-stage decimator composed of a CIC filter and an FIR filter: (a) Cascade implementation. (b) Single-stage equivalent.

323

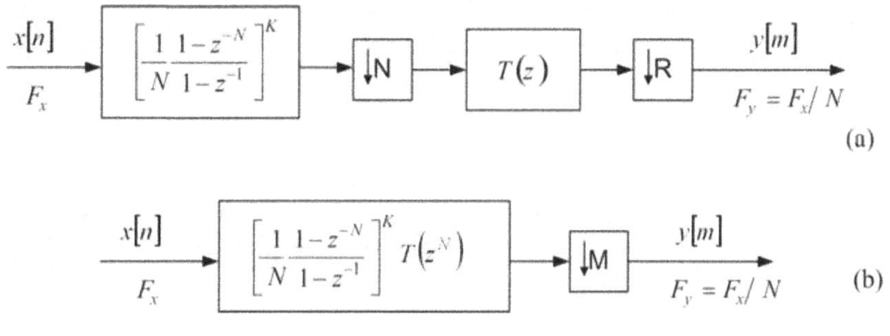

it might be beneficial to implement the second (last) stage as a CIC interpolator. The first stage is usually implemented as an FIR filter. Figure 11.8(a) depicts the two-stage interpolator consisting of the cascade of a factor-of-R FIR interpolator and a factor-of-N CIC interpolator. The corresponding single-stage equivalent is shown in Figure 11.8(b)

In the two-stage solutions of Figures 11. 7 and 11.8, the role of CIC decimator (interpolator) is to convert the sampling rate by the large conversion factor N, whereas the FIR filter $T(z)$ provides the desired transition band of the overall decimator (interpolator) and compensates the passband characteristic of the CIC filter. The example that follows illustrates the role and properties of each of the two filters.

Example 11.1

This example illustrates the two-stage decimator from Figure 11.7 (a), and computes the frequency response of the single-stage equivalent as given in Figure 11.7 (b).

Design specifications:
The overall decimation factor $M = 10$. The overall decimation filter $H(z)$ is specified by:

- Passband edge frequency $\omega_p = 0.05\pi$, and the deviations of the passband magnitude response are bounded to $a_p = \pm0.15$ dB.
- Stopband edge frequency $\omega_s = \pi/M = 0.1\pi$ with the requested minimal stopband attenuation $a_s = 50$ dB.
- The phase characteristic is linear.

Solution:
The overall factor-of-10 decimator is composed as the cascade of two decimators:

- For the 1st decimation stage we use the four-stage factor-of-5 comb decimator $G_C^4(z)$, i.e. $N = 5$ and $K = 4$. This filter has two natural nulls located at 0.4π and 0.8π. From equation (11.9), it follows that the selectivity factor, which determines the minimum suppression of aliasing, amounts here $\Phi_C^4 = 65.8753$ dB. On the other hand, according to (11.7), the passband droop is $d_C^4 = -0.8619$ dB.

Figure 11.8. Two-stage interpolator composed of an FIR filter in the first stage, and the CIC filter in the second stage: (a) Cascade implementation. (b) Single-stage equivalent.

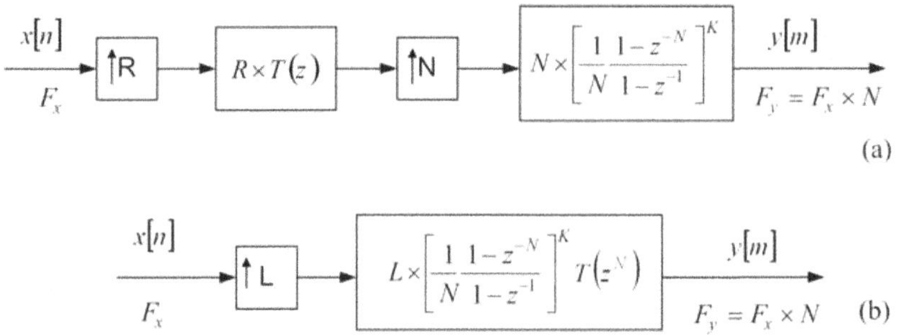

- 2nd decimation stage: factor-of-2 FIR decimator of the filter order $N_{ord} = 26$, which provides: stopband edge frequency of $\omega_s = \pi/M = 0.1\pi$ for the overall decimation filter; 50 dB stopband attenuation, and additionally, compensates the comb filter passband droop.

In the first step, we compute the frequency response of the comb decimation filter $G_C^4(z)$ whose decimation factor is $N = 5$.

```
N = 5; K = 4;                               % Setting the design parameters for G_K(z)
A = ([1,zeros(size(1:N-1)),-1]/N); B = [1,-1];   % Comb filter section G(z) = A(z)/B(z)
[G,f] = freqz(A,B,1024,2);                  % Computing the comb filter frequency response
GK = G.^K;                                  % Computing the frequency response of G(z)^K
```

Figure 11.9 displays the gain response of $G_C^4(z)$ with the indications of the signal band, the passband droop, and two aliasing bands.

In the second step, we set the design specifications for the factor-of-2 decimation filter and use the MATLAB function firpm to design a linear-phase FIR filter $T(z)$, and compute the corresponding frequency response:

```
Nord=26;                                    % Filter order for T(z)
Fo = [0,0.3,0.505,1]; Ao = [0.981,1.11,0,0];  %Setting the specifications for T(z)
t = firpm(Nord,Fo,Ao);                      % Computing the coefficients of T(z)
[T,f]=freqz(t,1,1024,2);                     % Computing the frequency response of T(z)
```

In the third step, we compute the frequency response of $T(z^N)$.

Figure 11.9. Gain response of the comb filter $G_C^4(z)$

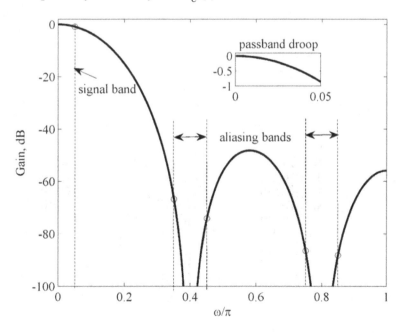

```
tN=upsample(t,N);                 % Computing the coefficients of T(z^N)
[TN,f]=freqz(tN,1,1024,2);        %  Computing the frequency response of T(z^N)
```

Figure 11.10 plots the gain responses of $G_C{}^4(z)$ and $T(z^N)$ and thus exposes the roles of both of them. Filter $T(z^N)$ ensures the desired transition band, compensates the passband droop of the comb filter of the first stage. The comb filter $G_C{}^4(z)$ has its two nulls just in the undesired passbands of the periodic filter $T(z^N)$ that ensure the requested stopband attenuation of the target two-stage decimator.

Finally, we compute the frequency response of the overall two-stage decimation filter, $H(z) = G_C{}^4(z)T(z^5)$,

```
H = TN.*GK;          % Computing the frequency response of (G(z)^K)*T(z^N)
```

Figure 11.11 displays the gain response of the resulting overall decimation filter $H(z)$.

Notice that the displayed frequency response meets the requested passband and the stopband specifications. The filter $H(z)$ exhibits a linear-phase characteristic since both filters $G_C{}^4(z)$ and $T(z)$ are linear-phase FIR filters.

Figure 11.12 depicts the implementation structure of the factor-of-10 decimator of *Example 11.1.* The cascade of $K = 4$ integrators operates before the down-sampling-by-5 takes place, and thus operates at the input (high rate) sampling frequency F_x. The comb section, composed of $K = 4$ differentiators, follows the down-sampler, and consequently, operates at the lower sampling frequency, $F_x/5$. Since the factor-of-5 decimator is based on the CIC filter, decimation-by-5 is performed without multiplications. In the second decimation stage, factor-of-two FIR decimator with the filter order $N_{ord} = 26$ is used. With the polyphase decomposition of $T(z)$ into two polyphase components, $T(z) = E_0(z^2) + z^{-1} E_1(z^2)$, FIR

Figure 11.10. Gain responses of the comb filter $G_C{}^4(z)$ (solid line), and that of periodic FIR filter $T(z^5)$ (dashed line)

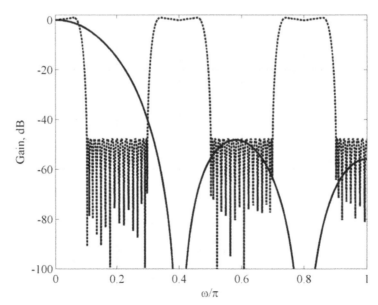

Figure 11.11. Gain response of the two-stage decimator implemented as a factor-of-5 comb decimator and a factor-of-two FIR decimator

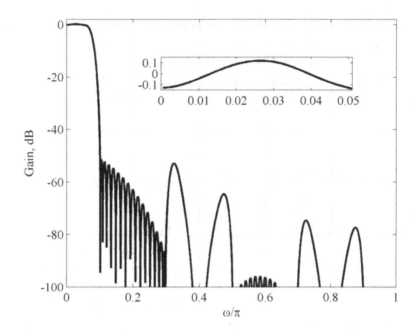

Figure 11.12. Implementation structure of the two-stage factor-of-10 decimator consisting of the cascade of factor-of-5 CIC decimator and factor-of-2 FIR decimator

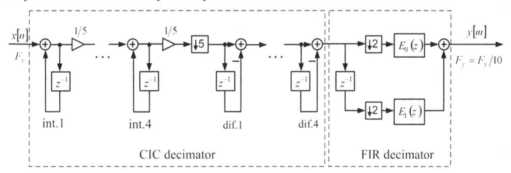

filtering is performed at the lowest possible sampling rate, $F_y = F_x/10$, as indicated in Figure 11.12, see Chapter IV, section Polyphase Implementation of Decimators and Interpolators.

The comb decimators and interpolators are extremely efficient due to their inherent multiplierless implementations. However, there are three main problems in the application of comb decimators and interpolators:

- The integrator sections operate at the maximum sampling rate and consequently the power consumption is very high.
- Register widths can become large for large sampling rate conversion factor N.

• The frequency response is fully determined by only two parameters N and K, and the desired frequency response cannot be met in the majority of practical applications.

In the last decade, various methods have been developed to solve the above mentioned problems. In the following sections, we will describe some of the recent advances that improve the implementation and performance aspects of the comb-based decimators and interpolators. In the next section, we present modifications in the implementation structure that is based on the polyphase decomposition of the comb filter transfer function. In the sequel, we consider methods for modifying and sharpening the frequency response.

POLYPHASE IMPLEMENTATION STRUCTURES

The comb decimators and interpolators displayed in Figures 11.4 and 11.5 use the recursive implementation technique. The major drawback of this technique is high power consumption in the integrator stage when operating at high frequencies. A typical example is a decimation filter in the oversampled $\Sigma\Delta$ analog-to-digital converters, where the integrator section operates at very high frequency before any decimation takes place.

Solutions for lowering the power consumption in comb-based filters were found in the applications of nonrecursive implementation techniques. It was shown that when applying the well known polyphase decomposition to the comb filter transfer function $G_C^K(z)$, the resulting filter achieves lower power consumption than the corresponding recursive structure (Aboushady, Dumonteix, & Mehrez, 2001).

Observe that the recursive implementation structure has been developed from the recursive form of the comb filter transfer function as given in equation (11.5). To develop the nonrecursive polyphase structure, we consider here the basic nonrecursive form of $G_C^K(z)$,

$$G_C^K(z) = \left[\frac{1}{N}\sum_{n=0}^{N-1} z^{-n}\right]^K. \qquad (11.12)$$

Recall that the filter length N is also the conversion factor of a comb factor-of-N decimator (interpolator). When the factor-of-N is expressible as a power-of-two, i.e., $N = 2^P$, where P is an integer, equation (11.12) can be factored as follows,

$$G_C^K(z) = 2^{-PK} \prod_{i=0}^{P-1}\left(1+z^{-2^i}\right)^K. \qquad (11.13)$$

For example, in the case of conversion factor $N = 16$, $G_C^K(z)$ is expressible in the form

$$G_C^K(z) = 2^{-16K}\left(1+z^{-1}\right)^K\left(1+z^{-2}\right)^K\left(1+z^{-4}\right)^K\left(1+z^{-8}\right)^K. \qquad (11.14)$$

It is straightforward to conclude from equations (11.13) and (11.14) that a decimator composed as the cascade of filter $G_C^K(z)$ and a factor-of-2^P down-sampler can be replaced with the cascade of P factor-of-

Figure 11.13. Cascade nonrecursive implementation structure for comb decimator: (a) Comb filter $G_C^K(z)$ and factor-of-2^P down-sampler. (b) Cascade of P factor-of-2 nonrecursive decimators.

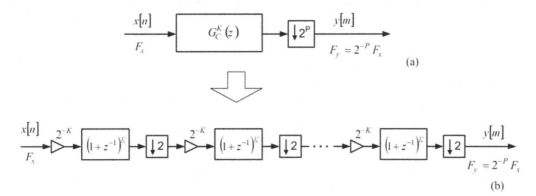

two decimators each consisting of a nonrecursive subfilter $(1 + z^{-1})^K$ and a factor-of-two down-sampler as indicated in Figure 11.13. This configuration is recognized as the cascade nonrecursive implementation structure.

Figure 11.13 illustrates that only the first decimation stage operates at the input sampling frequency. The operating frequency is successively reduced by two after each decimation stage. Moreover, the operating frequency can be further reduced by a factor-of-two when applying the polyphase decomposition to the FIR decimation subfilter $(1 + z^{-1})^K$,

$$\left(1 + z^{-1}\right)^K = E_0\left(z^2\right) + z^{-1} E_1\left(z^2\right), \tag{11.15}$$

where $E_0(z)$ and $E_1(z)$ are two polyphase components, see Chapter IV, section Polyphase Implementation of Decimators and Interpolators. We can now represent each subfilter in Figure 11.13 (b) in terms of the polyphase components $E_0(z)$ and $E_1(z)$, and then exploit the Third Identity to move the down-sampling-by-two before filtering, as indicated in Figure 11.14. The configuration of Figure 11.14 is called the polyphase implementation form. The advantage of the nonrecursive structures of Figures 11.13 and 11.14 is the absence of register overflow problems, and the word-length of each stage i for the input word-length of W_0 bits is limited to $(W_0 + K \cdot i)$ bits (Aboushady, Dumonteix, & Mehrez, 2001).

The new polyphase structure that additionally reduces the operation frequency was proposed by Aboushady, Dumonteix, & Mehrez (2001). The idea was to decimate as much as possible in the first decimation stage. To provide this, the overall decimator is decomposed into a first stage nonrecursive filter $H_1(z)$ with the decimation factor N_1, followed by a cascade of nonrecursive $(1 + z^{-1})^K$ filters with a decimation factor 2. Hence, the overall transfer function $G_C^K(z)$ is written in the form,

$$G_C^K(z) = H_1(z) H_2(z), \tag{11.16}$$

where,

$$H_1(z) = \left[\sum_{i=1}^{N_1-1} z^{-i}\right]^K, \tag{11.17a}$$

Figure 11.14. Polyphase implementation structure for comb decimator

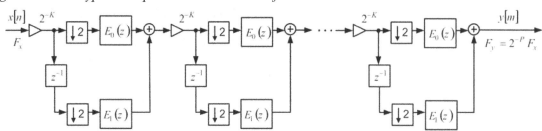

$$H_2(z) = \prod_{i=0}^{\log_2(N/N_1)-1} \left(1 + z^{-2^i}\right)^K.$$ (11.17b)

The expansion of $H_1(z)$ of equation (11.17a) results in an FIR filter of order $K(N_1 - 1)$

$$H_1(z) = \sum_{n=0}^{K(N_1-1)} h[n]z^{-n}.$$ (11.18)

The coefficients $h[n]$ are integers and are symmetrical, i.e. they satisfy the coefficient symmetry property,

$$h[n] = h[KN_1 - K - 1 - n].$$ (11.19)

According to the decompositions expressed by equations (11.16) – (11.18), the overall decimator can be implemented as a multistage system where the first stage is a factor-of-N_1 decimator with the filter $H_1(z)$, which is followed by the cascade of factor-of-two decimators that implement $H_2(z)$, see Figure 11.15(a).

The factor-of-N_1 decimator with the decimation filter $H_1(z)$ can be efficiently implemented with the use of polyphase form. Recall that the polyphase decomposition for FIR filters has been exposed in Chapter IV, section Polyphase Implementation of Decimators and Interpolators. Therefore, for the first decimation stage we can exploit the efficient polyphase realization of $H_1(z)$ and place the down-sampling-by-N_1 to the input of the overall system. The resulting polyphase implementation, which reduces by N_1 the operating frequency at the input of the system, is shown in Figure 11.15 (b).

It is to be pointed out that polyphase comb filters inevitably include multipliers. Unlike the recursive configuration, the polyphase structure contains integer-valued multiplication constants. The following example computes and displays the coefficients of $H_1(z)$ and the coefficients of the corresponding polyphase components $E_0(z)$, $E_1(z)$, ..., $E_{N1-1}(z)$ for the case $N_1 = 4$ and $K = 5$.

Example 11.2

In this example, we use equation (11.17a) to compute in MATLAB the impulse response coefficients $\{h_1[n]\}$ for $N_1 = 4$ and $K = 5$, and to decompose $\{h_1[n]\}$ into four polyphase components $\{e_0[n]\}$, $\{e_1[n]\}$, $\{e_2[n]\}$, $\{e_3[n]\}$.

Figure 11.15. Multistage polyphase implementation: (a) Cascade of FIR filter $H_1(z)$ and factor-of-N_1 down-sampler followed by the cascade of factor-of-two decimators. (b) Efficient polyphase implementation.

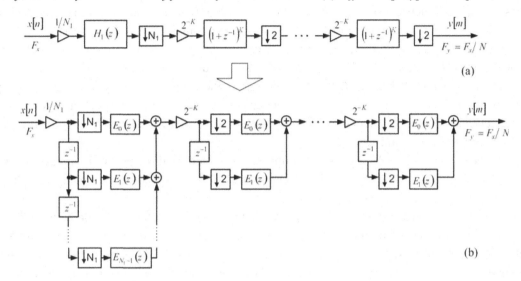

(a)

(b)

```
N1 = 4;
g = rectwin(N1); h1 = 1;
K = 5;
for I = 1:K
    h1 = conv(h1,g);
end

disp(h1') % Display the coefficients of the filter impulse response h1

1    5    15    35    65    101    135    155    155    135    101    65    35    15    5    1

% Polyphase components:
e0 = downsample(h1,4);
e1 = downsample(h1,4,1);
e2 = downsample(h1,4,2);
e3 = downsample(h1,4,3);

% Display the coefficients of polyphase components e0, e1, e2, e3
disp(e0'), disp(e1'), disp(e2'), disp(e3')

    1    65    155    35

    5    101    135    15

    15    135    101    5

    35    155    65    1
```

Example 11.2 illustrates clearly that reduction of the sampling rate by N_1 in the first decimation stage should be necessarily paid with multiplication operations. However, the coefficients of $\{h_1[n]\}$ are integers and are symmetric. For the implementation of the polyphase section with integer coefficients see the efficient configuration presented in (Aboushady, Dumonteix, & Mehrez, 2001).

When decreasing a sampling rate, we generally reduce the power consumption, but simultaneously, additional power dissipation should be produced due to multiplication operations. It was shown by Aboushady, Dumonteix, & Mehrez (2001) that, in spite of the additional multiplications, the polyphase configuration of Figure 11.15 (b) may achieve better performances in the respect of power consumption when compared with the recursive (CIC) implementation or with the nonrecursive configurations shown in Figures 11.13(b) and 11.14.

To achieve the savings in the power consumption, the selection of the proper first-stage decimation factor N_1 is crucial. The article of Aboushady, Dumonteix, & Mehrez (2001) presents the graphs useful for selecting the convenient value of N_1 in order to reduce the power dissipation.

In this section, we have exposed the polyphase implementation of the overall decimator when $H_1(z)$ and $H_2(z)$ in equation (11.16) are both comb filter transfer functions. In the papers of Abed and Narukar (Abed & Narukar, 2003; Narukar & Abed, 2006), a two-stage solution has been proposed, which implements the first stage as a comb decimator, and the second stage is composed of IIR decimators. For the comb decimator of the first stage, authors exploit the polyphase implementation of the comb filter as exposed in this section. The second stage is composed of a cascade of two IIR decimators where the first filter is a third-band IIR filter whereas the second is a halfband IIR filter. It is interesting to observe that both IIR filters belong to the class of EMQF (Elliptic Minimum Q-Factors) filters, which has been discussed in Chapter V of this book. It was demonstrated in (Narukar & Abed, 2006) that a solution based on the cascade of a polyphase comb decimator and the linear-phase EMQF filters achieves savings in power dissipation.

SHARPENED COMB FILTERS

Filter sharpening is a technique that improves passband/stopband filter performances in a manner that provides both: la smaller passband error and greater stopband attenuation. Several attractive approaches for the comb filter sharpening are based on the Kaiser-Hamming sharpened method that was introduced by Kaiser and Hamming (1977), see papers published by Kwentus, Jiang, & Willson (1997), Laddomada & Mondin (2004), and Jovanović-Doleček & Mitra (2005), Laddomada (2007a,b).

Kaiser and Hamming (1977) proposed a method to sharpen the magnitude response of a digital filter by using multiple realizations of a low-order basic filter. They introduced a family of sharpened filters $H_{nm}(f)$, which is expressible in the form

$$H_{nm}(f) = H_p^{n+1}(f) \cdot \sum_{k=0}^{m} \frac{(n+k)!}{n!\,k!} \left[1 - H_p(f)\right]^k, \tag{11.20}$$

where $H_p(f)$ is a low-order basic filter. The nonnegative integers n and m represent the number of the non-zero derivatives of $H_{nm}(f)$ at the points where $H_{nm}(f) = 0$ and $H_{nm}(f) = 1$, respectively. By choosing n and m, we select the order of tangency at the frequencies where $H_{nm}(f) = 0$ and $H_{nm}(f) = 1$.

In this chapter we utilize the simplest case $n = m = 1$, which leads to the filter

$$H_{11}(f) = H_p^2(f)\left[3 - 2H_p(f)\right]. \tag{11.21}$$

The Kaiser-Hamming sharpening technique is applicable to linear-phase FIR filters. For $H_p(z)$ being a causal linear-phase FIR filter with the group delay of D samples, the transfer function $H_{11}(z)$ is given by

$$H_{11}(z) = H_p^2(z)\left[3z^{-D} - 2H_p(z)\right]. \tag{11.22}$$

The usage of equation (11.22) produces the transfer function $H_{11}(z)$, which achieves the improved passband and stopband characteristics in comparison to that of the basic FIR filter $H_p(z)$. The term in the squared brackets in (11.21) is responsible for the passband droop reduction, whereas $H_p^2(z)$ can improve the stopband rejection.

The implementation of $H_{11}(z)$ requires three copies of $H_p(z)$, an integer multiplier of value 3, a trivial multiplier of value -2, an adder, and a delay line of D samples. The block diagram that implements the sharpened filter $H_{11}(z)$ is shown in Figure 11.16.

Kwentus, Jiang, and Willson (1997) introduced the comb filter transfer function $G_C^K(z)$ as the basic filter $H_p(z)$ in the structure of Figure 11.16, and developed technique suitable for constructing comb filters with sharpened magnitude characteristics. Figure 11.17 shows the implementation of a sharpened factor-of-N comb decimator for the case $K = 2$. Figure 11.17(a) presents the block diagram of Figure 11.16 where $H_p(z)$ is replaced with the original comb filter transfer function from (11.5) for $K = 2$. The expanded configuration of the overall decimator is shown in Figure 11.17(b). One observes that the cascade of integrators is placed at the input and before the down-sampler. Hence, the integrators operate at the high sampling rate before the sampling rate reduction takes place. On the other hand, differentiators operate at the reduced sampling rate. Therefore, the implementation scheme of the sharpened comb filter of Figure 11.17 suffers from the same drawback as the implementation scheme of the original comb decimator shown in Figure 11.4.

The computation of the magnitude response of the sharpened comb filter is straightforward. The expression for $|H_{11}(e^{j\omega})|$ results directly when applying expressions given in (11.4) to equation (11.22). Thereby, in the case $H_p(z) = G_C^K(z)$, one obtains for $|H_{11}(e^{j\omega})|$,

Figure 11.16. The block diagram of sharpened filter

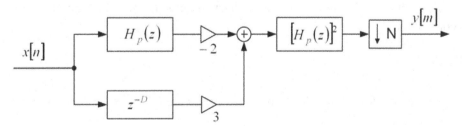

Figure 11.17. Implementation structure of the sharpened comb decimator. (a) Block diagram with comb filters transfer functions, K = 2. (b) Detailed implementation scheme. (c) Implementation of the integrator block. (d) Implementation of the differentiator block.

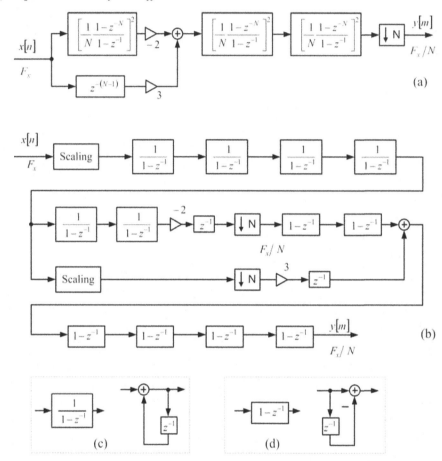

$$\left|H_{11}\left(e^{j\omega}\right)\right| = \left|3\left(\frac{1}{N}\frac{\sin\frac{\omega N}{2}}{\sin\frac{\omega}{2}}\right)^{2K} - 2\left(\frac{1}{N}\frac{\sin\frac{\omega N}{2}}{\sin\frac{\omega}{2}}\right)^{3K}\right|. \tag{11.23}$$

In the example that follows we show the effects of sharpening on the simple illustrative example of the single-stage comb filter.

Example 11.3

In this example, we compute and plot the gain responses for the following decimation filters:

(a)   Single-stage comb filter $G_C(z)$ with $N = 16$.

(b)   Sharpend filter $H_{11}(z)$ with the basic filter $H_p(z) = G_C(z)$.

We use the following code to compute the amplitude responses of $G_C(z)$ and $H_{11}(z)$:

```
N = 16;                          % Selecting the decimation factor
K = 1;                           % Selecting the number of stages
f = 0:0.00005:1;                 % Setting the vector of frequencies
G = sin(pi*f*N/2)./(N*sin(pi*f/2));   % Computing the single-comb filter amplitude response
H11 = 3*G.^(2*K)-2*G.^(3*K);     % Computing the sharpened filter amplitude response
```

Figure 11.8 displays the resulting gain responses for both filters in the low-frequency range $[0 - 0.35\pi]$ with markers at the passband edge 0.02π, and at the edges of the first aliasing band $[0.105\pi, 0.145\pi]$. The dashed line represents the single-stage comb filter whereas the sharpened filter is represented with the solid line. The effects of sharpening are visible in both the passband and in the stopband. It is to be noticed that with this sharpening technique the passband droop is practically eliminated.

TWO-STAGE SHARPENED COMB DECIMATOR

Many attractive methods have been advanced to improve the frequency responses of comb-based decimators and interpolators (Saramäki & Ritoniemi, 1997; Abu-Al-Saud & Stuber, 2003, 2006; Laddomada & Mondin, 2004; Kwentus, Jiang, & Willson, 1997; Lo Presti, 2000; Stephen & Stewart, 2004; Jovanović-Doleček & Mitra, 2005, 2006, 2007; Laddomada, 2007a&b).

In this section, we present the results achieved by Jovanović-Doleček and Mitra (2005) since their technique integrates the advantages of the methods presented in (Kwentus, Jiang, & Willson, 1997;

Figure 11.18. Example of filter sharpening: single-stage comb filter – dashed line, sharpened single-stage comb filter – solid line

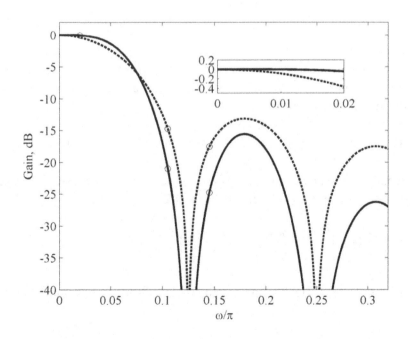

Aboushady, Dumonteix, & Mehrez; 2001), and introduces the modified sharpened comb filter that operates at a lower sampling rate than the sharpened comb filter shown in Figure 11.17.

Jovanović-Doleček and Mitra consider the case where the sampling rate conversion factor N is expressible as a product of two integers, $N = N_1 N_2$. In this case, the decimation filter transfer function $H(z)$ can be written as the product of two filter sections,

$$H(z) = \left[H_1\left(z^{N_1}\right) H_2(z) \right]^K,$$

(11.24)

where

$$H_1\left(z^{N_1}\right) = \frac{1}{N_2} \left[\frac{1 - z^{-N_1 N_2}}{1 - z^{-N_1}} \right],$$

(11.25)

$$H_2(z) = \frac{1}{N_1} \left[\frac{1 - z^{-N_1}}{1 - z^{-1}} \right].$$

(11.26)

Consequently, the amplitude responses $H_1(\omega N_1)$ and $H_2(\omega)$ are the following

$$H_1(\omega N_1) = \frac{1}{N_2} \left(\frac{\sin \frac{\omega N}{2}}{\sin \frac{\omega N_1}{2}} \right),$$

(11.27)

$$H_2(\omega) = \frac{1}{N_1} \left(\frac{\sin \frac{\omega N_1}{2}}{\sin \frac{\omega}{2}} \right).$$

(11.28)

Jovanović-Doleček and Mitra propose the specific roles for filters $H_1(z^{N_1})$ and $H_2(z)$ in the overall decimator. Filter $H_1(z^{N_1})$ is used to provide sharpening, and $H_2(z)$ to increase the stopband attenuation. Therefore, one can select the different number of stages for constructing the filter sections in (11.24), and obtain the modified transfer function $H_m(z)$ given by

$$H_m(z) = \left[H_1\left(z^{N_1}\right) \right]^K \left[H_2(z) \right]^L.$$

(11.29)

Applying the sharpening method from (11.22) to $H_1(z^{N_1})$, we arrive at the transfer function of the modified sharpened filter $H_{sh,m}(z)$ as proposed by Jovanović-Doleček and Mitra,

$$H_{sh,m}(z) = \left[H_2(z) \right]^L \left[H_1\left(z^{N_1}\right) \right]^{2K} \left\{ 3 z^{-N_1(N_2-1)K/2} - 2 \left[H_1\left(z^{N_1}\right) \right]^K \right\}.$$

(11.30)

Introducing expressions (11.27) – (11.28), one obtains the corresponding magnitude response for the modified sharpened filter

$$\left|H_{sh,m}\left(e^{j\omega}\right)\right| = \left|\left\{3\left(\frac{1}{N_2}\frac{\sin\frac{\omega N}{2}}{\sin\frac{\omega N_1}{2}}\right)^{2K} - 2\left(\frac{1}{N_2}\frac{\sin\frac{\omega N}{2}}{\sin\frac{\omega N_1}{2}}\right)^{3K}\right\}\left(\frac{1}{N_1}\frac{\sin\frac{\omega N_1}{2}}{\sin\frac{\omega}{2}}\right)^{L}\right|. \tag{11.31}$$

In the example that follows, we illustrate the properties of the magnitude responses of the modified sharpened filter and also those of the original comb filter, and of the comb filter sharpened according to (11.22). Notice that the investigations of Jovanović-Doleček and Mitra recommend the choice $L \geq 2K$.

Example 11.4

Compute and plot the magnitude response for the following decimation filters:

(a)   Magnitude response of the modified sharpened comb filter determined by: $N = 32$, $N_1 = 8$, $N_2 = 4$, $K = 2$, and $L = 5$.

(b)   Magnitude response of the original comb filter with $N = 32$, and $K = 2$.

(c)   Magnitude response of the sharpened comb filter (equations (11.22) and (11.23)) with $N = 32$, and $K = 2$.

The MATLAB code that computes and plots the requested characteristics is the following:

```
K = 2; N = 32; N1 = 8; N2 = 4; L = 5;          % Setting the design parameters
f = 0:0.001:1;                                 % Setting the frequency axis
omega = pi*f;
H1 = (sin(omega*N/2)./sin(omega*N1/2))/N2;     % Amplitude response of the CIC filter H1(z)
H2 = (sin(omega*N1/2)./sin(omega/2))/N1;       % Amplitude response of the CIC filter H2(z)
H_sh_m = abs((3*H1.^(2*K) - 2*H1.^(3*K)).*(H2.^L));  % Magnitude response of the modified sharpened
filter
figure (1)
plot(f,20*log10(H_sh_m))
axis([0,1,-250,5])
grid
hold on
G = (sin(omega*N/2)./sin(omega/2))/N;          % Amplitude response of the original CIC filter
GK = G.^K;                                     % Computing the amplitude response of G(z).^K
figure (1)
plot(f,20*log10(abs(GK)),'k','LineWidth',2)
H_sh = abs(3*G.^(2*K) - 2*G.^(3*K));           % Magnitude response of the original sharpened CIC filter
plot(f,20*log10(H_sh),'--')
```

Figure 11.19 displays the results. The overall magnitude responses are plotted in Figure 11.19(a), and the passband details are shown in Figure 11.19(b). The effects of sharpening techniques are evident in the augmented stopband attenuation and also in the decrease of the passband droop. We can notice that the passband characteristics of the sharpened filter, and of the modified sharpened filter are very similar in the low-frequency range, see Figure 11.19(b). Figure 11.19(a) shows that with the design parameters of this example, somewhat better stopband characteristics are achieved with the modified sharpened filter than with the sharpened filter. The reader is recommended to solve MATLAB Exercises 11.5 – 11.7 and compute the magnitude responses of the three filters for various values of N, N_1, N_2, K, and L.

Benefits of the modified sharpening technique proposed by Jovanović-Doleček and Mitra are found in the possibilities to exploit properly the two-stage decimation structure, and to reduce the sampling rate in the first decimation stage. The block diagrams of Figure 11.20 illustrate how the efficiency of the implementation is achieved. Figure 11.20(a) depicts the original structure of a sharpened factor-of-N decimator, see Figure 11.16. In Figure 11.20(b), the basic structure that implements the modified sharpened filter (11.30) proposed by Jovanović-Doleček and Mitra is shown. Here, filter $[H_2(z)]^L$ is placed at the input, whereas the sharpening technique is applied to the filter $[H_1(z^{N1})]^K$. Finally, Figure 11.20(c) presents the resulting two-stage implementation. The first stage performs the decimation-by-N_1 providing the sharpening section of the second stage to operate at the reduced sampling rate.

The first stage in Figure 11.20(c) is a factor-of-N_1 comb decimator which can be implemented in either recursive, or in a nonrecursive form. The advantage of nonrecursive realization is the possibility

Figure 11.19. Magnitude responses of the Modified Sharpened Filter (N = 32, N$_1$ = 8, N$_2$ = 4, K = 2, L = 5) – solid thick line; Sharpened Filter (N = 32, K = 2) – dashed thick line; comb filter (N = 32, K = 2) – thin line

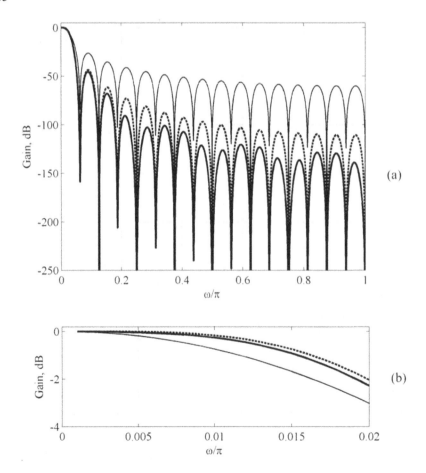

of down-sampling-by-N_1 at the input of the system. This is achieved by applying the polyphase decomposition to the filter $[H_2(z)]^L$. According to the procedure exposed in the section Polyphase Implementation Structures, we can decompose the filter $[H_2(z)]^L$ into N_1 polyphase components and move the down-sampling-by-N_1 at the input, see Figure 11.15(b). In this way, the first stage operates at the reduced sampling rate F_x/N_1, where F_x is the sampling frequency of the input signal.

For implementation of the second stage of Figure 11.20, the recursive configuration for sharpened comb filters can be used. This configuration is shown in Figure 11.17(b) for the case $K = 2$, and can be easily extended for the values $K > 2$. From Figures 11.17 (b) and 11.20(c), one observes that the operation rate for the integrator sections is F_x/N_1, and for differentiator sections the operation rate is equal to the output rate $F_y = F_x/(N_1 N_2) = F_x/N$.

An alternative approach for the two-stage sharpened decimator is proposed in (Jovanović-Doleček and Mitra, 2007), which introduces a sine-based compensation filter to decrease the passband droop of the original comb filter, and a cosine filter to improve the overall stopband characteristic.

In this section, we have concentrated on comb decimators and interpolators with the integer conversion factor. However, the comb-based filter architecture is also suitable for the rational conversion factors, see (Hentschel, 2002). An improvement of a rational sampling rate converter based on stepped triangular comb filter is proposed in (Jovanović-Doleček & Mitra, 2006).

Figure 11.20. Block diagram of the modified sharpened decimator: (a) Sharpened comb filter and down-sampler. (b) Structure of the modified sharpened comb decimator. (c) Two-stage implementation of the modified sharpened comb decimator.

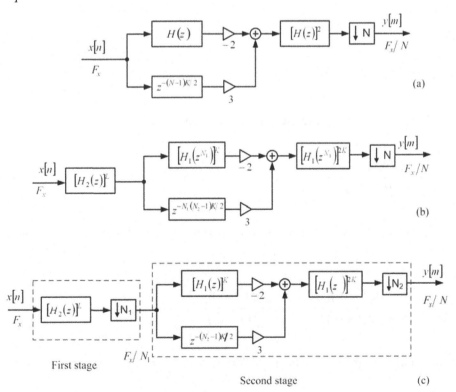

MODIFIED COMB DECIMATION FILTER: ZERO-ROTATION APPROACH

An attractive modification for comb filters developed for $\Sigma\Delta$ A/D converters was proposed by Lo Presti (2000) and extended by Laddomada (2007a,b). The new filter is aimed to improve the stopband attenuation in the aliasing bands in order to provide maximum suppression of the quantization noise in the first decimation stage. To achieve this, the concept of rotation of the natural nulls in the z-plane is applied to the comb filter sections. As a result, the new nulls placed in the intervals of the comb filter natural nulls are produced, and each pair of new nulls is located symmetrically around the natural comb filter nulls. In this way, a better distribution of the overall comb filter nulls may be achieved since they are not located one over the other any more. This approach provides the possibility to modify regularly the stopband attenuation in a more desirable form.

Consider a single-stage comb filter section $G_C(z)$ defined in (11.3),

$$G_C(z) = \frac{1}{N}\frac{1-z^{-N}}{1-z^{-1}},$$

(11.32)

whose z-plane zeros z_i, $i = 1, 2, (N-1)$, are uniformly distributed around the unit circle at the points $e^{j\pi/N}$, $e^{j2\pi/N}, \ldots, e^{j(N-1)\pi/N}$.

It is straightforward to show, see (Lo Presti, 2000), that the clockwise rotation by an angle α when applied to the zeros of $G_C(z)$ leads to the transfer function $G_{q+}(z)$ of the form

$$G_{q+}(z) = \frac{1}{N}\frac{1-z^{-N}e^{j\alpha N}}{1-z^{-1}e^{j\alpha}}.$$

(11.33)

The counter clockwise rotation by the same angle α applied to the zeros of $G_C(z)$ results in the transfer function $G_{q-}(z)$

$$G_{q-}(z) = \frac{1}{N}\frac{1-z^{-N}e^{-j\alpha N}}{1-z^{-1}e^{-j\alpha}}.$$

(11.34)

Multiplying (11.33) and (11.34) together, we obtain the modified filter cell $G_q(z)$,

$$G_q(z) = \frac{1}{N^2}\frac{1-2\cos(\alpha N)z^{-N}+z^{-2N}}{1-2\cos(\alpha)z^{-1}+z^{-2}}.$$

(11.35)

The resulting filter cell $G_q(z)$ is a linear-phase filter and has real coefficients. The frequency response $G_q(f)$ is given by

$$G_q(f) = \frac{e^{-j\pi f(N-1)}}{N^2} \cdot \frac{\sin\left[(\pi f + \alpha)N/2\right]}{\sin\left[(\pi f + \alpha)/2\right]} \cdot \frac{\sin\left[(\pi f - \alpha)N/2\right]}{\sin\left[(\pi f - \alpha)/2\right]}.$$

(11.36)

The zeros of $G_q(f)$ are placed at frequencies $(2i/N) \pm \alpha/\pi$, $i = 1, 2, \ldots, N/2$ for N even ($(N-1)/2$ for N odd).

Now, we compose the basic modified filter section $G_{MC3}(z)$ by multiplying the single-stage comb filter $G_C(z)$ with the modified filter cell $G_q(z)$ and obtain,

$$G_{MC3}(z) = G_C(z)G_q(z). \tag{11.37}$$

By introducing expressions for $G_C(z)$ and $G_q(z)$ from (11.32) and (11.35), respectively, we obtain $G_{MC3}(z)$ in the developed form,

$$G_{MC3}(z) = \frac{1}{N^3} \frac{1-z^{-N}}{1-z^{-1}} \frac{1-2\cos(\alpha N)z^{-N}+z^{-2N}}{1-2\cos(\alpha)z^{-1}+z^{-2}}. \tag{11.38}$$

The frequency response $G_{MC3}(f)$ is a product of $G_C(f)$ and $G_q(f)$, and therefore the product of the three sinc functions,

$$G_{MC3}(f) = \frac{1}{N^3} \frac{\sin(\pi Nf/2)}{\sin(\pi f/2)} \cdot \frac{\sin\left[(\pi f+\alpha)N/2\right]}{\sin\left[(\pi f+\alpha)/2\right]} \cdot \frac{\sin\left[(\pi f-\alpha)N/2\right]}{\sin\left[(\pi f-\alpha)/2\right]} \cdot e^{-j3\pi f(N-1)/2}. \tag{11.39}$$

The rotation angle α has to be chosen in such a way as to place the zeros of the transfer function $G_{MC3}(z)$ in the null intervals of the classical comb filter for the given value of N. Generally, the convenient value of α has to be selected in accordance with the signal bandwidth and with the bandwidth of the aliasing bands, i.e. the rotations of zeros for $\pm\alpha$ have to place nulls of $G_q(f)$ within the aliasing bands, see Figure 11.9. To this end, the value of α is expressible as

$$\alpha = q\pi f_m, \tag{11.40}$$

where q is a parameter, $q < 1$, and f_m is the highest frequency of the input signal. A detailed study on the optimal choice of q has been given in (Laddomada, 2007b).

Let us compare first by means of an example the distribution of the z-plane zeros of the modified filter cell $G_{MC3}(z)$ with the zero distribution of the three stage-comb filter $G_C^3(z)$ given by

$$G_C^3(z) = \left[\frac{1}{N} \frac{1-z^{-N}}{1-z}\right]^3. \tag{11.41}$$

Example 11.5

Figure 11.21 displays the zero-plots for two filters mentioned above for the case $N = 16$, $q = 0.78$, $f_m = 0.2$. Three-stage comb filter has triple zeros placed at the points $e^{j\pi/N}$, $e^{j2\pi/N}$,..., $e^{j(N-1)\pi/N}$ one over the other, whereas the modified filter $G_{MC3}(z)$ has single zeros placed at the points $e^{j(\pi/N-\alpha)}$, $e^{j\pi/N}$, $e^{j(\pi/N+\alpha)}$, $e^{j(2\pi/N-\alpha)}$, $e^{j2\pi/N}$, $e^{j(2\pi/N+\alpha)}$, ..., $e^{j[(N-1)\pi/N-\alpha]}$, $e^{j(N-1)\pi/N}$, $e^{j[(N-1)\pi/N+\alpha]}$, for $\alpha = 0.78\cdot0.2\cdot\pi$.

In the sequel, let us illustrate the effects of zero-rotation approach exposed above on the filter frequency response. Using the following MATLAB code we compute the frequency response of the classical three-stage comb filter and that of the modified comb filter.

```
N=16;                                         % Decimation factor
f = 0:0.00005:1;                              % Frequency scale
fm = 0.02;                                    % Maximum signal frequency
q = 0.78;                                     % Factor q
G=sin(pi*f*N/2)./(N*sin(pi*f/2));             % Comb filter frequency response
alpha = 0.78*pi*fm;                           % Rotating angle alpha
Gq = (sin((pi*f+alpha)*N/2)./sin((pi*f+alpha)/2));
Gq = Gq.*(sin((pi*f-alpha)*N/2)./sin((pi*f-alpha)/2));
Gq = Gq/(N^2);                                % Rotated filter section Gq(z)
GMC3 = Gq.*G;                                 % Modified comb filter section
```

The gain response of the two filters is shown in Figure 11.22, where the passband and the first aliasing band are emphasized. The markers are placed at the passband edge frequency, and at the edge frequencies of the first aliasing band. Figure 11.22 shows that the modified filter (solid line) provides 8 dB increase of selectivity in comparison with the classical three-stage comb filter (dashed line).

The various combinations of modified comb filters with classical comb filters can be used to construct the high-order filter structures. An exhaustive study of the modified comb filters including the optimal choice of parameters in order to construct the best decimation filter for the particular $\Sigma\Delta$ modulator is given in (Presti, 2000; Laddomada 2007 a & b).

The zero-rotation approach considerably improves the ability of the modified comb filter to suppress aliasing in decimation, but the passband droop is slightly increased. Naturally, the passband characteristics should be compensated.

The filter sharpening approach, aimed to improve the comb filter frequency response, which has been discussed earlier in this section, shows excellent results when applied to the modified filters with rotated zeros. The application of the sharpening technique has been exposed by Laddomada (2007 b), where the consequences of various combinations of the comb transfer functions and the modified comb transfer functions are examined in detail. It was demonstrated that attractive solutions can be achieved with respect to both: the stopband attenuation and the passband droop.

Figure 11.21. Z-plane zeros for: (a) Three-stage comb filter $G_C^{\,3}(z)$, (b) Modified filter $G_{MC3}(z)$

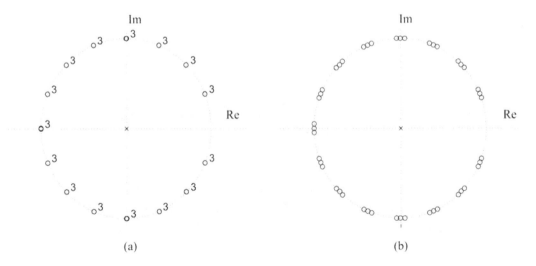

(a)　　　　　　　　　　　　　　(b)

Figure 11.22. Gain responses for: (a) Three-stage comb filter $G_C^3(z)$, (b) Modified filter $G_{MC3}(z)$

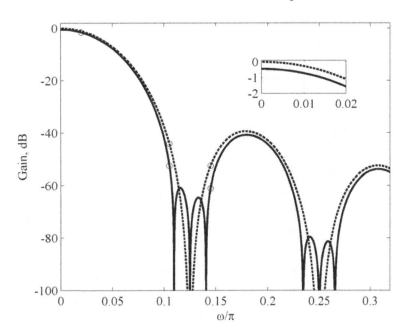

In this section, we show the effects of the simplest, 1st order shaping technique (Laddomada, 2007 b). The transfer function of the new filter $H_{11}(z)$ is defined by

$$H_{11}(z) = G_C^2(z)G_q(z)\left[3z^{-D} - 2G_C^2(z)\right] \tag{11.42}$$

where $G_C^2(z)$ is the two-stage classical comb filter and $G_q(z)$ is the modified filter cell as given in (11.35).

Let us apply the sharpening scheme of (11.42) to filter sections $G_C(z)$ and $G_q(z)$ obtained for the parameters $N = 16$, $q = 0.78$, $f_m = 0.2$ as already examined in *Example 11.5*. Using the frequency responses computed in *Example 11.5*, we easily compute the frequency response of the overall filter $H_{11}(z)$.

```
H11 = (G.^2).*Gq;
H11 = H11.*(3*ones(size(f))-2*(G.^2));
```

The gain response of the resulting filter $H_{11}(z)$ is shown in Figure 11.23. The attenuation in the first aliasing band exceeds 70 dB, and the passband droop is decreased to 0.2 dB. The good performance is a result of two techniques applied to the classical comb filter transfer function: zero-rotation and shaping technique. With higher order filters and higher order sharpening schemes very high stopband attenuations and small passband droops can be achieved, see (Laddomada 2007, b).

The excellent performances of this filter class have to be paid with an increase in the implementation complexity. The recursive realization structure of the filter is simple, but requires two multipliers, one in the recursive part of the structure, and the second in the nonrecursive part, see (Presti, 2000; Laddomada, 2007 a). The multiplier in the recursive part operates at the high sampling rate and, therefore, is particularly critical. The nonrecursive implementation schemes are investigated by Laddomada

Figure 11.23. Gain response of the sharpened modified comb filter with rotated zeros

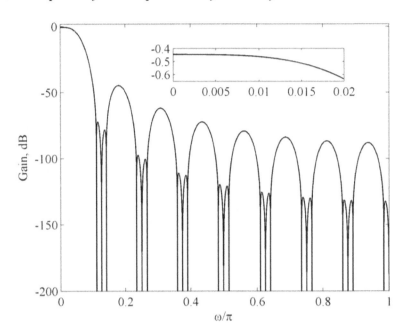

(2007 a), and a nonrecursive architecture suitable for power-of-2 decimation factors is proposed. The paper of Laddomada (2007 a) is recommended to the reader interested in the algorithms that lead to the multiplierless implementation of the nonrecursive implementation form.

MATLAB EXERCISES

11.1 Compute and plot the impulse response, magnitude response, and the pole-zero plot of the single-stage CIC filters $G(z)$ for $N = 8$, 16, and 32.

11.2 Compute and plot the impulse response, magnitude response, and the pole-zero plot of the K-stage CIC filters $G_K(z)$ for 16 and $K = 2$ and $K = 4$.

11.3 Modify specifications of *Example 11.1* to design a two-stage factor-of-16 decimator consisting of the CIC-based factor-of-8 decimator in the first stage, and the factor-of-two decimator in the second stage. Design CIC filter $G_K(z)$ and FIR filter $T(z)$. Display the magnitude responses and the pole-zero plots of $G_K(z)$ and $T(z)$. Display the magnitude response of the single-stage equivalent of the resulting overall decimator.

11.4 Consider the polyphase implementation of a CIC filter. Use equation (11.17a) and MATLAB program of *Example 11.2* to compute the coefficients of the FIR transfer function $H_1(z)$ for the following parameters: (i) $N_1 = 4$, $K = 4$; (ii) $N_1 = 8$, $K = 3$.

11.5 Modify MATLAB program of *Example 11.3* to compute and plot the magnitude response for the following decimation filters:

a) Magnitude response of the modified sharpened CIC filter determined by: $N = 32$, $N_1 = 4$, $N_2 = 8$, $K = 2$, and $L = 4$.

b) Magnitude response of the original CIC filter with $N = 32$, and $K = 2$.

c) Magnitude response of the sharpened CIC filter (equations (11.20) and (11.21)) with $N = 32$, and $K = 2$.

11.6 Using the procedure of Exercise 11.5, compute and plot the magnitude response for the following decimation filters:

a) Magnitude response of the modified sharpened CIC filter determined by: $N = 32$, $N_1 = 4$, $N_2 = 8$, $K = 2$, and $L = 5$.

b) Magnitude response of the original CIC filter with $N = 32$, and $K = 2$.

c) Magnitude response of the sharpened CIC filter (equations (11.20) and (11.21)) with $N = 32$, and $K = 2$.

11.7 Display the magnitude response and pole-zero plot of the CIC filter $G_K(z)$ determined by $N = 32$ and $K = 2$. Compose the CIC subfilters $[H_1(z^{N1})]^K$ and $[H_2(z)]^L$ of the modified sharpened filter as given in equations (11.23) and (11.24). Compute and plot the magnitude responses and pole-zero plots of $[H_1(z^{N1})]^K$ and $[H_2(z)]^L$ for the parameters $N = 32$, $N_1 = 4$, $N_2 = 8$, $K = 2$, and $L = 5$.

REFERENCES

Abed, K.H., & Nerurkar, S.B. (2003). Implementation of a low power decimation filter using 1/3-band IIR filter. *IEEE Wireless Communications and Networking, WCNC 2003.* Volume 1, 460-465.

Aboushady, H.; Dumonteix, Y., Louerat, M.-M., & Mehrez, H. (2001). Efficient polyphase decomposition of comb decimation filters in ΣΔ analog-to-digital converters. *IEEE Transactions on Circuits and Systems II: Analog and Digital Signal Processing*, *48*(10), 898-903.

Abu-Al-Saud, W.A. ,& Stuber, G.L. (2003). Modified CIC filter for sample rate conversion in software radio systems. *IEEE Signal Processing Letters*, *10*(5), 152 -154.

Abu-Al-Saud, W.A. & Stuber, G.L. (2006). Efficient sample rate conversion for software radio systems. *IEEE Transactions on Signal Processing,* Volume *54*(3), 932 – 939.

Filter design toolbox for use with MATLAB. User's guide. Version 6. (2006). Natick: MathWorks.

Hentchel, T. (2002). *Sample rate conversion in software configurable radios*. Morwood, MA: Artech House.

Hogenauer, E.B. (1981). An economical class of digital filters for decimation and interpolation. *IEEE Trans. Acoustics, Speech, and signal processing*, *29*(2), 155-162.

Jovanović-Doleček, G., & Mitra, S. K. (2005). A new two-stage sharpened comb decimator. *IEEE Transactions on Circuits and Systems – I: Regular Papers*, *52*(7), 1414-1420.

Jovanović-Doleček, G., & Mitra, S. K. (2006). CIC filter for rational sample rate conversion. *Proc. 2006 IEEE Asia Pacific Conference on Circuits and Systems – APCCAS*, 918-921.

Jovanović-Doleček, G., & Mitra, S. K. (2007). A new two-stage CIC-based decimator filter. *Proc. 5ᵗʰ International Symposium on Image and Signal Processing and Analysis – ISPA*, 218-223.

Kaiser, J. F., & Hamming, R. W. (1977). Sharpening the response of a symmetric nonrecursive filter. *IEEE Trans. Acoustics, Speech, and Signal Processing*, *25*(5), 415-422.

Kwentus, A.,Jiang, Z. & Willson A. Jr. (1997). Application of filter sharpening to cascaded integrator-comb decimation filter. *IEEE Transactions on Signal Processing*, *45*(2), 457-467.

Laddomada, M. (2007a). Generalized Comb Decimation Filters for $\Sigma\Delta$ A/D Converters: Analysis and Design. *IEEE Transactions on Circuits and Systems I: Regular Papers*, 54, 994-1005.

Laddomada, M. (2007b). Comb-Based Decimation Filters for $\Sigma\Delta$ A/D Converters: Novel Schemes and Comparisons. *IEEE Transactions on Signal Processing*, *55*(5), Part 1, 1769-1779.

Laddomada, M., & Mondin, M. (2004). Decimation schemes for $\Sigma\Delta$ A/D converters based on Kaiser and Hamming sharpened filters. *IEE Proceedings - Vision, Image and Signal Processing*, 151(4), 287-296.

Lo Presti, L. (2000). Efficient modified-sinc filters for sigma-delta A/D converters. *IEEE Trans. Circuits Syst. II, Analog and Digital Signal Processing*, 47(11), 1204-1213.

Meyer-Baese, U.; Rao, S.; Ramirez, J., & Garcia, A. (2005). Cost-effective Hogenauer cascaded integrator comb decimator filter design for custom ICs. *Electronics Letters*, *41*(3), 158-160.

Mitra, S. K. (2006). *Digital signal processing: A computer based approach*. 3rd edition. New York: The McGraw-Hill Companies, Inc.

Nerurkar, S.B., & Abed, K.H. (2006). Low-power decimator design using approximated linear-phase N-band IIR filter. *IEEE Transactions on Signal processing*, *54*(4), 1550-1553.

Stephen, G., & Stewart, R.W. (2004). High-speed sharpening of decimating CIC filter. *Electronics Letters*, *40*(21), 1383-1384.

Saramäki T., & Ritoniemi, T. (1997). A modified comb filter structure for decimation. *Proc. IEEE International Symp. Circuits and Systems – ISCAS*, 2353-2356.

Ze Tao, & Signell, S. (2006). Multi-Standard Delta-Sigma Decimation Filter Design. *IEEE Asia Pacific Conference on Circuits and Systems – APCCAS 2006*, 1212-1215.

Chapter XII
Examples of Multirate Filter Banks

INTRODUCTION

The purpose of this chapter is to illustrate by means of examples the construction of the analysis and synthesis filter banks with the use of FIR and IIR two-channel filter banks as the basic building blocks. In Chapter VIII, we have discussed the design and properties of several types of complementary filter pairs, and in Chapters IX and X we have shown how those filter pairs are used in the synthesis of digital filters with sharp spectral constraints. In this chapter, we demonstrate the application of the complementary filter pairs as two-channel filter banks used to decompose the original signal into two channel signals and to reconstruct the original signal from the channel signals. Signal decomposition is referred to as the signal analysis, whereas the signal reconstruction is referred to as the signal synthesis. Thereby, the filter bank used for the signal decomposition is called the analysis filter bank, and the bank used for signal reconstruction is called the synthesis filter bank.

The two-channel filter bank is usually composed of a pair of lowpass and highpass halfband filters, which satisfy some complementary properties. The bandwidth that occupies each of two channel signals is a half of the original signal bandwidth. Hence, the channel signals can be processed with the sampling rate which is a half of the original signal sampling rate. At the output of the analysis bank, the channel signals are down-sampled-by-two and then processed at the lower sampling rate. For the signal reconstruction, each of two channel signals has to be up-sampled-by-two first, and then fed into the synthesis bank.

The sampling rate alteration in the two-channel filter bank causes the unwanted effects: the down-sampling produces aliasing, and the up-sampling produces imaging. The essential feature of the two-channel filter bank is that the aliasing produced in the analysis side can be compensated in the synthesis side. This is achieved by choosing the proper combination of filters in the analysis and synthesis banks. The elimination of aliasing opens the possibility of the perfect (and nearly perfect) reconstruction of the original signal. The perfect reconstruction means that the signal at the output of the cascade connection of the analysis and synthesis bank is a delayed replica of the original input signal. Constructing perfect reconstruction and nearly perfect reconstruction analysis/synthesis filter banks is an unbounded area of research.

An important and widely used application of the two-channel filter banks is the construction of multichannel filter banks based on the tree-structures where the two-channel filter bank is used as a building block. In this way, a multilevel multichannel filter bank can be obtained with either uniform or nonuniform separation between the channels. The two-channel filter banks are particularly useful in generating octave filter banks.

Depending on applications, the filter bank can be requested to provide frequency-selective separation between the channels, or to preserve the original waveform of the signal. The example applications of the frequency-selective filter banks are audio and telecommunication applications. The importance of preserving the original waveform is related with the images. In the case of the discrete-time wavelet banks, the frequency-selectivity is less important. The main goal is to preserve the waveform of the signal.

The purpose of this chapter is to illustrate by means of MATLAB examples the signal analysis and synthesis based on the two-channel filter banks. We give first a brief review of the properties of the two-channel filter banks with the conditions for aliasing elimination. We discuss the perfect reconstruction and nearly perfect reconstruction properties and show the solutions based on FIR and IIR QMF banks and the orthogonal two-channel filter banks. In the sequel, the tree-structured multichannel filter banks are considered. The process of signal decomposition and reconstruction is illustrated by means of examples.

TWO-CHANNEL FILTER BANKS

The block diagram representing the analysis/synthesis two-channel filter bank with the processing unit between the analysis and synthesis parts is shown in Figure 12.1.

In the analysis bank, the original signal $x[n]$ is filtered using the lowpass/highpass filter pair $[H_0(z),$ $H_1(z)]$, and the lowpass and highpass channel signals $x_0[n]$ and $x_1[n]$ are obtained. Therefore, their z-transforms $X_0(z)$ and $X_1(z)$ are given by

$$X_0(z) = H_0(z)X(z), \text{ and } X_1(z) = H_1(z)X(z). \tag{12.1}$$

Figure 12.2 illustrates typical magnitude frequency responses of $H_0(z)$ and $H_1(z)$.

The spectra of the filtered signals $x_0[n]$ and $x_1[n]$ occupy a half of the baseband of the original signal $x[n]$, and according to this, $x_0[n]$ and $x_1[n]$ can be further processed at the half of the input sampling rate. The filtered signals $x_0[n]$ and $x_1[n]$ are then down-sampled by-a-factor-of-two, and subband signals $v_0[n]$ and $v_1[n]$ are obtained. If the sampling rate at the input is F_0, the subband signal components $v_0[n]$ and $v_1[n]$ are sampled at the rate $F_0/2$.

Figure 12.1. The two-channel analysis/synthesis filter bank

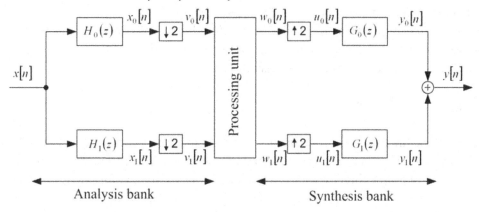

Figure 12.2. Typical magnitude responses of the lowpass filter $H_0(z)$, and the highpass filter $H_1(z)$

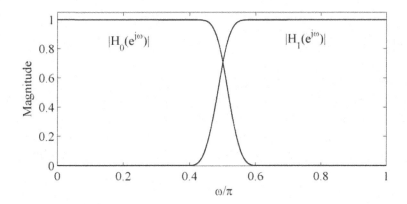

Since $v_0[n]$ and $v_1[n]$ are the down-sampled versions of $x_0[n]$ and $x_1[n]$, the z-transforms $V_1(z)$ and $V_2(z)$ are expressible in terms of the z-transforms $X_0(z)$ and $X_1(z)$. We simply apply equation (2.17) to the z-transforms $X_0(z)$ and $X_1(z)$ with the down-sampling factor $M = 2$. Additionally, by introducing relations (12.1), we express $V_1(z)$ and $V_2(z)$ in terms of the z-transform of the input signal $X(z)$,

$$V_0(z) = \frac{1}{2}\left[H_0\left(z^{1/2}\right)X\left(z^{1/2}\right) + H_0\left(-z^{1/2}\right)X\left(-z^{1/2}\right)\right], \qquad (12.2a)$$

$$V_1(z) = \frac{1}{2}\left[H_1\left(z^{1/2}\right)X\left(z^{1/2}\right) + H_1\left(-z^{1/2}\right)X\left(-z^{1/2}\right)\right]. \qquad (12.2b)$$

The first terms in equations (12.2a – b) represent the z-transforms of the desired decimated signal components, whereas the second terms represent the aliasing components that overlap in the basebands of the decimated signals.

The signals $v_0[n]$ and $v_1[n]$ are processed in the processing unit, usually coded and compressed. Before inputting to the synthesis bank, signals are usually decoded and decompressed. The coding and quantization errors in the processing unit may cause derogation of the signals. In that case the resulting signals $w_0[n]$ and $w_1[n]$ differ from the original signals $v_0[n]$ and $v_1[n]$.

In order to examine the performances of the analysis/synthesis bank, the errors that may be produced in the processing unit are neglected. Hence, the future considerations of the analysis/synthesis filter bank in this Chapter are evaluated under the assumption that

$$w_0[n] = v_1[n], \quad \text{and} \quad w_1[n] = v_1[n]. \tag{12.3}$$

In the synthesis bank, the two signal components are up-sampled-by-two first, then filtered by $G_0(z)$ and $G_1(z)$, and finally added together to compose the output signal $y[n]$. The z-transforms of the up-sampled signals $u_0[n]$ and $u_1[n]$ follow directly from equation (2.22) when applied for the up-sampling factor $L = 2$. With the assumption given in (12.3), i.e., $w_0[n] = v_0[n]$ and $w_1[n] = v_1[n]$, the z-transforms $U_0(z)$ and $U_1(z)$ become expressible in terms of $X_0(z)$ and $X_1(z)$,

$$U_0(z) = V_0(z^2) = \frac{1}{2}\left[X_0(z) + X_0(-z)\right], \tag{12.4a}$$

$$U_1(z) = V_1(z^2) = \frac{1}{2}\left[X_1(z) + X_1(-z)\right]. \tag{12.4b}$$

With the up-sampling operation, the sampling rate increases from $F_0/2$ to F_0. In equations (12.4a – b), the terms $X_0(z)$ and $X_1(z)$ represent the desired signal component without aliasing, and the terms $X_0(-z)$ and $X_1(-z)$ represent the unwanted aliased signal components.

In the second step, signals $u_0[n]$ and $u_1[n]$ are processed by the lowpass/highpass synthesis filter pair $G_0(z)$ and $G_1(z)$. Thereby, the z-transforms $Y_0(z)$ and $Y_1(z)$ of the output signals $y_0[n]$ and $y_1[n]$ are given by

$$Y_0(z) = G_0(z)V_0(z^2) = \frac{1}{2}\left[G_0(z)X_0(z) + G_0(z)X_0(-z)\right], \tag{12.5a}$$

$$Y_1(z) = G_1(z)V_1(z^2) = \frac{1}{2}\left[G_1(z)X_1(z) + G_1(z)X_1(-z)\right]. \tag{12.5b}$$

Finally, in the third step the filtered signals $y_0[n]$ and $y_1[n]$ are added together to yield the output $y[n]$. Accordingly, the z-transform $Y(z)$ is the sum of $Y_0(z)$ and $Y_1(z)$,

$$Y(z) = Y_0(z) + Y_1(z). \tag{12.6}$$

Introducing the substitutions from equations (12.2), (12.4) and (12.5) into equation (12.6), we arrive to the expression, which relates the z-transform of the output $Y(z)$ with the z-transform of the input $X(z)$ and with its aliased component $X(-z)$,

$$Y(z) = \frac{1}{2}\left[H_0(z)G_0(z) + H_1(z)G_1(z)\right]X(z) + \frac{1}{2}\left[H_0(-z)G_0(z) + H_1(-z)G_1(z)\right]X(-z) \tag{12.7}$$

The first term in equation (12.7) represents the input/output relation of the overall analysis synthesis filter bank without aliasing and imaging effects. The second term represents the effects of aliasing and imaging.

This second term has to be eliminated by the proper combination of the transfer functions $H_0(z)$, $H_1(z)$, $G_0(z)$ and $G_1(z)$. Conditions for the alias-free filter bank will be considered in the next subsection.

Alias-Free Filter Banks

Technically, the alias-free design of the two-channel filter bank means the elimination of the second term in equation (12.7).

The solution is based on the supposition that the aliasing produced by the decimation in the analysis part could be cancelled by the images produced by interpolation in the synthesis part. This can be achieved with the proper combination of the transfer functions of the filters in the analysis and synthesis parts.

Equation (12.7) is usually expressed in the form

$$Y(z) = T(z)X(z) + A(z)X(-z)$$

(12.8)

where

$$T(z) = \frac{1}{2}\left[H_0(z)G_0(z) + H_1(z)G_1(z)\right]$$

(12.9)

is the *distortion transfer function*, and

$$A(z) = \frac{1}{2}\left[H_0(-z)G_0(z) + H_1(-z)G_1(z)\right]$$

(12.10)

is called the *aliasing transfer function*.

The alias-free transfer function could be generated if the aliasing transfer function is cancelled, i.e., if we make $A(z) \equiv 0$. A simple way to achieve this is to choose the synthesis filters $G_0(z)$ and $G_1(z)$ as follows,

$$G_0(z) = 2H_1(-z)$$

(12.11a)

$$G_1(z) = -2H_0(-z)$$

(12.11b)

With this choice, the aliasing term $A(z)$ disappears, and equation (10.8) becomes

$$Y(z) = T(z)X(z)$$

(12.12)

where

$$T(z) = H_0(z)H_1(-z) - H_0(-z)H_1(z)$$

(12.13)

The $T(z)$, which is called the distortion transfer function, represents the distortion of the overall analysis/synthesis filter bank. In the frequency domain, we consider the frequency characteristics of $T(e^{j\omega})$,

$$T(e^{j\omega}) = \left|T(e^{j\omega})\right|e^{j\arg\left[T(e^{j\omega})\right]}$$

(12.14)

where $|T(e^{j\omega})|$ represents the *amplitude distortion*, and $\arg[T(e^{j\omega})]$ is the *phase distortion* of the analysis/synthesis filter bank.

Perfect-Reconstruction and Nearly Perfect-Reconstruction Filter Banks

In this subsection, we consider the conditions for the perfect-reconstruction and nearly perfect-reconstruction filter banks. The *perfect-reconstruction property* means that the signal at the output of the analysis/synthesis filter bank $y[n]$ is a delayed version of the original signal $x[n]$,

$$y[n] = x[n - K].$$

(12.15)

In the majority of applications, the perfect-reconstruction of the original signal is a desirable, but not indispensable property. More economical solutions with the reduced computational complexity can be obtained if the filter bank achieves the nearly perfect-reconstruction property defined by

$$y[n] \approx x[n - K].$$

(12.16)

The distortion transfer function of the two-channel analysis/synthesis filter bank satisfying the perfect-reconstruction property is a pure delay,

$$T(z) = z^{-K}.$$

(12.17)

For the *nearly-perfect reconstruction property*, the distortion transfer function is an approximation of the ideal solution of equation (12.16).

If $T(z)$ is an all-pass transfer function,

$$\left| T\left(e^{j\omega}\right) \right| = \text{constant, for all } \omega,$$

(12.18)

the filter bank satisfies the *magnitude-preserving property*. In the case that $T(z)$ is a linear phase transfer function, i.e.,

$$\arg\left[T\left(e^{j\omega}\right) \right] = K\omega,$$

(12.19)

there are no phase distortions in the analysis/synthesis filter bank.

In the subsections that follow, we describe some typical two-channel filter banks satisfying the perfect-reconstruction and nearly perfect-reconstruction properties.

Quadrature Mirror (QMF) Filter Banks

The term *quadrature mirror (QMF) filter bank* denotes the quadrature mirror symmetry of the lowpass/highpass halfband filter pair $H_0(z)$ and $H_1(z)$. Hence, we choose

$$H_1(z) = H_0(-z),$$

(12.20)

and according to (12.11a – b) the synthesis filters $G_0(z)$ and $G_1(z)$ are given by

$$G_0(z) = 2H_0(z), \quad \text{and} \quad G_1(z) = -2H_1(z). \tag{12.21}$$

In this way, the aliasing in the analysis/synthesis bank is easily eliminated. The distortion transfer function $T(z)$ is expressible in terms of the lowpass filter $H_0(z)$,

$$T(z) = H_0^2(z) - H_1^2(z) = H_0^2(z) - H_0^2(-z). \tag{12.22}$$

It is apparent that the amplitude and phase distortions of the overall QMF bank depend on the performances of the lowpass filter $H_0(z)$. If $H_0(z)$ has a linear phase, the phase distortion of the overall filter bank is eliminated. It is shown (Mitra, 2006; Vaidianathan, 1987 and 1993) that for $H_0(z)$ being a linear-phase FIR filter of the length N, the overall frequency response $T(e^{j\omega})$ can be expressed in the following manner

$$T(e^{j\omega}) = e^{-j\omega(N-1)} \left[\left| H_0(e^{j\omega}) \right|^2 - (-1)^{(N-1)} \left| H_1(e^{j\omega}) \right|^2 \right]. \tag{12.23}$$

Since the filter pair $[H_0(z), H_1(z)]$ is a halfband filter pair, their magnitude responses at the crossover frequency $\omega_c = \pi/2$ are equal, i.e., $|H_0(e^{j\pi/2})| = |H_1(e^{j\pi/2})|$. This implies that for odd values of N, $T(e^{j\omega})$ may have severe amplitude distortions in the vicinity of $\omega = \pi/2$. Therefore, when using the linear-phase FIR filters for a QMF bank, the filter length N should be an even number.

When the filter length N is an even number, equation (12.23) reduces to

$$T(e^{j\omega}) = e^{-j\omega(N-1)} \left[\left| H_0(e^{j\omega}) \right|^2 + \left| H_1(e^{j\omega}) \right|^2 \right]. \tag{12.24}$$

It follows from the above equation that the FIR QMF filter bank with linear-phase filters in the analysis and synthesis parts would satisfy the condition of a perfect reconstruction if

$$\left| H_0(e^{j\omega}) \right|^2 + \left| H_1(e^{j\omega}) \right|^2 = 1. \tag{12.25}$$

Thereby, the perfect reconstruction may be achieved when the linear-phase analysis filters are power-complementary.

It was shown that a QMF bank with linear-phase filters can achieve a perfect reconstruction property only when $H_0(z)$ and $H_1(z)$ are trivial first-order transfer functions (Mitra, 2006; Vaidianathan, 1987 and 1993).

According to equation (12.24), the QMF bank with linear phase filters has no phase distortion, but the amplitude distortion will always exist except for the trivial first-order case. It is an important approximation problem to adjust the coefficients of the transfer function $H_0(z)$ that simultaneously provide the selective frequency responses of the analysis/synthesis filters and guarantee a small reconstruction error. A computer-aided optimization method can be employed, which iteratively adjusts the coefficients of $H_0(z)$ to achieve

$$\left| H_0(e^{j\omega}) \right|^2 + \left| H_1(e^{j\omega}) \right|^2 \cong 1 \tag{12.26}$$

for all values of ω. This method has been applied by Johnston (1980). The tabulated values of the optimized FIR filter coefficients can also be found in (Ansary & Liu, 1993).

Since the appearance of the Johnston's work, the optimization of the linear-phase FIR QMF filter bank was the subject of permanent research. A design based on neural networks containing also the relevant references has been published recently by Yue-Dar Jou (May, 2007).

An efficient realization of an FIR alias-free QMF bank is obtained when using the polyphase decompositions. The lowpass/highpass analysis filters $H_0(z)$ and $H_1(z)$ are expressible in terms of two polyphase components $E_0(z)$ and $E_1(z)$,

$$H_0(z) = E_0(z^2) + z^{-1}E_1(z^2) \quad \text{and} \quad H_1(z) = E_0(z^2) - z^{-1}E_1(z^2), \tag{12.27}$$

and according to equation (12.21), the synthesis filters $G_0(z)$ and $G_1(z)$ are given by

$$G_0(z) = 2E_0(z^2) + 2E_1(z^2) \quad \text{and} \quad G_1(z) = -2E_0(z^2) + 2z^{-1}E_1(z^2). \tag{12.28}$$

Starting from the above polyphase representations of the analysis/synthesis filter pairs $[H_0(z), H_1(z)]$ and $[G_0(z), G_1(z)]$, a computationally efficient realization structure for a linear-phase FIR QMF bank can be obtained. Figure 12.3 depicts the solution for the overall filter bank where the analysis (synthesis) filter pair is implemented at the cost of a single filter, and moreover, all the arithmetic operations in the polyphase components are to be evaluated at the half of the input/output sampling rate

We illustrate the characteristics of an FIR linear-phase QMF bank, and its application in signal decomposition and reconstruction, by means of the following example.

Example 12.1

In this example, we choose the Johnston's FIR filter of the length $N = 12$ to examine the performances of a linear-phase FIR QMF bank. We determine first the analysis filters $H_0(z)$ and $H_1(z)$, and plot their characteristics in the time and frequency domains. In the following, we compose the polyphase components $E_0(z)$ and $E_1(z)$. Finally, we use the structure of Figure 12.3 to perform the decomposition and reconstruction of the test signal. In this example, we use the test signal of a rectangular shape.

Vector B1, given below, contains the first half of the linear-phase FIR filter coefficients (Ansari & Liu, 1993),

Figure 12.3. Computationally efficient realization of the two-channel FIR QMF bank

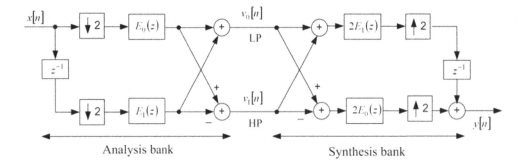

354

B1 = [-0.006443977,0.02745539,-0.00758164,-0.0913825,0.09808522,0.4807962];

We use the given coefficients and exploit the coefficient symmetry property to generate the impulse response of the linear-phase lowpass analysis filter $H_0(z)$,

```
h0 = [B1,fliplr(B1)];        % Generating the lowpass filter H0(z)
```

The highpass analysis filter $H_1(z)$ is defined by equation (12.20), and consequently its impulse response is determined by

```
for k = 1:length(h0)
   h1(k) = ((-1)^k)*h0(k);   % Generating the highpass filter H1(z)
end
```

Figure 12.4 plots the impulse responses of the filter pair $[H_0(z), H_1(z)]$, and Figure 12.5 (upper subfigure) shows the gain responses of two filters. The bottom subfigure of Figure 12.5 shows the resulting amplitude distortion of the overall filter bank.

Following the structure of Figure 12.3, we select the polyphase components $E_0(z)$ and $E_1(z)$,

```
e0 = h0(1:2:length(h0));     % Polyphase component E_0(z)
e1 = h0(2:2:length(h0));     % Polyphase component E_1(z)
```

The rectangular test signal $x[n]$ is determined by

Figure 12.4. Two-channel FIR QMF bank. Impulse responses of the lowpass filter $H_0(z)$ and the highpass filter $H_1(z)$

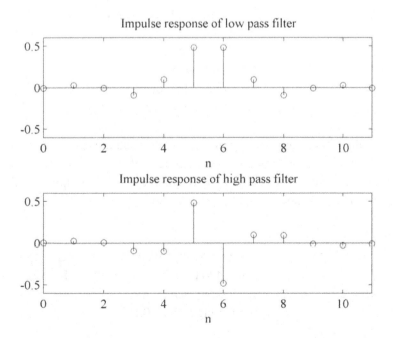

Figure 12.5. Two-channel FIR QMF bank: amplitude responses

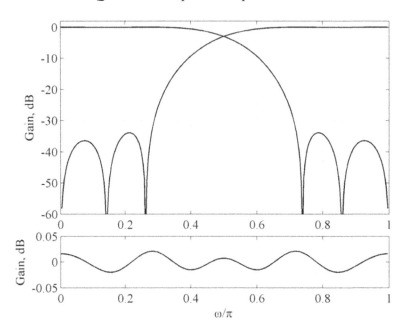

```
x = [zeros(size(1:100)),ones(size(101:250)),zeros(size(251:511))]; % Generating the test signal
```

First, we decompose the original signal stored in vector x into two signal components $v_0[n]$ and $v_1[n]$ using the analysis part of the two-channel analysis/synthesis bank presented in Figure 12.3. The MATLAB code that implements the two-channel analysis bank is the following,

```
% Analysis part
x0 = x(1:2:length(x))/2;            % Down-sampling, branch E_0
x1 = [0,x(2:2:length(x)-1)]/2;      % Down-sampling, branch E_1
v00 = filter(e0,1,x0);              % Filtering with E_0(z)
v11 = filter(e1,1,x1);              % Filtering with E_1(z)
v0 = v00 + v11;                     % Output of the lowpass filter
v1 = v00 - v11;                     % Output of the highpass filter
```

Secondly, we use the signal components $v_0[n]$ and $v_1[n]$ stored in vectors v0 and v1 to reconstruct the original signal. The signal reconstruction is based on the synthesis part of the analysis/synthesis bank of Figure 12.3 where the signal components $v_0[n]$ and $v_1[n]$ are the two input signals. The two-channel synthesis bank can be implemented in MATLAB as follows,

```
% Synthesis part
w00 = v0 + v1;                      % Input to polyphase component E_1(z), synthesis part
w11 = v0 - v1;                      % Input to polyphase component E_0(z), synthesis part
u00 = 2*filter(e1,1,w00);          % Filtering with E_1(z)
u11 = 2*filter(e0,1,w11);          % Filtering with E_0(z)
```

```
y0 = zeros(size(1:512));
y0(1:2:512) = u00;                    % Upsampling, branch E_0
y1 = zeros(size(1:512));
y1(1:2:512) = u11;                    % Upsampling, branch E_1
y0 = [0,y0(1:511)];                   % Inserting the delay z^(-1)
y = 2*(y0 + y1);                      % Reconstructed signal
```

Figure 12.6 presents the original signal $x[n]$, signal components $v_0[n]$ and $v_1[n]$, and the reconstructed signal $y[n]$. We can observe that the reconstructed signal is practically a delayed replica of the original signal. Although Figure 12.5 shows that this QMF bank produces some minor amplitude distortion, the effects on this particular test signal are negligible.

Alias-Free Two-Channel IIR QMF Banks with Magnitude-Preserving Property

An IIR lowpass/highpass halfband filter pair whose transfer functions are expressible as a sum and difference of two stable all-pass functions $A_0^{HB}(z)$ and $A_1^{HB}(z)$ can be used to construct the alias-free two-channel QMF bank.

The transfer functions of the analysis filter pair are given by

$$H_0(z) = \frac{1}{2}\left(A_0^{HB}(z^2) + z^{-1}A_1^{HB}(z^2)\right),$$ (12.29a)

$$H_1(z) = \frac{1}{2}\left(A_0^{HB}(z^2) - z^{-1}A_1^{HB}(z^2)\right).$$ (12.29b)

Figure 12.6. Two-channel FIR QMF bank: signal decomposition and reconstruction

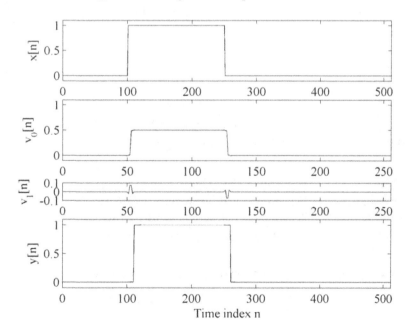

Recall that this class of lowpass/highpass transfer functions have been considered earlier in Chapter VIII as a *Class I* complementary filter pair. This filter class was shown to satisfy the power complementary property, and therefore, to satisfy the conditions for the magnitude-preserving property given by equation (12.18).

The alias-free property of the overall analysis/synthesis filter bank is achieved when the synthesis filters $G_0(z)$ and $G_1(z)$ are chosen according to the conditions defined by equations (12.21),

$$G_0(z)=2H_0(z)=A_0^{HB}(z^2)+z^{-1}A_1^{HB}(z^2),$$ (12.30a)

$$G_1(z)=-2H_1(z)=-A_0^{HB}(z^2)+z^{-1}A_1^{HB}(z^2).$$ (12.30b)

Introducing the expressions for $H_0(z)$, $H_1(z)$, $G_0(z)$ and $G_1(z)$ from (12.29) and (12.30) into equation (12.9), we express the distortion transfer function in terms of all-pass functions $A_0^{HB}(z)$ and $A_1^{HB}(z)$,

$$T(z)=z^{-1}A_0^{HB}(z^2)A_1^{HB}(z^2).$$ (12.31)

It is seen from equation (12.31) that the distortion transfer function $T(z)$ is an IIR all-pass transfer function, and therefore, the analysis/synthesis filter bank satisfies the magnitude-preserving property. However, this filter bank has a nonlinear phase response.

The computationally efficient realization structure of the two-channel alias-free IIR QMF bank is shown in Figure 12.7.

Example 12.2

In this example, we use the 5th order Butterworth filter to examine the performances of a magnitude-preserving IIR QMF bank. We determine first the analysis filters $H_0(z)$ and $H_1(z)$, and plot their magnitude responses. In the following, we compose the allpass subfilters $A_0(z)$ and $A_1(z)$. Finally, we use the structure of Figure 12.7 to perform the decomposition and reconstruction of the test signal. In this example, we use the test signal of a rectangular shape.

Firstly, we design the 5th order Butterworth filter pair $[H_0(z), H_1(z)]$ and verify the power-complementary property of the filter pair.

```
N = 5;                          % Selecting the Butterworth filter order
[z,p,k] = butter(N,0.5);        % Butterworth filter design

B = k*poly(z); A=poly(p);       % Lowpass filter H0(z) = B(z)/A(z)
[H0,f] = freqz(B,A,256,2);      % Lowpass filter frequency response
BB = k*poly(-z);   AA = A;      % Highpass filter H1(z) = BB(z)/AA(Z)
[H1,f] = freqz(BB,AA,256,2);    % Highpass filter frequency response

ver = abs(H0).^2+abs(H1).^2;    % Verification of the power-complementary property
```

Figure 12.7. Computationally efficient realization of two-channel IIR QMF bank

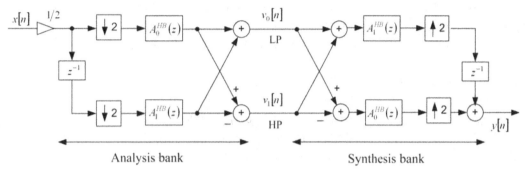

Analysis bank   Synthesis bank

Figure 12.8. Magnitude responses of the lowpass/highpass Butterworth filter pair, and verification of the power-complementary property

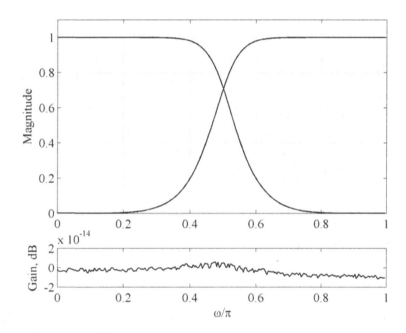

The upper subfigure of Figure 11.8 displays the magnitude responses of $H_0(z)$ and $H_1(z)$, and the bottom subfigure displays the magnitude distortion of the analysis/synthesis filter bank in dB, computed as 10*log10(ver). It is to be observed that the very small magnitude-distortion shown in Figure 12.8 is caused by a numerical error.

Using the filter poles stored in the vector p, we compute the constants for the all-pass subfilters $A_0^{HB}(z)$ and $A_1^{HB}(z)$, see Chapter VIII, subsection *Class I*: Power-Complementary and All-Pass Complementary Filter Pairs Implemented as a Parallel Connection of Two All-Pass Subfilters, particularly equations $(8.31 - 32)$.

```
c = abs(p).^2;                         % Computing the squared moduli of the filter poles
g = sort(c(1:2:length(c)));            % Sorting the squared moduli of the filter poles in ascending order
bet0 = g(2:2:length(g));               % Selecting the beta coefficients for A0(z)
bet1 = g(3:2:length(g));               % Selecting the beta coefficients for A1(z)


% Composing the allpass branches A_0(z) and A_1(z)
P0 = poly(-bet0);                      % Allpass branch A_0(z)=Q0(z)/P0(z)
Q0 = fliplr(P0);                       % Allpass branch A_0(z)=Q0(z)/P0(z)
P1 = poly(-bet1);                      % Allpass branch A_1(z)=Q1(z)/P1(z)
Q1 = fliplr(P1);                       % Allpass branch A_1(z)=Q1(z)/P1(z)
```

For the 5th-order Butterworth filter, the following all-pass functions $A_0^{HB}(z)$ and $A_1^{HB}(z)$ have been obtained:

$$A_0^{HB}(z) = \frac{0.1056 + z^{-1}}{1 + 0.1056z^{-1}}, \quad A_1^{HB}(z) = \frac{0.5279 + z^{-1}}{1 + 0.5279z^{-1}}$$

The overall analysis/synthesis IIR filter bank of Figure 12.7 needs only 4 multiplication constants in implementation: two constants for the analysis filter pair $[H_0(z), H_1(z)]$, and two constants for the synthesis filter pair $[G_0(z), G_1(z)]$.

In order to examine the phase nonlinearity of the analysis/synthesis IIR QMF bank, we compute the group delay characteristic of the distortion transfer function $T(z)$. In equation (12.31) the distortion transfer function is expressed in terms of the all-pass functions $A_0^{HB}(z)$ and $A_1^{HB}(z)$. Based on expression (12.31), we write the following code for computing the group-delay characteristic.

```
TB = conv(upsample(Q0,2),upsample(Q1,2));
TB = [0,TB];
TA = conv(upsample(P0,2),upsample(P1,2));
[gd,f] = grpdelay(TB,TA,512,2);
```

Figure 12.9 plots the resulting group delay characteristic.

We examine the IIR QMF bank of Figure 12.7 with the rectangular test signal $x[n]$, which is the same test signal already used in *Example 12.1*,

```
x = [zeros(size(1:100)),ones(size(101:250)),zeros(size(251:511))]; % generating the test signal
```

First, we perform the decomposition of the original signal stored in vector x into two signal components $v_0[n]$ and $v_1[n]$ using the analysis part of the two-channel analysis/synthesis bank of Figure 12.7.

```
% Analysis bank
xx = x/2;                              % Scaling the input signal by 1/2
x0 = xx(1:2:length(xx));               % Down-sampling, branch A0
x1 = [0,xx(2:2:length(xx))];           % Down-sampling, branch A1
v00 = filter(Q0,P0,x0);                % Filtering with A_0(z)
```

Figure 12.9. Group-delay characteristic of the distortion transfer function of the IIR two-channel filter bank composed of 5th-order Butterworth filter

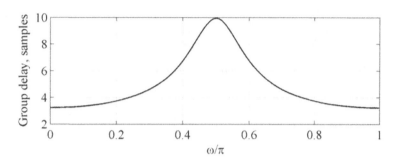

```
v11 = filter(Q1,P1,x1);        % Filtering with A_1(z)
v0 = v00 + v11;                % Lowpass output
v1 = v00 - v11;                % Highpass output
```

Secondly, we reconstruct the original signal from the signal components $v_0[n]$ and $v_1[n]$, which are stored in vectors v0 and v1. The signal reconstruction is based on the synthesis part of the analysis/synthesis bank of Figure 12.7.

```
% Synthesis bank
w00 = v0 + v1;                 % Input to A_0(z)
w11 = v0 - v1;                 % Input to B_0(z)
u00 = filter(Q1,P1,w00);       % Filtering with A_1(z)
u11 = filter(Q0,P0,w11);       % Filtering with A_0(z)
y0 = zeros(size(1:512));
y0(1:2:512) = u00;             % Up-sampling, branch A_1
y1 = zeros(size(1:512));
y1(1:2:512) = u11;             % Up-sampling, branch B_1
y0 = [0,y0(1:511)];            %  Inserting delay z^(-1)

y = y0 + y1;                   % Reconstructed signal
```

Figure 12.10 shows the original signal $x[n]$, signal components $v_0[n]$ and $v_1[n]$, and the reconstructed signal $y[n]$. The shape of the reconstructed signal shows the difference to that of the original signal, especially edges. This is the consequence of the nonlinear phase characteristic of the IIR QMF bank.

Example 12.2 illustrates the computational efficiency of an IIR QMF filter bank, and also its main disadvantage, the phase nonlinearity. Due to the nonlinear phase characteristic, this filter bank is not suitable for the applications where the shape of the signal waveform is the main requirement.

The solutions for a linear-phase IIR QMF bank can be achieved with anticausal filters in the synthesis part. In that case, the analysis filters $H_0(z)$ and $H_1(z)$ are the same as defined by equations (12.29a – b), whereas the synthesis filters $G_0(z)$ and $G_1(z)$ are the anticausal versions of the solutions given in equations (12.30a – b),

Figure 12.10. Two-channel IIR QMF bank: signal decomposition and reconstruction

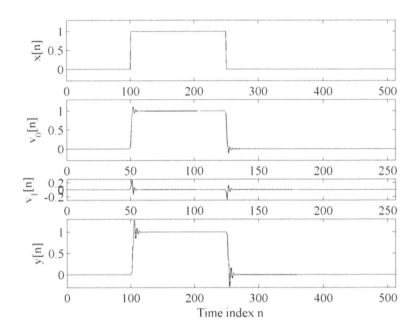

$$G_0(z) = 2H_0(z^{-1}) = A_0^{HB}(z^{-2}) + z^{-1}A_1^{HB}(z^{-2}),$$

(12.32a)

$$G_1(z) = -2H_1(z^{-1}) = -A_0^{HB}(z^{-2}) + z^{-1}A_1^{HB}(z^{-2}).$$

(12.32b)

Therefore, an IIR QMF filter bank is a perfect-reconstruction yielding $Y(z) \equiv X(z)$, when the analysis filters are causal filters constructed according to equations (12.29a – b), and the synthesis filters are anticausal constructed according to equations (12.32a – b). The main problem in implementation of this perfect-reconstruction filter bank is the implementation of the anticausal synthesis filters $G_0(z)$ and $G_1(z)$. Solutions for the causal implementation of the anti-causal transfer functions for the finite length signals and also for the infinite length signals exist in the literature, see for example (Saramäki & Bregovic, 2002) for the references.

Orthogonal Two-Channel FIR Filter Banks

We consider in this subsection the orthogonal filter bank, which satisfies the perfect-reconstruction property as defined by expression (12.15). The analysis/synthesis filters are nonlinear phase halfband FIR filters whose transfer functions $H_0(z)$, $H_1(z)$, $G_0(z)$ and $G_1(z)$ are related in such a manner that the overall analysis/synthesis bank is an alias-free filter bank satisfying the perfect-reconstruction property.

The transfer functions of the lowpass filters $H_0(z)$ and $G_0(z)$ [highpass filters $H_1(z)$ and $G_1(z)$] satisfy the mirror-image symmetry,

$$G_0(z) = 2z^{-(N-1)}H_0(z^{-1}), \quad G_1(z) = 2z^{-(N-1)}H_1(z^{-1}).$$

(12.33)

Here, N is the filter length - an even number. With the real-coefficients transfer functions, the above equation implies that the magnitude responses of the synthesis filters are the scaled-by-two versions of the magnitude responses of the analysis filters, whereas the phase responses are of the opposite sign,

$$G_0\left(e^{j\omega}\right)=2H_0\left(e^{-j\omega}\right), \quad G_1\left(e^{j\omega}\right)=2H_1\left(e^{-j\omega}\right).$$
(12.34)

The transfer functions of the lowpass/highpass analysis filters $[H_0(z), H_1(z)]$ in the orthogonal bank are related by

$$H_1\left(z\right)=-z^{-(N-1)}H_0\left(-z^{-1}\right).$$
(12.35)

In summary, the analysis/synthesis filters are related as follows,

$$G_0\left(z\right)=2H_1\left(-z\right)=2z^{-(N-1)}H_0\left(z^{-1}\right),$$
(12.36a)

$$G_1\left(z\right)=-2H_0\left(-z\right)=2z^{-(N-1)}H_1\left(z^{-1}\right).$$
(12.36b)

The overall filter bank is alias-free. This can be verified when introducing relations from (12.36) into the expression for the aliasing transfer function defined by equation (12.10).

The perfect-reconstruction property $y[n] = x[n–K]$ (defined in equation (12.15)) is achieved when the filters $H_0(z)$ and $G_0(z)$ are the spectral factors of the linear-phase halfband transfer function $E(z)$ of the order $2N–2$,

$$E\left(z\right)=H_0\left(z\right)G_0\left(z\right)=\sum_{n=0}^{2N-2}e[n]z^{-n}$$
(12.37)

whose impulse-response coefficients satisfy

$$e[n]=\begin{cases}0.5, & n=K\\ 0, & n \text{ is odd, and } n\neq K\end{cases}$$
(12.38)

The proof follows from the definition of the distortion transfer function (12.9),

$$T\left(z\right)=\frac{1}{2}\left[H_0\left(z\right)G_0\left(z\right)+H_1\left(z\right)G_1\left(z\right)\right].$$
(12.39)

Let us define the highpass transfer function

$$\hat{E}\left(z\right)=H_1\left(z\right)G_1\left(z\right)=\sum_{n=0}^{2N-2}\hat{e}[n]z^{-n}$$
(12.40)

According to the lowpass/highpass complementary properties, the impulse-response coefficients satisfy

$$\hat{e}[n] = \begin{cases} -e[n], & \text{for } n \text{ even} \\ e[n], & \text{for } n \text{ odd} \end{cases}. \tag{12.41}$$

Following equations (12.37), (12.39) and (12.40), the distortion transfer functions is expressible by

$$T(z) = E(z) + \hat{E}(z) = \sum_{n=0}^{2N-2} (e[n] + \hat{e}[n]) z^{-n}. \tag{12.42}$$

Consequently, the impulse response of $T(z)$ satisfies

$$t[n] = e[n] + \hat{e}[n] = \begin{cases} 1, & \text{for } n = K \\ 0, & \text{for } n \neq K \end{cases}, \tag{12.43}$$

and the distortion transfer function becomes

$$T(z) = z^{-K} \tag{12.44}$$

In the case where the analysis/synthesis filters are of the same length N, the overall delay of the bank amounts to $K = N-1$ samples.

Notice that the analysis/synthesis filters $H_0(z)$, $H_1(z)$, $G_0(z)$ and $G_1(z)$ are causal FIR filters. When the impulse response $h_0[n]$ is known, it is straightforward to determine the impulse responses $h_1[n]$, $g_0[n]$ and $g_1[n]$, see equations (12.35) and (12.36a – b). Figure 12.11 illustrates their mutual relations for the filters designed in *Example 12.3*.

The design problem of the perfect-reconstruction two-channel orthogonal filter bank can be considered as a design problem of the low-pass filter $H_0(z)$. The remaining transfer functions $H_1(z)$, $G_0(z)$ and $G_1(z)$ are related to $H_0(z)$ according to equations (12.35) and (12.36a – b). The key problem is to design a linear-phase separable (factorizable) FIR transfer function $E(z)$ of the order $2N–2$ and to extract the spectral factor $H_0(z)$. Notice that $2N–2$ should be two-times an odd integer. The transfer function $H_0(z)$ can be selected as a minimum-phase spectral factor of $E(z)$. In that case $H_1(z)$ is the maximum-phase transfer function. Between those solutions, one can select for $H_0(z)$ a mixed-phase spectral factor in order to achieve a better phase characteristics in channel filters.

In Chapter VIII, subsection Power-Complementary Filter Pairs, we have considered the design problem of the separable linear-phase FIR filters and the extraction of the minimum phase spectral factor. In this Chapter, we use MATLAB function to design directly the analysis/synthesis filters of the orthogonal FIR filter bank.

The MATLAB function firpr2chfb from the *Filter Design Toolbox* can be used to compute the impulse-response coefficients of the analysis/synthesis filters $H_0(z)$, $H_1(z)$, $G_0(z)$ and $G_1(z)$ of the orthogonal filter bank. In the following example, we show the impulse responses of the four filters, the magnitude response of the lowpass/highpass filter pair, and verification of the perfect-reconstruction property of the overall orthogonal filter bank.

Example 12.3

In this example, we design the analysis filter pair $[H_0(z), H_1(z)]$ and the synthesis filter pair $[G_0(z), G_1(z)]$. The filter length is $N = 18$, and the passband edge frequency of the lowpass halfband filter $H_0(z)$ $[G_0(z)]$ is $\omega_p = 0.4\pi$.

The MATLAB function firpr2chfb returns the impulse responses $h_0[n]$, $h_1[n]$, $g_0[n]$ and $g_1[n]$ for the input parameters:

```
N = 18; fp = 0.4;                    % Seting the input parameters for firpr2chfb
[h0,h1,g0,g1] = firpr2chfb(N-1,fp);  % Generating four filters for the orthogonal FIR filter bank
```

The impulse responses, which are stored in vectors h0, h1, g0, g1 are shown in Figure 12.11. The gain characteristics of the lowpass/highpass halfband filter pair are given in the upper subfigure of Figure 12.12. The bottom subfigure of Figure 12.12 illustrates the perfect-reconstruction property of the overall bank determined by using the code fragment

```
t = (conv(h0,g0)+conv(h1,g1))/2     % Computing the impulse response of the distortion transfer
function
stem(0:2*(N-1),t)
```

Figure 12.12 shows that the impulse response of $T(z)$, the sequence $t[n]$, results in delayed-by-$(N-1)$ version of the unit sequence,

$$t[n] = \delta[n - N + 1].$$

Figure 12.11. Impulse responses $h_0[n]$, $h_1[n]$, $g_0[n]$ and $g_1[n]$

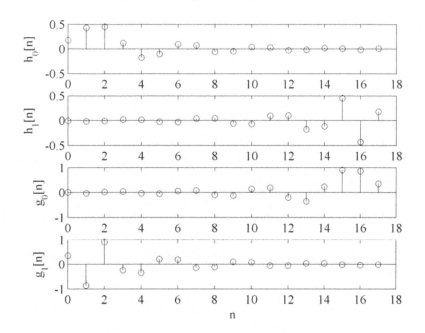

Figure 12.12. The gain responses of the lowpass/highpass filter pair, and the verification of the perfect-reconstruction property of the two-channel analysis/synthesis bank

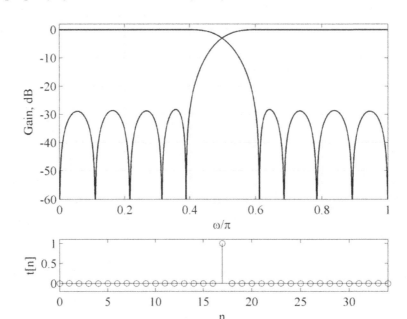

The orthogonal two-channel filter bank demonstrated in *Example 12.3* is composed of FIR filters designed to provide a maximal stopband attenuation for the given filter order. In many applications, such as wavelet filter banks, it is requested to provide the maximal order of flatness at $\omega = \pi$ for the lowpass filter, and at $\omega = 0$ for the highpass filter. In this case, the lowpass transfer function $H_0(z)$ should have the multiple zero at the point $z = -1$.

We use the *Wavelet Toolbox* to illustrate the performances of a two-channel wavelet filter bank composed of the orthogonal filters with the regularity order greater than one. In the following example, we show the design and performances of the two-channel orthogonal filter bank based on the Daubechies' db5 wavelet.

Example 12.4

In this example, we use the function wfilters from the *Wavelet Toolbox* to design four filters of the analysis/synthesis bank. We specify 'db5' for the wavelet name. Here, for the analysis/synthesis filters we use the notations from the *Wavelet Toolbox*:

- Lo_D, Hi_D is the lowpass/highpass analysis (decomposition) filter pair denoted earlier in this chapter as $H_0(z)$ and $H_1(z)$.
- Lo_R, Hi_R is the lowpass/highpass synthesis (reconstruction) filter pair denoted earlier in this chapter as $G_0(z)$ and $G_1(z)$.

We use the following code to design the analysis and synthesis filters:

```
wname = 'db5'; % Set wavelet name.
% Compute the four filters associated with wavelet name given by the input string wname.
[Lo_D,Hi_D,Lo_R,Hi_R] = wfilters(wname);
```

Figure 12.13 shows the impulse responses of the analysis (decomposition) and synthesis (reconstruction) filters.

In the sequel, we display the pole-zero plots of the four filters using the code

```
subplot(2,2,1), zplane(Lo_D,1);
subplot(2,2,2), zplane(Hi_D,1);
subplot(2,2,3), zplane(Lo_R,1);
subplot(2,2,4), zplane(Hi_R,1);
```

The results shown in Figure 12.14 illustrate the following transfer functions' properties:

- The 5th-order zero at the point $z = -1$ for lowpass filters (left-hand side of Figure 12.4), and the 5th-order zero at the point $z = +1$ for high pass filters (right-hand side of Figure 12.4).
- The analysis filter pair is composed of the maximum-phase lowpass filter and the minimum-phase highpass filter (two upper subfigures).
- The synthesis filter pair is composed of the minimum-phase lowpass filter and the maximum-phase highpass filter (two bottom subfigures).

Figure 12.13. Impulse responses of the analysis/synthesis db5 filters

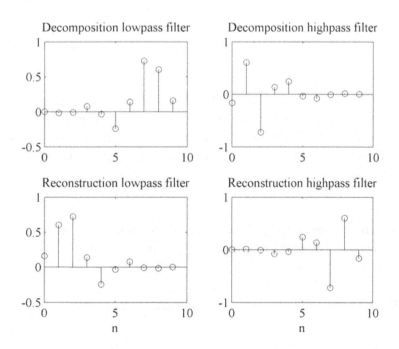

Figure 12.14. The z-plane locations of the analysis/synthesis db5 filters

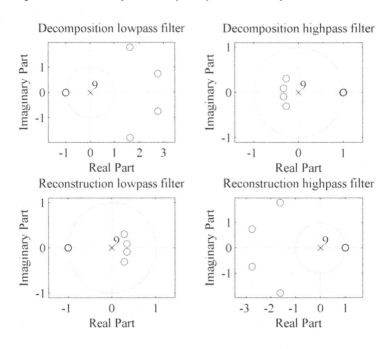

Figure 12.15 plots the magnitude responses of the lowpass/highpass filter pair (upper subfigure). The displayed magnitude response indicates √2 scaling factor commonly associated with the wavelet analysis and synthesis filter banks. The bottom subfigure demonstrates the perfect-reconstruction property of the overall analysis/synthesis filter bank verified by computing the impulse response of the distortion transfer function $t[n]$,

```
t = (0:2*(N-1),conv(Lo_D,Lo_R)+conv(Hi_D,Hi_R));
stem(0:2*(N-1),t)
```

The resulting sequence $t[n]$ is a scaled and delayed version of the unit sample sequence $\delta[n]$ thus confirming the perfect reconstruction property of the analysis/synthesis bank of this example.

The two-channel orthogonal FIR filter bank can be implemented in the direct form using the general structure of Figure 12.1. For the impulse responses $h_0[n]$, $h_1[n]$, $g_0[n]$, $g_1[n]$, and an impute signal $x[n]$ we compute in MATLAB,

```
% Analysis part
x0 = filter(h0,1,x); v0 = downsample(x0,2);  % Lowpass signal component
x1 = filter(h1,1,x); v1 = downsample(x1,2);  % Highpass signal component

% Synthesis part
u0 = upsample(v0,2); y0 = filter(g0,1,u0);     % Lowpass signal component
```

Figure 12.15. Magnitude responses of the lowpass/highpass db5 filter pair, and verification of the perfect-reconstruction property of the two-channel analysis/synthesis bank

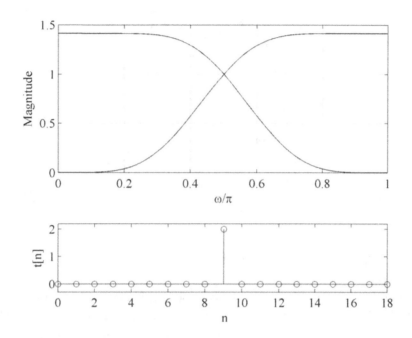

```
u1 = upsample(v1,2); y1 = filter(g1,1,u1);    % Highpass signal component
y = y1 + y2;                                    % Reconstructed signal
```

The orthogonal filter banks can be realized efficiently using the lattice form, see (Fliege 1994; Saramäki and Bregovic, 2002; Vaidyanathan, 1993). The solution for the analysis filter pair is shown in Chapter VIII, Figure 8.12.

TREE-STRUCTURED MULTICHANNEL FILTER BANKS

An approach for constructing a multichanel filter bank is based on a tree-structure, which uses the two-channel filter banks as building blocks. With the use of this approach, the multichannel filter banks with uniform and nonuniform separation between the channels can be generated. If the two-channel filter banks satisfy the perfect-reconstruction (nearly perfect-reconstruction) property, the overall tree-structure filter bank also satisfies the perfect- reconstruction (nearly perfect-reconstruction) property.

In this section, we show examples of uniform and nonuniform filter banks built on the basis of the two-channel filter banks. For the sake of simplicity, we use in this section the symbolic representations of the analysis and synthesis two-channel filter banks as shown in Figure 12.16. The single-input/two-output device $A^{(k)}(z)$ symbolizes a two-channel analysis filter bank, and two-input/single-output device $S^{(k)}(z)$ symbolizes a two-channel synthesis bank.

Figure 12.16. Symbolic representation of two-channel filter bank: (a) Analysis bank. (b) Synthesis bank

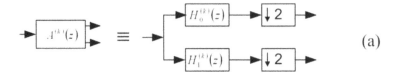

Filter Banks with Equal Passband Widths

A two-level four-channel analysis/synthesis filter bank is indicated in Figure 12.17. This figure illustrates how the tree-structured filter banks are developed. Let us consider the analysis part. The first level consists of the two-channel bank indicated as $A^{(1)}(z)$, which is the starting point for developing the multichannel bank. The two outputs of $A^{(1)}(z)$, $v_0^{(1)}[n]$ and $v_1^{(1)}[n]$, are inputted into the two banks of the second level indicated as $A^{(2)}(z)$. In the second level, the four outputs $v_0^{(2)}[n]$, $v_1^{(2)}[n]$, $v_2^{(2)}[n]$ and $v_3^{(2)}[n]$, are produced. The process can be continued in the same manner until arriving up to the desired number of channels. We say that the multichannel filter bank is developed by iterating the two-channel filter bank.

On the synthesis side, the four input signals $w_0^{(2)}[n]$, $w_1^{(2)}[n]$, $w_2^{(2)}[n]$ and $w_3^{(2)}[n]$, are used as the inputs to the synthesis banks denoted by $S^{(2)}(z)$. The two outputs, $w_0^{(1)}[n]$ and $w_1^{(1)}[n]$, are inputted to the synthesis bank $S^{(1)}(z)$. The output $y[n]$ represents the reconstructed signal. The reconstructed signal $y[n]$ is a delayed replica of the input signal $x[n]$ under the following conditions:

- The processing unit is omitted from the structure of Figure 12.17.
- The analysis and synthesis two-channel filter banks $A^{(k)}(z)$ and $S^{(k)}(z)$ satisfy the perfect-reconstruction property.

The equivalent single-stage structure for the four-channel filter bank of Figure 12.17 can be developed by making use of the *Third* and *Sixth Identities*, see Chapter II. The equivalent structure shown in Figure 12.18 represents the analysis filter bank as a parallel connection of four filters cascaded with factor-of-four down-samplers, and the synthesis bank is represented as the cascade of factor-of-four up-samplers and the four filters. Figure 12.18 shows the expressions for the equivalent transfer functions of the corresponding channel filters in the single-stage representation.

The tree-structured filter bank can be composed of identical two-channel building blocks. However, we may combine several different building blocks for constructing the multilevel multichannel filter bank. The reconstruction property of the overall bank is achieved if the individual two-channel building blocks retain the reconstruction properties that are requested for the overall bank.

Figure 12.17. Two-level four-channel filter bank

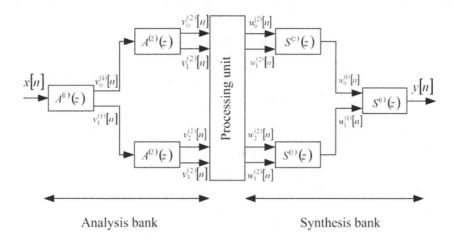

Figure 12.18. Single-stage equivalent of the two-level four-channel filter bank

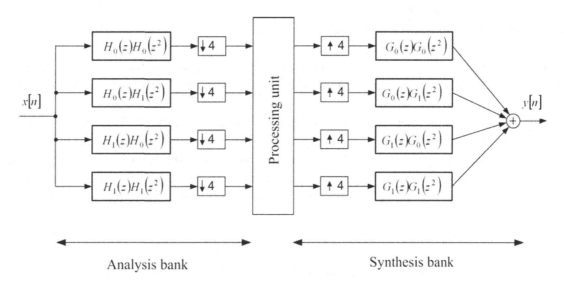

We illustrate the characteristics of a multichannel filter bank with equal passband widths by means of example. We construct the tree-structured four-channel filter bank of Figure 12.17 and examine the magnitude responses of the individual channels and the reconstruction property of the overall filter bank.

Example 12.5

In this example, we use the two-channel orthogonal filter bank of *Example11.3* to construct the four-channel filter bank using the tree-structure of Figure 12.17. Hence, the MATLAB function firpr2chfb computes the coefficients for the analysis/synthesis filters of the two-channel filter banks.

```
N=18;                              %Setting the filter length
[h0,h1,g0,g1]=firpr2chfb(N-1,0.4); % MATLAB function for the analysis/synthesis orthogonal bank
```

The gain responses for the analysis (synthesis) filter pair are shown in Figure 12.13, together with the verification of the perfect-reconstruction property of this filter bank.

We compute the equivalent impulse responses for the four-channel analysis bank, and in the sequel, we compute the equivalent frequency responses.

```
% Analysis filters, impulse responses
hh0 = conv(h0,upsample(h0,2));  % Impulse response, 2nd level
hh1 = conv(h0,upsample(h1,2));  % Impulse response, 2nd level
hh2 = conv(h1,upsample(h0,2));  % Impulse response, 2nd level
hh3 = conv(h1,upsample(h1,2));  % Impulse response, 2nd level
```

```
% Analysis filters, frequency responses
[HH0,f] = freqz(hh0,1,1024,2);    % Frequency response, 2nd level
[HH1,f] = freqz(hh1,1,1024,2);    % Frequency response, 2nd level
[HH2,f] = freqz(hh2,1,1024,2);    % Frequency response, 2nd level
[HH3,f] = freqz(hh3,1,1024,2);    % Frequency response, 2nd level
```

The gain responses of the four analysis filters are shown in the upper subfigure of Figure 12.19. One observes that all filters have equal passband widths defined according to the 3-dB crossover frequencies.

Since the two-channel filter banks are linear-phase FIR filters, the overall bank is also of a linear phase. To verify the perfect-reconstruction property, we verify whether the overall filter bank satisfies the magnitude-preserving property (12.18). Since the analysis and synthesis filters of the building-block two-channel banks exhibit equal magnitude responses, it is sufficient to examine the power-complementary property of the analysis part of the overall four-channel filter bank. Therefore, we compute

```
ver= abs(HH0).^2+abs(HH1).^2+abs(HH2).^2+abs(HH3).^2;
```

The vector ver returns the frequency characteristic of the squared magnitude distortion function $|T(e^{j\omega})|^2$ of the overall four-channel filter bank. The bottom subfigure of Figure 12.19 shows that $|T(e^{j\omega})|$ is a constant and therefore, confirms the magnitude-preserving property of the overall analysis/synthesis four-channel bank.

Octave Filter Banks

An octave filter bank is a multilevel filter bank generated of the two-channel filter bank. Development process of an analysis bank is indicated in Figure 12.20. The process starts from the two-channel filter bank of Figure 12.16 (a). In the second level, the filtered and decimated lowpass signal is used as an input to the next two-channel filter bank, which is identical to the original bank. The process can be continued in the same manner to generate the multilevel filter bank with the desired number of channels. Figure

Figure 12.19. Gain responses of four analysis filters and magnitude response of the distortion transfer function of Example 12.5

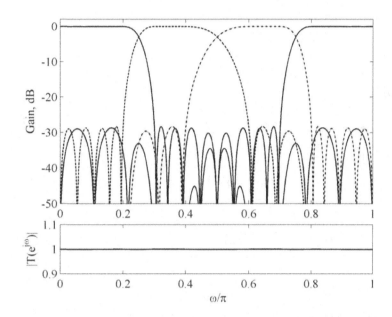

Figure 12.20. Generation of four-level analysis octave filter bank

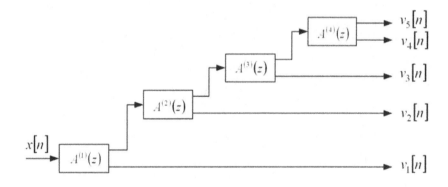

12.20 shows the four-level five-channel analysis octave bank. The equivalent representation useful for the analysis purposes is shown in Figure 12.21.

To illustrate a typical magnitude response of an octave filter bank, we construct the five-channel bank of Figures 11.20 and 11.21 by using the orthogonal two-channel filter bank of *Example 12.3*. The resulting magnitude responses are shown in Figure 12.22.

The structure of the synthesis octave bank is given in Figure 12.23 for the case of four-level five-channel filter bank. For constructing this bank, the two-channel synthesis filter bank of Figure 12.16 (b) is used. One can easily develop the equivalent representation for the synthesis bank, as shown in Figure 12.21 for the analysis bank, see (Mitra, 2006; Saramäki, & Bregović, 2002).

The term *binary tree* is used to denote the tree-structure octave banks of Figures 11.20 and 11.23 when the two-channel filter banks of Figures 11.16 (a) – (b) are used as building blocks.

Figure 12.21. Equivalent representation of the analysis octave filter bank

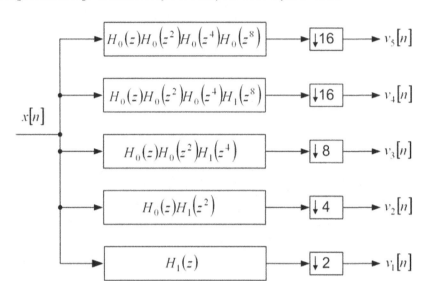

Figure 12.22. Magnitude responses of the example five-channel octave filter bank

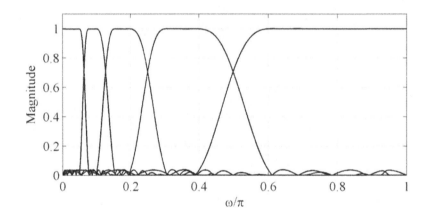

Figure 12.23. Generation of four-level synthesis octave filter bank

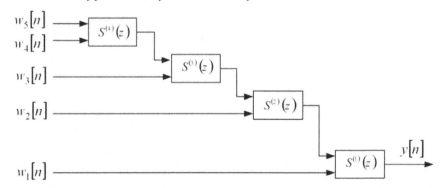

An analysis/synthesis octave bank retains the reconstruction properties of the building-block analysis/synthesis two-channel filter bank. If the analysis/synthesis bank $[A^{(k)}(z), S^{(k)}(z)]$ satisfies the perfect-reconstruction (nearly perfect-reconstruction) property, the same is true for the overall analysis/synthesis multi-channel multi-level bank.

When using the synthesis bank for reconstructing the original signal, the signal components need a pre-processing before inputted to the synthesis filters. Let us consider the analysis/synthesis four-level filter bank presented in Figure 12.24. In the fourth level, the two outputs of the analysis bank $A^{(4)}(z)$, $v_4[n]$ and $v_5[n]$, are inputted directly to the synthesis bank $S^{(4)}(z)$, and they form the distortion transfer function $T^{(4)}(z)$. If $A^{(4)}(z)$ and $S^{(4)}(z)$ satisfy the perfect-reconstruction property, we have $T^{(4)}(z) = z^{-(N-1)}$, where N is the length of the analysis (synthesis) filters. Therefore, in the 3rd level, the signal component $v_3[n]$ has to be delayed by $N-1$ samples before inputted to the synthesis bank $S^{(3)}(z)$. The same approach has to be continued in the lower levels as indicated in Figure 12.24.

In two examples that follow, we demonstrate how the process of signal decomposition and reconstruction is developed through the four-level analysis/synthesis octave bank of Figure 12.24. Our goal is threefold. Firstly, we intend to demonstrate in MATLAB the computations process in the analysis and synthesis sections of the bank. Secondly, we intend to demonstrate the perfect reconstruction of the test signal in the orthogonal multichannel bank. Thirdly, we intend to show the importance of the type of the filter transfer function for the analysis/synthesis process.

In *Examples 12.6* and *12.7* that follow, the computations are based on the four level analysis bank, and on the four level synthesis bank, as given in Figure 12.24. The analysis/synthesis filter pairs $[A^{(k)}(z), S^{(k)}(z)]$ are orthogonal two-channel filter banks in both examples. In *Example 12.6*, the transfer function of the lowpass analysis filters $H_0(z)$ is a minimum-phase halfband FIR filter with the equiripple magnitude characteristic in the pass and stop bands, whereas in *Example 12.7* we use the Daubechies db9 filter for $H_0(z)$.

Example 12.6

In this example, we use the orthogonal two-channel filter bank of *Example 12.3* to construct the four-level five-channel analysis/synthesis octave bank according to the tree-structure of Figure 12.24.

We compute first the coefficients of the analysis and synthesis filters, $H_0(z)$, $H_1(z)$, $G_0(z)$ and $G_1(z)$,

```
N = 18;                      %Setting the filter length
[h0,h1,g0,g1] = firpr2chfb(N-1,0.4);   % MATLAB function for the analysis/synthesis orthogonal bank
```

Figure 12.24. Analysis/synthesis four-level octave bank

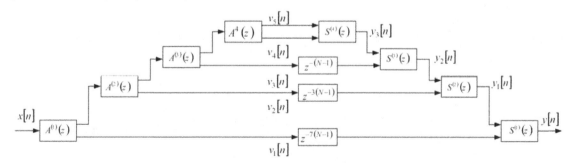

The impulse responses of the analysis and synthesis filters are already shown in Figure 12.11, and the gain responses for the analysis (synthesis) filter pair are plotted in the upper subfigure of Figure 12.12. The magnitude responses of the five channel filters of the four-level octave bank have been considered in *Example 12.5* with the plots displayed in Figure 12.22.

In this example, we utilize the above mentioned octave bank to demonstrate the decomposition and reconstruction of the rectangular-shape test signal defined by the code

```
x=[zeros(size(1:20)),ones(size(21:50)),zeros(size(51:511))];   % Generating the test signal
```

and shown in Figure 12.25.

The test signal is used as an input to the four-level octave bank of Figure 12.24. In the analysis part of the bank we perform the signal decomposition by computing five signal components $v_1[n]$, $v_2[n]$, $v_3[n]$, $v_4[n]$, $v_5[n]$. The process starts at the first level of the analysis bank where the test signal is decomposed into the lowpass and highpass signal components. The highpass component of the first level represents the signal component $v_1[n]$, whereas the lowpass component is inputted to the second level for the next decomposition. The highpass component computed in the second level represents the signal component $v_2[n]$. The lowpass component of the second level is the input for the third level. The process ends up after computing the lowpass and highpass components of the fourth level.

We use the following MATLAB code to implement the four-level analysis bank of Figure 12.24, and to evaluate the computations described above.

```
% Analysis part
% Level 1
x0 = filter(h0,1,x);            % Lowpass filtering
x1 = filter(h1,1,x);            % Highpass filtering
v0 = downsample(x0,2);          % Down-sampling, signal component v_0[n]
v1 = downsample(x1,2);          % Down-sampling, signal component v_1[n]

% Level 2
x2 = v0;                        % Selecting the lowpass output from Level 1 for the input to Level 2
x02 = filter(h0,1,x2);          % Lowpass filtering
x12 = filter(h1,1,x2);          % Lowpass filtering
v02 = downsample(x02,2);        % Down-sampling
```

Figure 12.25. The rectangular-shape test signal

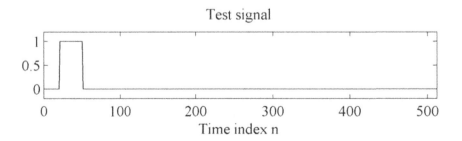

```
v12 = downsample(x12,2);          % Down-sampling
v2 = v12;                         % Signal component v_2[n]


% Level 3
x3 = v02;                         % Selecting the lowpass output from Level 2 for the input to Level 3
x03 = filter(h0,1,x3);            % Lowpass filtering
x13 = filter(h1,1,x3);            % Highpass filtering
v03 = downsample(x03,2);          % Down-sampling
v13 = downsample(x13,2);          % Down-sampling
v3 = v13;                         % Signal component v_3[n]


% Level 4
x4 = v03;                         % Selecting the lowpass output from Level 3 for the input to Level 4
x04 = filter(h0,1,x4);            % Lowpass filtering
x14 = filter(h1,1,x4);            % Highpass filtering
v04 = downsample(x04,2);          % Down-sample
v14 = downsample(x14,2);          % Down-sample
v4 = v14;                         % Signal component v_4[n]
v5 = v04;                         % Signal component v_5[n]
```

We have obtained here five signal components $v_1[m]$, $v_2[m]$, $v_3[m]$, $v_4[m]$, $v_5[m]$, which are computed as the five outputs of the analysis part of the bank. In this way, the rectangular test signal $x[n]$ is decomposed into five signal components. Figure 12.26 displays the results.

The signal reconstruction has been performed by making use of the synthesis part of the four-channel octave bank. According to Figure 12.24, signals $v_4[m]$ and $v_5[m]$ are directly inputted into the two-channel synthesis bank $S^{(4)}(z)$, and the output intermediate signal $y_3[n]$ is computed. In the next level, the signal $v_3[m]$ has been delayed for $N-1$ samples first, and then together with $y_3[n]$ inputted into the synthesis bank $S^{(3)}(z)$. The process has been continued in the same manner until obtaining the output $y[n]$. In the following, we give the MATLAB description of the reconstruction process.

```
% Synthesis part
% Level 4
w04=upsample(v5,2);                              % Up-sampling
w14=upsample(v4,2);                              % Up-sampling
y3=filter(g0,1,w04) + filter(g1,1,w14);          % Filtering and addition, signal y_3[n]


% Level 3
w13 = [zeros(size(1:N-1)),v3(1:length(v3)-(N-1))];   % Inserting the delay z^(-(N-1)), Figure 12.24
w13 = upsample(w13,2);                           % Up-sampling
yu3 = upsample(y3,2);                            % Up-sampling
y2 = filter(g0,1,yu3) + filter(g1,1,w13);        % Filtering and addition, signal y_2[n]


% Level 2
w12 = [zeros(size(1:3*(N-1))),v2(1:length(v2)-3*(N-1))];   % Inserting the delay z^(-3(N-1)), Figure 12.24
```

Figure 12.26. Decomposition of the rectangular-shape test signal with the five-channel octave bank based on the FIR filters exhibiting the equiripple magnitude responses

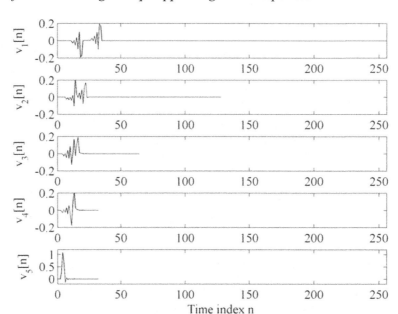

Time index n

```
w12 = upsample(w12,2);                          % Up-sampling
yu2 = upsample(y2,2);                           % Up-sampling
y1 = filter(g0,1,yu2) + filter(g1,1,w12);       % Filtering and addition, signal y_1[n]


% Level 1
w11 = [zeros(size(1:7*(N-1))),v1(1:length(v1)-7*(N-1))];   % Inserting the delay z^(-7(N-1)), Figure 12.24
w11 = upsample(w11,2);
yu1 = upsample(y1,2);
y = filter(g0,1,yu1) + filter(g1,1,w11);        % Filtering and addition, reconstructed signal y[n]
```

Figure 12.27 plots the intermediate signals $y_3[n]$, $y_2[n]$, $y_1[n]$ and the result of the synthesis process, signal $y[n]$. So, we can follow how the reconstruction process advances through the synthesis bank.

Signal $y[n]$ shows the result of the decomposition and reconstruction process based on the two-channel orthogonal filter banks. Since the building-block two-channel filter bank exhibits the perfect-reconstruction property, the same is evidently true for the overall multi-level octave bank. Namely, the reconstructed signal $y[n]$ is the delayed replica of the test signal $x[n]$, compare Figure 12.25 with the bottom subfigure of Figure 12.27.

Example 12.6 demonstrates the ability of the octave bank to reconstruct perfectly the original signal when the building-block two-channel filter bank satisfies the perfect-reconstruction property.

The signal decomposition and reconstruction in *Example 12.6* has been performed by making use of minimum-phase FIR filters having equiripple magnitude characteristics in the pass and stopbands.

Figure 12.27. Reconstruction of the rectangular-shape test signal with five-channel octave bank based on the FIR filters exhibiting the equiripple magnitude responses

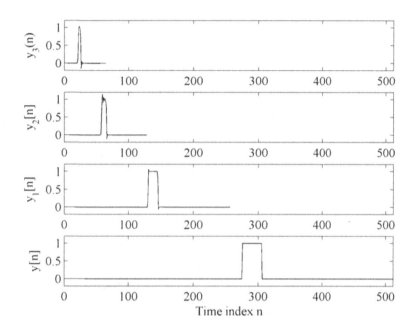

Instead of this filter class, we may prefer a transfer function with different performances. We may need, for example, the smooth magnitude response which can be achieved with maximally-flat FIR filters. In the following example, we consider the process of signal analysis and synthesis based on Daubechies dB9 filter. It is interesting to compare the filter characteristics and the analysis/synthesis process with that of *Example 12.6*.

Example 12.7

In this example, we design four filters of the orthogonal two-channel filter bank using the Daubechies db9 wavelet, and compose the four-level five-channel analysis/synthesis octave bank. We use this filter bank to perform the decomposition and reconstruction of the rectangular shape signal of Figure 12.25, generated in *Example 12.6*.

The db9 filter is of the same complexity as the FIR filter of *Example 12.6*. Namely the length of db9 is also 18 (2×9).

First, we compute the coefficients of the analysis and synthesis filters $H_0(z)$, $H_1(z)$, $G_0(z)$, $G_1(z)$ using the function wfilters from the *Wavelet Toolbox*. The code fragment for filter design is the following

```
wname = 'db9';                  % Set wavelet name.
% Compute the four filters associated with wavelet name given by the input string wname.
[h0,h1,g0,g1] = wfilters(wname);   % Designing the four db9 filters
N = 18;                         % The length of db9 filter
```

Figure 12.28. Magnitude responses of the analysis filters: Solid line db9 filter pair. Dashed line minimum-phase FIR filter pair from Example 12.6.

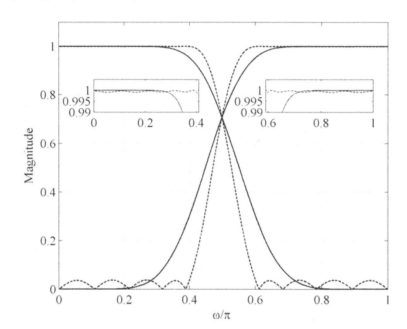

Vectors h0, h1, g0, g1 return the coefficients for $H_0(z)$, $H_1(z)$, $G_0(z)$, $G_1(z)$. Here, the values of coefficient are multiplied by $\sqrt{2}$ according to the scaling factors associated with the analysis and synthesis wavelet banks.

In Figure 12.28, we display the magnitude responses of the db5 filter pair along with the minimum-phase FIR filter pair with equiripple magnitude response from *Example 12.6*. The magnitude responses of db5 are normalized with $1/\sqrt{2}$ in order to provide the same level for all filters represented in Figure 12.28.

In this example, we perform the signal analysis and synthesis by utilizing the same MATLAB code already presented in *Example 12.6*. The difference is in the type of the filter transfer function.

Results of the signal decomposition based on db9 filters are displayed in Figure 12.29, whereas Figure 12.30 illustrates the corresponding reconstruction process. It is apparent that the analysis synthesis octave bank based on db9 filters provides the perfect reconstruction of the input test signal. Namely, signal $y[n]$ shown in the bottom subfigure of Figure 12.30 is the delayed replica of the original test signal as given in Figure 12.25.

It is interesting to summarize the results of the signal analysis obtained with two different types of the filter transfer functions exposed in *Examples 11.6* and *11.7*. Since both filter banks are octave banks satisfying the perfect-reconstruction property, they provide both the perfect reconstruction of the input signal. However, the significant differences can be observed in the signal components $v_1[n]$, $v_2[n]$, $v_3[n]$, $v_4[n]$ and $v_5[n]$ obtained at the output of the analysis bank, which are displayed in Figures 11.26 and 11.29. Different signal components are obtained since the different transfer functions are used in *Examples 11.6*

Figure 12.29. Decomposition of the rectangular-shape test signal with the five-channel octave bank based on db9 filters

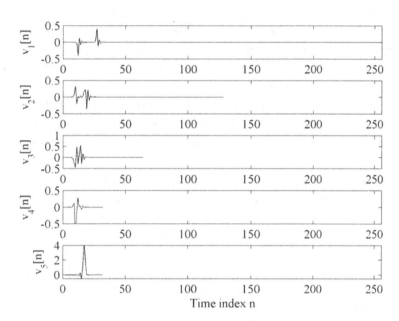

Figure 12.30. Reconstruction of the rectangular-shape test signal with five-channel octave bank based on db9 filters

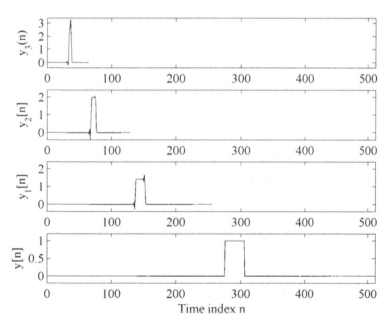

and *11.7*. It is apparent that changing the type of the transfer functions of the analysis bank we change shapes of the signal components. Comparing components $v_1[n]$ in Figures 11.26 and 11.29 we can notice that db9 filter sharply detects the discontinuity of the rectangular test signal.

In this chapter, we have mainly considered examples based on FIR filters. However, IIR filter banks become desirable solutions in applications where the computational efficiency is of greatest importance. The reader is recommended to develop the program for the analysis/synthesis IIR multichannel filter bank by solving MATLAB Exercise 12.8.

The octave filter banks can be designed either to provide a sharp channel-selectivity or to preserve the waveform of the original signal. In some applications, such as speech and telecommunication applications, channel selectivity is very important. In the case of images, for example, the waveform of the signal is of greatest interest.

Octave filter banks are also used in computing the discrete wavelet transform (DWT) and the inverse discrete wavelet transform (IDWT). The analysis multilevel octave bank is used as an algorithm for computing DWT, whereas the synthesis bank is used for computing IDWT. Some of the built-in functions in the MATLAB *Wavelet Toolbox* are based on the analysis and synthesis octave banks. An example is wavedec, which computes the DWT coefficients. Another example is the function waverec, which computes the coefficients of IDWT. Octave banks used for computing DWT and IDWT are called also the wavelet filter banks. The wavelet banks are distinguished by the special classes of FIR filters developed for the purposes of wavelet analysis.

MATLAB EXERCISES

12.1 Repeat the computations of *Example 12.1*. The impulse response of the linear-phase lowpass FIR filter $H_0(z)$ is specified by vector B1, which contains the first half of the filter coefficients,

B1=[0.002329266, -0.005182978, -0.002273145, 0.01354012, -0.0006504669, -0.02755195, 0.01004621, 0.05088162, -0.03464143, -0.09987885, 0.12464520, 0.4686479];

Generate and plot the new versions of Figures 12.4 – 12.6. Comment on the results.

12.2 Repeat the computations of *Example 12.2* using the 7th order Butterworth filter. Generate and plot the new versions of Figures 12.8 – 12.10. Comment on the results.

12.3 Repeat the computations of *Example 12.2* using the 5th order elliptic filter with the passband edge frequency $\omega_p = 0.4\pi$. For the elliptic half-band IIR filter design use the MATLAB program halfbandiir given in Appendix A. Generate and plot the new versions of Figures 12.8 – 12.10. Comment on the results.

12.4 Design the analysis/synthesis two-channel orthogonal filter bank using the MATLAB function firpr2chfb. The filter length is $N = 32$, and the lowpass passband edge frequency $\omega_p = 0.43\pi$. Plot the impulse responses of the analysis and synthesis filters: $h_0[n]$, $h_1[n]$, $g_0[n]$ and $g_1[n]$. Compute and plot the poles and zeros of the transfer functions $H_0(z)$, $H_1(z)$, $G_0(z)$ and $G_1(z)$. Compute and plot the magnitude and group delay responses of the analysis filters $H_0(z)$ and $H_1(z)$. Compute and plot the impulse response of the distortion transfer function $t[n]$. Comment on the results.

12.5 Compose the eight-channel tree-structured analysis/synthesis filter bank. As a basic building block use the orthogonal two-channel filter bank of *Example 12.3* and construct the analysis and synthesis banks. Compute and plot the magnitude responses of the resulting eight analysis filters. Verify the magnitude-preserving property of the overall bank. Comment on the phase characteristic of the overall analysis/synthesis bank.

12.6 Design an orthogonal two-channel FIR filter bank with filter lengths of $N = 32$. Build the five-channel orthogonal analysis filter bank. Develop the MATLAB program which computes the magnitude responses of the bank.

12.7 Develop the MATLAB program for signal decomposition and reconstruction:
(a) Using MATLAB function firpr2chfb design the analysis/synthesis two-channel orthogonal filter bank for $N = 22$ and $f_p = 0.4$.
(b) Generate the triangular-shape test signal.
(c) Using the two-channel filter bank from (a), construct the analysis four-level octave bank.
(d) Perform the signal decomposition using the four-level analysis filter bank.
(e) Using the two-channel filter bank from (a), construct the synthesis four-level octave bank.
(f) Using the signal components from (d) reconstruct the original signal.
(g) Comment on the results.

12.8 Develop the MATLAB program for signal decomposition and reconstruction:
(a) Design the two-channel analysis/synthesis bank using db5 wavelet filter.
(b) Generate the triangular-shape test signal.
(c) Using the two-channel filter bank from (a), construct the analysis three-level octave bank.
(d) Perform the signal decomposition using the three-level analysis filter bank.
(e) Using the two-channel filter bank from (a), construct the synthesis three-level octave bank.
(f) Using the signal components from (d) reconstruct the original signal.
(g) Comment on the results.

12.9 Develop the MATLAB program for signal decomposition and reconstruction based on the IIR two-channel filter banks:
(a) Utilize the two-channel Butterworth filter bank of *Example 12.2* to construct the two-level three-channel IIR octave analysis bank.
(b) Compute and plot the magnitude responses of the bank.
(c) Utilize MATLAB program fir2 to generate the test signal with triangular shape spectrum.
(d) Compute the signal components of the test signal as the outputs of the three-channel analysis bank
(e) Construct the two-level three-channel synthesis bank and reconstruct the original test signal.
(f) Comment on the results.

REFERENCES

Ansari,R., & Liu,B., (1993). Multirate signal processing. In Sanjit. K. Mitra and James F. Kaiser (ed.), *Handbook for Digital Signal Processing*. New York: John Wiley-Interscience, 981-1084.

Filter design toolbox for use with MATLAB. User's guide. Version 6. (2006). Natick: MathWorks.

Fliege, N. J. (1994). *Multirate digital signal processing*. New York, NY: John Wiley.

Johnston, J.D. (March 1980). A filter family designed for use in quadrature mirror filter banks. *Proceedings of the IEEE International Conference Acoustics, Speech, and Signal Processing*, 291–294.

Mitra, S. K. (2006). *Digital signal processing: A computer based approach*. 3rd edition. New York, NY: The McGraw-Hill Companies, Inc.

Saramäki, T. *Multirate Signal Processing*. (2001). Lecture notes for a graduate course, the Institute of Signal Processing, Tampere University of Technology, Finland.

Saramäki, T., & Bregovic, R. (2002). Multirate systems and filter banks. In Gordana Jovanović-Doleček, (ed.), *Multirate Systems: Design & Applications*. Hershey, PA: Idea Group Publishing, 27-85.

Signal processing toolbox for use with MATLAB. User's guide. Version 6. (2006). Natick: Math-Works.

Strang, G., & Nguyen, T. (1996). *Wavelets and Filter Banks*.Wellesley, MA: Wellesley-Cambridge Press.

Vaidyanathan, P.P. (1987). Quadrature mirror filter banks, M-band extensions and perfect-reconstruction techniques. IEEE ASSP Magazine, *4*(3), 4-20.

Vaidyanathan, P.P. (1993). *Multirate systems and filter banks*. Englewood Cliffs, NJ: Prentice Hall.

Wavelet toolbox for use with MATLAB. User's guide. Version 3. (2006). Natick: MathWorks.

Vetterli, M., & Kovačević, J.(1995). *Wavelets and Subband Codding*. Englewood Cliffs, N.J.: Prentice Hall.

Yue-Dar Jou. (May 2007). Design of two-channel linear-phase quadrature mirror filter banks based on neural networks. *Signal Processing*, *87*(5), 1031-1044.

Appendix A

This Appendix contains three MATLAB programs, which are used through the book.

1. Program halfbandiir.m computes the coefficients and poles and zeros of the IIR halfband filters.
2. Program qmflattice.m for computing the lattice coefficients of the power-complementary minimum-phase/maximum-phase FIR filter pair.
3. Program minphase.m for extracting minimum-phase factor from the separable linear-phase FIR filter.

1. halfbandiir: **program for IIR halfband filter design**

References

Lutovac, M. D., & Tošić, D. V., & Evans, B. L. (2000). *Filter design for signal processing using MATLAB and Mathematica*. Upper Saddle River, N J: Prentice Hall.

Milić, L. D., & Lutovac, M.D. (2002). Efficient multirate filtering. In Gordana Jovanović-Doleček, (ed.), *Multirate Systems: Design & Applications*. Hershey, PA: Idea Group Publishing, 105-142.

Milić, L. D., & Lutovac, M.D. (2003). Efficient algorithm for the design of high-speed elliptic IIR filters. *AEÜ Int. J. Electron. Commun., 57*(4), 255-262.

```
function [b,a,z,p,k] = halfbandiir(N,fp,varargin)
%
% HALFBANDIIR  Halfband IIR filter design.
% [B,A,Z,P,K] = HALFBANDIIR(N,Fp) designs a lowpass N-th order
% halfband IIR filter with an equiripple characteristic.
%
% The filter order, N, must be selected such that N is an odd integer.
% Fp determines the passband edge frequency that must satisfy
% 0 < Fp < 1/2 where 1/2 corresponds to pi/2 [rad/sample].
%
% [B,A,Z,P,K] = HALFBANDIIR('minorder',Fp,Dev) designs the minimum
% order IIR filter, with passband edge Fp and ripple Dev.
```

```
%   Dev is a passband ripple that must satisfy 0 < Dev (linear) < 0.29289
%   or stopband attenuation that must satisfy Dev (dB) > 3.1
%
%   The last three left-hand arguments are the zeros and poles returned in
%   length N column vectors Z and P, and the gain in scalar K.
%
%   [B,A,Z,P,K] = HALFBANDIIR(...'high') returns a highpass halfband filter.
%
%   EXAMPLE: Design a minimum order halfband filter with given max ripple
%      [b,a,z,p,k]=halfbandiir('minorder',.45,0.0001);
%

error(nargchk(2,4,nargin));

[minOrderFlag,lowpassFlag,msg] = validateParseInput(N,fp,varargin{:});
error(msg);

omegap = fp*pi;
omegas = (1-fp)*pi;
if minOrderFlag,
  if varargin{1} <1
    deltap = varargin{1};
    deltas = sqrt(deltap*(2-deltap));
    Rp = -20*log10(1-deltap);
    Rs = -20*log10(deltas);
  else
    Rs = varargin{1};
    deltas = 10^(-Rs/20);
    deltap = 1-sqrt(1-deltas^2);
    Rp = -20*log10(1-deltap);
  end
  N = estimateOrder(omegap,omegas,Rp,Rs);
end

if N < 3, N = 3; end

fa = (1-fp)/2;
x = 1/(tan(pi*fp/2))^2;
if x >= sqrt(2)
   t = .5*(1-(1-1/x^2)^(1/4))/(1+(1-1/x^2)^(1/4));
   q = t+2*t^5+15*t^9+150*t^13;
else
   t = .5*(1-1/sqrt(x))/(1+1/sqrt(x));
   qp = t+2*t^5+15*t^9+150*t^13;
   q = exp(pi^2/log(qp));
end
L = (1/4)*sqrt(1/q^N-1);
Rp = 10*log10(1+1/L);
Rs = 10*log10(1+L);

if N == 3
```

```
  beta = nfp2beta(N,fp/2);
  p = [0;i*sqrt(beta);-i*sqrt(beta)];
  z = [-1
      (beta-1 + i*sqrt(3*beta^2 + 2*beta-1))/(2*beta)
      (beta-1 - i*sqrt(3*beta^2 + 2*beta-1))/(2*beta)];
  k =  beta/2;
  b = [beta 1 1 beta]/2;
  a = [1 0 beta 0];
elseif N == 5
  [beta,zeroi,select] = nfp2beta(N,fp/2);
  k = 1/2;
  p = 0;
  z = -1;
  for ind = 2:2:N
    ind2 = ind/2;
    p = [p;i*sqrt(beta(ind2));-i*sqrt(beta(ind2))];
    z = [z;(zeroi(ind2)^2-select+i*2*sqrt(select)*zeroi(ind2))/(select+zeroi(ind2)^2)];
    z = [z;(zeroi(ind2)^2-select-i*2*sqrt(select)*zeroi(ind2))/(select+zeroi(ind2)^2)];
    k = k*(1+beta(ind2))*(1+zeroi(ind2)^2/select)/4;
  end
  aOdd = 1;  bOdd = 1;
  for ind = 1:2:length(beta)
    aOdd = conv(aOdd,[1 0 beta(ind)]);
    bOdd = conv(bOdd,[beta(ind) 0 1]);
  end
  aEven = 1; bEven = 1;
  for ind = 2:2:length(beta)
    aEven = conv(aEven,[1 0 beta(ind)]);
    bEven = conv(bEven,[beta(ind) 0 1]);
  end
  a = conv(conv(aOdd,aEven),[1 0]);
  b = (conv(conv(aOdd,bEven),[0 1]) + conv(conv(bOdd,aEven),[1 0]))/2;
else
  %  [z,p,k] = ellip(N,Rp,Rs,omegap/pi)
  %  [b,a] = ellip(N,Rp,Rs,omegap/pi)
  [beta,zeroi,select] = nfp2beta(N,fp/2);
  k = 1/2;
  p = 0;
  z = -1;
  for ind = 2:2:N
    ind2 = ind/2;
    p = [p;i*sqrt(beta(ind2));-i*sqrt(beta(ind2))];
    z = [z;(zeroi(ind2)^2-select+i*2*sqrt(select)*zeroi(ind2))/(select+zeroi(ind2)^2)];
    z = [z;(zeroi(ind2)^2-select-i*2*sqrt(select)*zeroi(ind2))/(select+zeroi(ind2)^2)];
    k = k*(1+beta(ind2))*(1+zeroi(ind2)^2/select)/4;
  end
  aOdd = 1;  bOdd = 1;
  for ind = 1:2:length(beta)
    aOdd = conv(aOdd,[1 0 beta(ind)]);
    bOdd = conv(bOdd,[beta(ind) 0 1]);
  end
```

```
  aEven = 1; bEven = 1;
 for ind = 2:2:length(beta)
   aEven = conv(aEven,[1 0 beta(ind)]);
   bEven = conv(bEven,[beta(ind) 0 1]);
 end
 a = conv(conv(aOdd,aEven),[1 0]);
 b = (conv(conv(aOdd,bEven),[0 1]) + conv(conv(bOdd,aEven),[1 0]))/2;
end

if ~lowpassFlag,
  z = -z;
  b = b.*((-(ones(size(b)))).^(1:length(b)));
end


%------------------------------------------------------------
function N = estimateOrder(omegap,omegas,Rp,Rs)
 N = ellipord(omegap/pi,omegas/pi,Rp,Rs);
 N = adjustOrder(N);

%------------------------------------------------------------
function N = adjustOrder(N)
 if (N+1) ~= 2*fix((N+1)/2),
   N = N + 1;
 end

%------------------------------------------------------------
function [minOrderFlag,lowpassFlag,msg] = validateParseInput(N,fp,varargin)
 msg = '';
 minOrderFlag = 0;
 lowpassFlag = 1;
 if nargin > 2 & ischar(varargin{end}),
   stringOpts = {'low','high'};
   lpindx = strmatch(lower(varargin{end}),stringOpts);
   if ~isempty(lpindx) & lpindx == 2,
     lowpassFlag = 0;
   end
 end
 if ischar(N),
   ordindx = strmatch(lower(N),'minorder');
   if ~isempty(ordindx),
     minOrderFlag = 1;
     if nargin < 3,
       msg = 'Passband ripple, Dev, must be specified for minimum order design.';
       return
     end
     if ~isValidScalar(varargin{1}),
       msg = 'Passband ripple must be a scalar.';
       return
     elseif varargin{1} <= 0 | ((varargin{1} >= 0.29289)&(varargin{1} <= 3.1)) ,
       msg = ['Dev=' num2str(varargin{1}) ', it must be 0<Dev(linear)<0.29289, or Dev (dB) >3.1'];
       return
```

```
      end
    else
      msg = 'Specified unrecognized order.';
      return
    end
  elseif ~isValidScalar(N),
    msg = 'Specified unrecognized order.';
    return
  else
    if (N+1) ~= 2*fix((N+1)/2),
      msg = ['N=' num2str(N) ', order must be an odd integer.'];
      return
    end
    if nargin > 2 & ~ischar(varargin{1}),
      msg = 'Passband ripple, Dev, can be specified for minimum order design, only.';
      return
    end
  end
  if length(fp) ~= 1,
    msg = ['Length of Fp = ' num2str(length(fp)) ', length must be 1.'];
    return
  else,
    if ~isValidScalar(fp),
      msg = 'Passband edge frequency must be a scalar, 0<Fp<1/2.';
      return
    end
    if fp <= 0 | fp >= 0.5,
      msg = ['Fp=' num2str(fp) ', passband edge frequency must satisfy 0<Fp<1/2.'];
      return
    end
  end

%------------------------------------------------------------------------
function bol = isValidScalar(a)
  bol = 1;
  if ~isnumeric(a) | isnan(a) | isinf(a) | isempty(a) | length(a) > 1,
    bol = 0;
  end


%------------------------------------------------------------------------
function [beta,xx,a] = nfp2beta(n,fp)
  a = 1/tan(pi*fp)^2;
  for ind = 1:(n-1)/2
    x = ellipj( ((2*ind-1)/n + 1)*ellipke(1/a^2), 1/a^2);
    b(ind) = ((a+x^2) - sqrt((1-x^2)*(a^2-x^2)) )/((a+x^2) + sqrt((1-x^2)*(a^2-x^2)));
    xx(ind) = x;
  end
  beta = sort(b);

%------- [EOF] ----------------------------------------------------
```

2. qmflattice: **program for computing lattice coefficients of FIR complementary filter pair**

References

Fliege, N. J. (1994). *Multirate digital signal processing.* New York, NY: John Wiley.

Mitra, S. K. (2006). *Digital signal processing: A computer based approach.* New York, NY: The Mc-Graw-Hill Companies, Inc.

Vaidyanathan, P.P., (1993). *Multirate systems and filter banks.* Englewood Cliffs, NJ: Prentice Hall.

```
function k = qmflattice(h);
% h ? impulse-response coefficients of a minimum-phase FIR filter
% k ? lattice coefficients
% Author: Jelena Certic
hnorm = h/h(1);
k(length(hnorm)-1) = -hnorm(length(hnorm));
for n = length(hnorm)-3:-2:1
   gnorm = fliplr(hnorm);
   gnorm(1:2:length(gnorm)) = -gnorm(1:2:length(gnorm));
   hnorm = (hnorm+k(n+2)*gnorm)/(1+k(n+2)^2);
   hnorm = hnorm(1:n+1);
   k(n) = -hnorm(n+1);
end;

%------- [EOF] ------------------------------------------------
```

3. minphase: **program for extracting minimum-phase factor from the separable linear-phase FIR filter**[1]

Reference

Orchard, H. J., & Wilson, A. N. (2003, March). On the computation of a minimum phase spectral factor. *IEEE Trans. Circuits and Systems-I: Fundamental Theory and Application, 50*(3), 365–375.

```
function [y,ssp,iter] = minphase(g);
% Extracts the minimum phase factor y from a linear-phase filter.
% Input: g = [g(0) g(1) . . . g(N)] (row vector) where
% the g vector is the right-half of a linear-phase FIR filter.
% Assumption: Unit-circle zeros of g are of even multiplicity.
% Copyright (c) January 2002 by H.J. Orchard and A.N. Willson, Jr.
% Initialize y (poly. with all zeros at z = 0)
y = [1 zeros(1,length(g)-1)];
ssp = realmax;% A large number (for previous norm)
ss = ssp/2;   % Smaller large no. (for current norm)
iter = 0; e = 0;  % Initialize iter. counter & correction vector
```

```
while ( ss < ssp)
   y = y + e'; ssp = ss;% Update y and move old norm(e) value
   iter = iter + 1; % Increment the iteration count
   AR = toeplitz([y(1),zeros(1, length(y)-1)], y);
   M = toeplitz([y(length(g)),zeros(1,length(g)-1)],fliplr(y));
   AL = fliplr(M);
   A = AL + AR;% Create the A matrix
   b = g' - AL*y'; % and the b vector
   e = A\b;% Solve Ae = b for the correction vector e
   ss = norm(e);   % Get norm to see if it is still decreasing
end;

%------- [EOF] -------------------------------------------------------
```

ENDNOTE

[1]   Reproduced with permission of IEEE.

About the Author

Ljiljana D. Milić was born in Čačak, Serbia, in 1939. She received the Dipl.-Eng., MSc and DSc degrees from the University of Belgrade in 1962, 1973 and 1978, respectively, all in electrical engineering. She spent the academic year 1974-1975 at Swiss Federal School of Technology, Lausanne, Switzerland, where she worked as a research assistant.

Since 1962 she has been with Mihajlo Pupin Institute, Belgrade. She worked on computer applications to filter design and on the sensitivity and tolerances of electrical filters. Since 1979 she has dealt with digital signal processing, primarily in the domain of digital filters. She also treated the problems of real-time spectrum analysis and analog and digital filter banks. Dr. Milić was leader of numerous R&D projects and gave practical solutions that went into production. For several years she served as assistant director and from 1987-1990 as director of Telecommunication department of Mihajlo Pupin Institute. She was appointed associate professor (1991) and professor (2003) in Digital Signal Processing at the School of Electrical Engineering, University of Belgrade. She spent summers in 2002, 2003 and 2005 as a visiting professor at the Tampere International Center for Signal Processing (TICSP), Tampere University of Technology, Finland. Presently, she is a chief scientist in Mihajlo Pupin Institute and professor at the School of Electrical Engineering of Belgrade University.

Dr. Milic is the author of more than 140 scientific papers mainly in the field of digital signal processing, and the author of two text books in Serbian: *Recursive Digital Filters* (1982) and *Introduction to Digital Signal Processing* (1999). She is also the author of the book chapters and the editor for several scientific publications. Dr. Milić is a senior member of IEEE and a corresponding member of Academy of Engineering Sciences of Serbia.

Her current research interest is in digital filters, multirate signal processing, and DSP in communications.

For more details visit http://kondor.etf.rs/~milic/

Index